Tsunamis in the Pacific Ocean

Tsunamis in the Pacific Ocean

Proceedings of
The International Symposium on Tsunamis
and Tsunami Research
Jointly Sponsored by
The International Union of Geodesy
and Geophysics Committee on Tsunamis
and the East-West Center
at the University of Hawaii
Honolulu, Hawaii
October 7-10, 1969

Edited by William Mansfield Adams
Secretary, Tsunami Committee
International Union of Geodesy and Geophysics

East-West Center Press — Honolulu

Financial support for this volume has been
provided, in part, by the United Nations Educational,
Scientific, and Cultural Organization and by
Environmental Science Services Administration.

Copyright © 1970 by East-West Center Press
University of Hawaii
All Rights Reserved
International Standard Book Number: 0-8248-0095-8
Library of Congress Catalog Card Number: 77-124716
Printed in the United States of America
First Edition

Contents

SEISMIC SOURCE AND ENERGY TRANSFER
Convener: Dr. *S. L. Soloviev*
Co-chairman: *L. M. Murphy*

1. The Generation of Tsunamis
    and the Focal Mechanism
    of Earthquakes, *K. Iida* — 3

2. Generation of the Tsunami Wave
    by the Earthquake, *G. S. Podyapolsky* — 19

3. A Model Experiment on the Generation
    of the Tsunami of March 28, 1964
    in Alaska, *W. G. Van Dorn* — 33

4. Relationship of Tsunami Generation
    and Earthquake Mechanism
    in the Northwestern Pacific,
    L. M. Balakina — 47

5. Features of Tsunamigenic Earthquakes,
    W. M. Adams and A. S. Furumoto — 57

6. Dimensions and Geographic Distribution
    of Tsunami Sources near Japan, *T. Hatori* — 69

7. The Tsunami Accompanying the
    Tokachi-oki Earthquake, 1968, *Z. Suzuki* — 85

8. Statistical Studies of Tsunami Sources
    and Tsunamigenic Earthquakes
    Occurring in and near Japan, *H. Watanabe* — 99

9. Ionospheric Recordings of Rayleigh Waves
    for Estimating Source Mechanisms,
    A. S. Furumoto — 119

10. Identification of Source Region
    from a Single Seismic Record,
    T. Sokolowski and G. Miller — 135

11. Recurrence of Tsunamis
    in the Pacific, *S. L. Soloviev* — 149

12. The Tsunami in the Alberni Inlet
    Caused by the Alaska Earthquake
    of March 1964, *T. S. Murty and L. Boilard* — 165

TSUNAMI INSTRUMENTATION
Convener: *Dr. M. Vitousek*
Co-chairman: *Dr. G. L. Pickard*

13. Tide-Gauge Data Telemetry
    between the Tsunami Warning Center
    at Honolulu, Hawaii
    and Selected Stations in Canada,
    G. C. Dohler — 191

14. Recent Advances in Tsunami Instrumentation
    in Japan, *K. Terada* — 207

15. Bourdon-Tube, Deep-Sea Tide Gauges,
    J. H. Filloux — 223

16. An Instrumentation System
    for Measuring Tsunamis
    in the Deep Ocean,
    M. Vitousek and G. Miller — 239

17. Estimating Earthquake Rupture Length
    from *T* Waves, *R. H. Johnson* — 253

18. Developments and Plans
    for the Pacific Tsunami Warning System,
    L. M. Murphy and R. A. Eppley — 261

19. Problems of the Tsunami Warning Service
    in the U.S.S.R., *Z. K. Abouziyarov* — 271

TSUNAMI PROPAGATION AND RUN-UP
Convener: *Dr. K. Kajiura*
Co-chairman: *J. W. Brodie*

20. Tsunami Propagation over Large Distances, *R. D. Braddock* — 285

21. Some Hydrodynamic Models of Nonstationary Wave Motions of Tsunami Waves, *S. S. Voit and I. Sebekin* — 305

22. Some Problems of Hydrodynamics of Tsunami Waves, *L. V. Cherkesov* — 319

23. Transformation of Tsunamis on the Continental Shelf, *A. V. Nekrasov* — 337

24. Experimental Investigations of Tsunami Waves, *M. I. Krivoshey* — 351

25. A Laboratory Model of a Double-humped Wave Impingent on a Plane, Sloping Beach, *J. A. Williams and J. M. Jordaan, Jr.* — 367

26. Response of Narrow-mouthed Harbors to Tsunamis, *G. F. Carrier and R. P. Shaw* — 377

27. An Inverse Tsunami Problem, *R. O. Reid and C. E. Knowles* — 399

28. Experimental Investigations of Wave Run-up under the Influence of Local Geometry, *L. Hwang and A. C. Lin* — 407

29. A Model Study of Wave Run-up at San Diego, California, *R. W. Whalin, D. R. Bucci, and J. N. Strange* — 427

30. The Numerical Simulation of Long Water Waves: Progress on Two Fronts, *R. L. Street, R. K. C. Chan, and J. E. Fromm* — 453

GENERAL REPORTS

31. Initiating an IBM System/360 Document
    Processing System for Tsunami Research
    Using an Existing KWIC Index Data Base,
    *J. M. Walling, D. Freeman, and
    W. M. Adams* 477

32. Report of the International Union
    of Geodesy and Geophysics
    Tsunami Committee,
    Chairman: *B. D. Zetler* 485

33. Ecumenical Tsunamigaku, *D. Cox* 489

34. Names and Addresses of Authors
    and Conference Participants 495

    Index 501

Preface

The International Symposium on Tsunamis and Tsunami Research was held in Honolulu, Hawaii, from 7 to 10 October 1969. These proceedings document the scientific findings reported at the Symposium.

Australia, Canada, Germany, Japan, New Zealand, Union of Soviet Socialist Republics, and the United States of America sent representatives.

There were three seminars:

1. "Seismic Source and Energy Transfer," convened by Dr. S. Soloviev.
2. "Tsunami Instrumentation," convened by Dr. Vitousek.
3. "Tsunami Propagation and Run-up," convened by Dr. K. Kajiura.

Many of the scientific findings were of considerable significance. With respect to tsunami generation, papers by Dr. Iida, Dr. Watanabe, and Dr. Hatori provide vital statistical correlations of tsunami sources and tsunamigenic earthquakes. The theoretical dependence of gravity-wave parameters on source parameters is reported by Dr. G. Podyapolsky.

In the discussion on instrumentation, an automatic, remote, vocal type gauge is described by G. Dohler. Dr. K. Terada reports on the progress in Japan for both remote analog and remote digital tide gauges. Four submerged long wave recorders were used to study seiche in Ofunato Bay which was severely damaged by the 1960 Chilean tsunami. An offshore ocean-wave meter has been modified to report for the tsunami period range. Design, development, and utilization of a Bourdon-tube, deep-sea, tide gauge and of a Snodgrass vibrotron system repackaged in glass spheres are meticulously described by Drs. J. Filloux and M. Vitousek, respectively.

The plans for the Pacific Tsunami Warning System are reviewed by L. Murphy and R. Eppley. Dr. R. Johnson proposes a method for using a hydrophone array to record T-Phase data suitable for estimating fault characteristics. The suggested method is designed to eliminate from consideration those events for which no significant tsunami has been generated.

The propagation of tsunamis is analyzed both analytically and numerically. Analytical studies of point-source generation, significance of Coriolis effect, the effects of heterogeneity and viscosity of the fluid medium are presented by Drs. S. Voit and B. Sebekin. Dr. A. Nekrasov analyzes the transformation of waves by a step model of the continental shelf. The tsunami response of a uniform-depth model harbor connected to an ocean by a channel is considered by Drs. G. Carrier and R. Shaw. The numerical studies are also represented by a dynamic-programming approach for improving prediction of arrival times devised by Dr. R. Braddock. Professor R. Reid and Dr. C. Knowles describe the estimation of the deep-water tsunami form by an inverse transformation of a marigram obtained near the island. Several laboratory model studies are reported: local geometry effects by L. Hwang and A. Lin; wave intrusion into a harbor by R. Whalin and D. Bucci; and a verification of the Carrier-Greenspan transform theory, as applied to a double-humped wave, by Dr. J. Williams and Dr. J. Jordaan, Jr. An unusually small, distorted scale model of 1:65,000 and 1:12,500 was used to quantitatively study deep-ocean tsunami propagation by Dr. M. Krivoshey. The tsunami inundation was evaluated as to extent and depth by larger scale, distorted models of 1:5,000 and 1:350.

Recent progress towards solving the problem of storing and quickly retrieving relevant tsunami facts or documents is reported by J. Walling. A digital computer is used for analysis of indexed properties of documents stored in microfiche form on a five-second, random-access, display file.

Chairman of the Tsunami Committee of the International Union of Geodesy and Geophysics, Bernard Zetler, selected the conveners of the morning sessions and the chairmen of the afternoon sessions. The conveners invited the speakers for the morning sessions, with the afternoon sessions reserved for the contributed papers. However, due to the excellent response from the invitees, some invited papers were shifted from a morning session to an afternoon session without due notice. The tolerance of all involved was commendable.

Sponsorship of the meetings has been provided by the International Union of Geodesy and Geophysics (IUGG) Tsunami Committee, and the East-West Center. These organizations and the United Nations Educational, Scientific, and Cultural Organization have provided partial funding for the travel and other expenses incurred by the attendees. Financial assistance in the preparation of this manuscript has been provided by the Environmental Science Services Administration. A set of the papers, before editing, was provided in draft form to the attendees and authors. This was funded by the State of Hawaii allotment to the Tsunami Research Program. The assistance of all these cooperating organizations is gratefully acknowledged.

The idea for this Symposium was first presented at the one-day Tsunami Symposium held at Berne, Switzerland, in conjunction with the IUGG Triennial meeting on August, 1966.

Programs, housing, and social events were arranged by Joyce Terada, with the assistance of Jack Durham. Judy Loomis, Lois McAllister, and Joan Fukumoto performed registration and Message Center duties. Projection equipment was operated by Edith Fujikawa, Roy Nishioka, Joe Malina, or gracious registrants. Preparation of the "camera-ready copy" of these *Proceedings* has involved Lois McAllister as assistant editor, Joan Fukumoto as managing typist, with Judy Loomis as lay-out artist. Mahalo to all of them.

I accept full responsibility for technical editing: I have attempted to assure understanding by those with a minimal capability in English, yet retain the style of the author.

William Mansfield Adams

1 April 1970

SEISMIC SOURCE AND ENERGY TRANSFER

1. The Generation of Tsunamis and the Focal Mechanism of Earthquakes

KUMIZI IIDA
Nagoya University
Nagoya, Japan

ABSTRACT

The generation of tsunamis is discussed in connection with the type of faulting associated with the focal mechanism of tsunamigenic earthquakes. The focal mechanism of tsunamigenic earthquakes and of earthquakes not accompanied by tsunamis, occurring in or near Japan during the period 1926 to 1968, were investigated on the basis of radiation pattern of P-waves and for some, of S-waves. Only earthquakes having a magnitude greater than 6.3 were taken into consideration.

The faulting inferred from the focal mechanism of tsunamigenic earthquakes is mostly the dip-slip type; the generation of large-magnitude tsunamis is especially associated with dip-slip type. The earthquakes associated with strike-slip faulting are almost never accompanied by tsunamis, even though the magnitude of the earthquake is nearly seven. Also, most aftershocks are not associated with the generation of tsunamis.

The generation of tsunamis is closely related to the type and dimension of dislocation in the source, being associated with characteristics of the earthquake, such as magnitude, focal depth, and water depth at the epicenter.

INTRODUCTION

The generation of tsunamis associated with the focal mechanism of tsunamigenic earthquakes occurring in or near

Japan was discussed by the present author (Iida, 1967) at the eleventh Pacific Science Congress at Tokyo in 1966. The generation of tsunamis has been found to be greatly dependent on characteristics of an earthquake, such as magnitude, focal depth, focal mechanism, aftershock activity, and water depth at the epicenter. This paper is a progress report. The present investigation is concerned with tsunamigenic earthquakes during the period 1900 to 1968. However, since the more accurate data are those from 1926 to 1968, during which the focal mechanism of all earthquakes having a magnitude greater than 6.3, even those not accompanied by tsunamis, was investigated, this time interval was specially studied.

CATALOGUE OF TSUNAMIGENIC EARTHQUAKES AND TSUNAMI MAGNITUDE

A catalogue of tsunamigenic earthquakes, compiled from various reports and papers and from mareographic records of tidal stations distributed along the coasts of Japan, is presented in Table 1 along with tsunami magnitudes. Tsunami magnitudes m were determined by the following formula

$$m = \log_2 h \text{ or } h = 2.0^m, \qquad (1)$$

in which h is the maximum height, in meters, measured at a coast 10 to 300 km. from the tsunami origin.

The geographic distribution of epicenters of tsunamigenic earthquakes from 1900 to 1968 is shown in Fig. 1, classified according to tsunami magnitudes. From Fig. 1, it is apparent that most of these tsunami sources are located on the Pacific side in northeastern Japan. Figure 2 shows the relationship between the earthquake magnitude and focal depth of an earthquake accompanied by a tsunami. The magnitude of a tsunami is shown in round numbers in Figure 2. Earthquakes occurring off the coast accompanied by tsunamis and not accompanied by tsunamis are depicted by filled-in circles and by plain circles, respectively. Data for the magnitudes and focal depths of the earthquakes were taken mainly from the "Catalogue of Major Earthquakes Which Occurred In and Near Japan" published by the Japan Meteorological Agency (1958, 1966).

Figure 2 shows that there is an approximate linear boundary between earthquakes accompanied by tsunamis and those that

Table 1: Catalogue of Earthquakes Accompanied by Tsunamis In and Near Japan, 1900 to 1968.

Date	Time d. h. m. (G.C.T.)	Epicenter Lat. °N	Long. °E	Location	Depth (km)	M	m	Type
1901 June	24 07 04	28.3	129.3	Off Amami-Oshima	shallow	7.9	-1	
1901 Aug.	9 09 24	40.5	141.5	Off Hachinohe	"	7.7	0	
1901 Aug.	9 18 34	40.5	141.5	"	"	7.8	0	
1911 June	15 14 25	29.0	129.0	Off Amami-Oshima		8.2	1	
1912 June	8 04 42	39.3	143.3	Off Iwate	"		0	
1914 Jan.	12 09 29	31.6	130.7	Kagoshima Bay	"	6.1	0	
1915 Nov.	1 07 25	38.9	143.1	Off Sanriku	"	7.5	1	
1918 Sep.	7 17 16	45.0	153.0	Off Uruppu	"	8.2	3.6	
1918 Nov.	8 04 42	41.0	146.5	SE off Hokkaido	"	7.4	1	
1923 June	1 17 25	36.0	141.4	Kashimanada	0	6.3	-1.5	
1923 Sep.	1 02 58	35.3	139.3	Kanto	20	7.9	3.6	
1923 Sep.	2 02 46	35.1	140.4	Off Katsuura	20	7.4	0	
1927 Mar.	7 09 27	35.6	135.1	Tango	10	7.5	0.4	SS
1927 Aug.	5 21 13	38.0	142.0	Off Miyagi	20	6.9	-2	DS
1927 Aug.	18 19 28	34.0	142.0	Off Boso	shallow	6.4	-2	UN
1928 May	27 09 50	40.0	143.2	Off Sanriku	0-10	7.0	-2	DS
1931 Mar.	9 03 49	41.2	142.5	Off Aomori	0	7.6	-2	DS
1931 Nov.	2 10 03	32.2	132.1	Hiuganada	20	6.6	-0.3	UN
1933 Mar.	2 17 31	39.1	144.7	Off Sanriku	0-20	8.3	4.8	*DS
1933 June	18 21 37	38.1	142.2	Off Miyagi	20	7.1	-2	DS
1935 July	19 00 50	36.7	141.3	Off Ibaraki	0	6.5	-2	DS
1935 Oct.	12 16 45	40.0	143.6	Off Sanriku	40	7.2	-2	UN
1935 Oct.	18 00 12	40.3	144.2	"	20-40	7.1	-2	UN
1936 Nov.	2 20 46	38.2	142.2	Off Miyagi	50-60	7.7	-1.6	DS
1938 May	23 07 18	36.7	141.4	Off Ibaraki	10	7.1	-1.3	DS
1938 June	10 09 53	25.3	125.2	Off Miyakojima	10	6.7	0	UN
1938 Nov.	5 08 43	37.1	141.7	Off Fukushima	20	7.7	-0.8	DS
1938 Nov.	5 10 50	37.2	141.7	"	15	7.6	-0.8	DS
1938 Nov.	6 08 54	37.5	141.8	"	0	7.5	-0.6	DS
1938 Nov.	6 21 39	37.0	141.7	"	0	7.1	-0.7	DS
1938 Nov.	13 22 31	37.0	141.5	"	40	6.0	-1.5	UN
1938 Nov.	22 01 14	37.0	141.8	"	10	6.7	-2	DS
1938 Nov.	30 02 30	37.0	141.8	"	5	7.0	-2	DS
1939 Mar.	20 03 22	32.3	131.7	Off Miyazaki	10	6.6	-0.3	DS
1939 May	1 06 00	40.0	139.8	Oga Pen.	0	7.0	-2	SS
1940 Aug.	1 15 08	44.1	139.5	Off Hokkaido	0-20	7.0	1.8	DS
1941 Nov.	18 16 46	32.6	132.1	Hiuganada	0-20	7.4	0	UN
1943 June	13 05 12	41.1	142.7	Off Sanriku	20	7.1	-1.8	UN
1944 Dec.	7 04 36	33.7	136.2	Off Tonankai	0-30	8.0	2.9	DS
1945 Jan.	12 18 38	34.7	137.2	Mikawa Bay	0	7.1	-0.7	SS
1945 Feb.	10 04 58	40.9	142.1	Off Aomori	30	7.3	-1.5	DS
1946 Dec.	20 19 19	33.0	135.6	Nankaido	30	8.1	2.4	DS
1947 Nov.	4 00 09	43.8	141.5	Off Hokkaido	0-30	7.0	1	UN
1948 Apr.	17 16 11	33.1	135.6	Off Shionomisaki	40	7.2	-1	DS
1952 Mar.	4 01 23	42.2	143.9	Off Tokachi	45	8.1	2	*DS
1952 Mar.	9 17 04	41.7	143.5	Off Hokkaido	0-20	7.0	-2.5	DS
1952 Sep.	16 02 50	31.9	140.0	Off Hachijo Is.	0		-3	UN
1952 Sep.	23 04 20	31.9	140.0	"	0		-3	UN
1952 Sep.	24 03 23	31.9	140.0	"	0		-3	UN
1952 Sep.	26 03 33	31.9	140.0	"	0		-3	UN
1953 Mar.	11-25	31.9	140.0	"	0		-3	UN
1953 Nov.	25 17 48	34.3	141.8	Off Boso	40-60	7.5	1.6	UN

(continued)

Table 1. (continued)

Date	Time d. h. m. (G.C.T.)	Epicenter Lat. °N	Long. °E	Location	Depth (km)	M	m	Type
1956 Mar.	5 23 29	44.3	144.1	Off Hokkaido	0-20	5.8	-2	UN
1956 Aug.	12 16 59	33.8	138.8	Off Izu Pen.	40-60	6.5	-2	SS
1958 Nov.	6 22 58	44.3	148.5	Off Iturup Is.	80	8.2	2	SS
1959 Jan.	21 20 10	37.6	142.3	Off Fukushima	30	6.8	-3	DS,SS
1959 Oct.	26 07 35	37.6	143.2	"	20	6.7	-3	DS,SS
1960 Mar.	20 17 07	39.8	143.5	Off Sanriku	20	7.5	-1	*DS
1960 Mar.	23 00 23	39.3	143.8	"	20	6.7	-3	DS
1960 July	29 17 31	40.2	142.6	Off Iwate	30	6.7	-3	DS
1961 Jan.	16 07 20	36.0	142.3	Off Ibaraki	40	6.8	-2	*DS
1961 Jan.	16 11 19	36.0	141.9	"	20	6.4	-2	DS
1961 Jan.	16 12 12	36.2	142.0	".	20	6.5	-2	DS
1961 Feb.	26 18 10	31.6	131.9	Hiuganada	40	7.0	-0.8	DS
1961 July	18 14 04	29.6	131.8	Yakushima	60	6.6	-3	*DS
1961 Aug.	11 15 51	42.9	145.6	Off Kushiro	80	7.3	-3	DS,SS
1961 Nov.	15 07 17	42.7	145.6	"	60	6.9	-3	DS
1962 Apr.	12 00 53	38.0	142.8	Off Miyagi	40	6.8	-1	DS,SS
1962 Apr.	23 05 58	42.2	143.6	Off Hokkaido	60	7.0	-1	*DS,SS
1963 Oct.	12 11 26	44.5	148.9	Off Iturup Is.	0	6.3	-2	DS
1963 Oct.	13 05 17	43.8	150.0	"	20	8.1	2	DS
1963 Oct.	20 00 53	44.1	150.1	"	0	6.7	3.5	*DS,SS
1964 May	7 07 58	40.3	139.0	Off Akita	0	6.9	-3	DS
1964 June	16 04 01	38.4	139.2	Off Niigata	40	7.5	2.5	DS
1964 Dec.	10 15 11	40.4	138.9	Off Akita	40	6.3	-3	DS
1968 Jan.	29 10 19	43.2	147.0	Off Hokkaido	30	6.9	-3	UN
1968 Apr.	1 00 42	32.3	132.5	Hiuganada	30	7.5	1	DS
1968 May	16 00 49	40.7	143.6	Off Sanriku	0	7.9	2.3	*DS,SS
1968 May	16 10 39	41.4	142.9	Off Aomori	40	7.5	-2	*DS,SS
1968 June	12 13 41	39.4	143.1	Off Sanriku	0	7.2	0	DS,SS

*   Normal Faulting
DS: Dip-slip fault type
SS: Strike-slip fault type
M: Earthquake magnitude

m: Tsunami magnitude
DS,SS: Dip-slip or strike-slip fault type
UN: Earthquake mechanism not determined. (unknown)

are not accompanied by tsunamis. This boundary is considered to be a limiting magnitude for earthquakes; below this magnitude tsunamis do not occur, as reported in previous papers (Iida, 1958, 1963, 1965). This limiting magnitude for tsunamigenic earthquakes may be generally expressed by the straight line as

$$M = 6.3 + 0.005 H, \qquad (2)$$

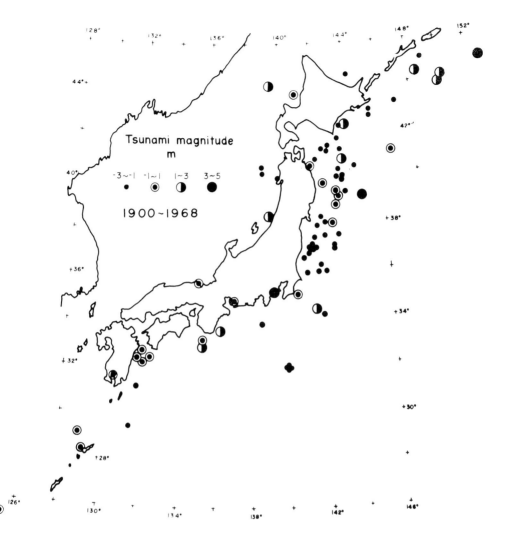

Fig. 1: Geographic Distribution of the Epicenters of Tsunamigenic Earthquakes Classified According to Tsunami Magnitude.

where the focal depth H is in kilometers. This straight line is shown as the full line in Figure 2. The coefficient of H in equation 2 is somewhat different from that in previous papers (Iida, 1958, 1963). From Fig. 2, it can be seen that there are three tsunamigenic earthquakes located on the left side of the straight line expressing the limit of equation 2. Taking those three into consideration, the limit may be expressed by

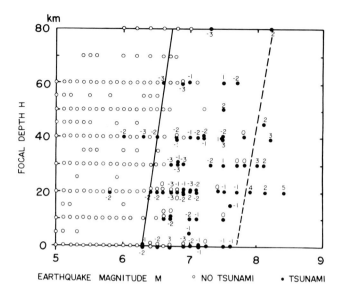

Fig. 2: Relationship Between Earthquake Magnitude and Focal Depth of Submarine Earthquakes During the Period 1926 to 1968. The numeral outside of the circle is the tsunami magnitude in round numbers.

$$M = 5.6 + 0.01 \, H. \tag{3}$$

This relation, expressing the smallest magnitude of a tsunamigenic earthquake, is the same as before.

The limit for disastrous tsunamis, having a magnitude of more than 2, may also be determined by

$$M = 7.7 + 0.005 \, H, \tag{4}$$

as shown by the broken line in Figure 2. Thus the earthquake accompanied by tsunamis having magnitude of less than 2 may generally be located between the full and broken lines in Figure 2. It is, however, noticed that four tsunamis (1940 Hokkaido, 1953 Boso, 1963 Iturup, and 1964 Niigata) having a magnitude of 2 or 3 are located on the left side of the broken line of equation 4. It is noted that the magnitude of the 1963 Iturup tsunami was especially great compared with its earthquake magnitude.

As seen in Fig. 2, even though the magnitude is greater than that located on the right side of the line representing equation 2, there are a number of earthquakes not accompanied by tsunamis. The number of these earthquakes for the period from 1926 to 1968 totaled 79; the epicenters are shown in Figure 3. These earthquakes are classified into three groups: (1) the aftershocks, 32 in number, (2) the relatively deep-focus shocks, 15 in number, and (3) others, 32 in number. Groups (1) and (2) of these earthquakes are generally considered not to generate tsunamis. Therefore, the reason group (3) is not accompanied by tsunamis must be explained. For this investigation, the mechanism of earthquakes is considered.

MECHANISM OF EARTHQUAKES AND TYPES OF FAULTING

To see the difference between tsunamigenic earthquakes and non-tsunamigenic earthquakes with magnitude greater than 6.3, focal mechanism is investigated. There are a number of papers concerning the focal mechanism of earthquakes. Most of the data concerned were taken from the publications of the Dominion Observatory, Ottawa by Wickens and Hodgson (1965, 1967) and publications by Stauder and Bollinger (1964, 1965). Further, some of the data were taken from the technical reports of the Japan Meteorological Agency, especially from the report on individual earthquakes (1969) and from the papers by Ichikawa (1966) and others (Hirasawa, 1965).

Some examples of focal-plane solutions of tsunamigenic and non-tsunamigenic earthquakes are shown in Fig. 4, which shows the distribution patterns of compression and dilatation of P initial motions. Some radiation patterns of S-waves were also considered, together with those of P-waves, especially in the results by Stauder and Bollinger (1964, 1965). From these radiation patterns, not only the nodal planes and their dips can be determined, but also the strike and inclination of the nodal planes. Assuming that the faulting occurs along the nodal plane, the dip and strike components of the unit vector of the movement due to faulting can be calculated. When the strike component thus obtained is greater than the dip component, the faulting is called strike-slip type; when the dip component is greater than the strike component, the

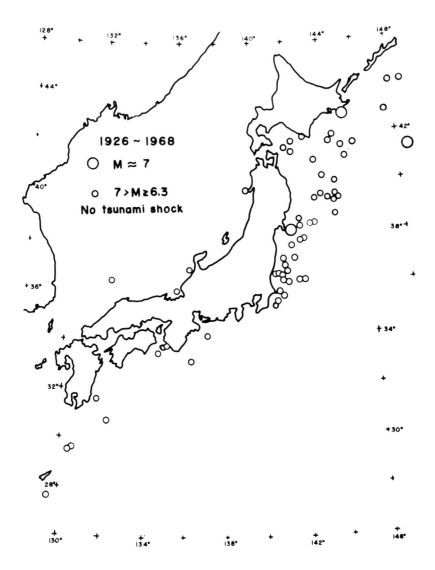

Fig. 3: Geographic Distribution of the Epicenters of Non-Tsunamigenic Earthquakes, Classified According to Magnitude.

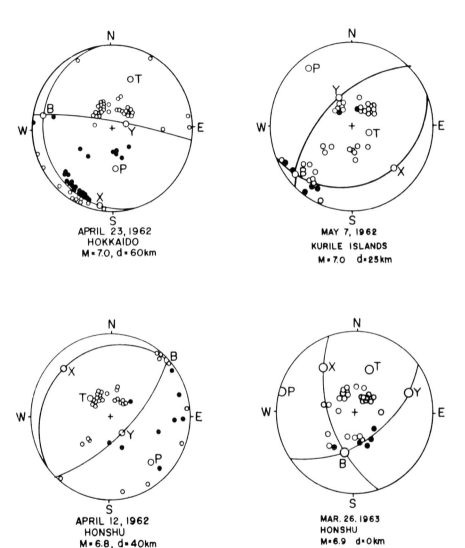

Fig. 4: Some Examples of Focal Plane Solutions of Earthquakes, Redetermined After the Data by Stauder and Bollinger (1964, 1965). o Compression. • Dilitation, P,T,B: Axes. X,Y: Nodal Plane Poles.
Left side: Tsunamigenic Earthquake.
Right side: Non-Tsunamigenic Earthquake.

Table 2. Examples of Non-Tsunamigenic Earthquakes and Types of Faulting.

Date	Time d. h. m. (G.C.T.)	Epicenter Lat. °N	Long. °E	Location	Depth (km)	M	Plane A SC	DC	Plane B SC	DC	Type
1935 Sep. 11	14 04	42.7	145.1	Off Hokkaido	60	7.1	0.89	0.49	0.77	0.65	SS
1938 Jan. 11	15 12	33.7	135.2	Kii-suido	20	6.7	1.00	0	1.00	0	SS
1950 Sep. 10	03 21	35.3	140.5	Off Chiba	30	6.5	0.83	0.55	0.71	0.70	SS
1950 June 27	15 41	42.7	138.7	Off Hokkaido	80	6.4	0.95	0.26	0.93	0.38	SS
1950 Sep. 10	03 21	35.3	140.5	Off Chiba	40	6.5	0.82	0.58	0.38	0.92	SS,
1950 Nov. 5	17 37	33.5	134.9	Kii-suido	30	6.9	0.45	0.93	0.95	0.20	*SS,*D
1951 Oct. 18	08 26	41.4	142.1	Off Hokkaido	40	6.5	0.07	1.00	0.24	0.97	DS
1952 Mar. 7	07 32	36.5	136.2	Off Noto	20	6.8	0.99	0.07	0.98	0.19	SS
1954 July 18	09 07	35.5	141.1	Off Chiba	40	6.4	0.57	0.83	0.42	0.90	*DS
1962 Jan. 4	04 33	33.6	135.2	Off Shikoku	40	6.4	1.00	0	1.00	0	SS
1962 May 7	17 39	45.3	146.7	Off Iturup	25	7.0	0.70	0.70	0.74	0.67	SS
1963 Mar. 26	21 34	35.8	135.8	Off Fukui	0	6.9	1.00	0.40	0.96	0.29	SS

SC: Strike component
DC: Dip component
* Normal faulting

faulting is called dip-slip type. In this way the type of faulting was investigated for each of the earthquakes mentioned above. The results are listed in Tables 1 and 2. It can be seen that more than 60 percent of the tsunamigenic earthquakes are from faulting of dip-slip type.

The geographic distribution of the epicenters of tsunamigenic earthquakes classified according to the type of faulting is shown in Figure 5. From this figure, it can be determined that there are few tsunamigenic earthquakes (only 8 percent) produced by faults of strike-slip type. Further, there are a number of earthquakes on the northern Pacific Ocean side for which the types of faulting are not determined. Figure 6, which corresponds to Fig. 3, shows the geographic distribution of the epicenters of submarine earthquakes not accompanied by tsunamis, classified according to the type of faulting. From Fig. 6 it is seen that most earthquakes show the strike-slip type fault, whereas few earthquakes are from dip-slip type faulting.

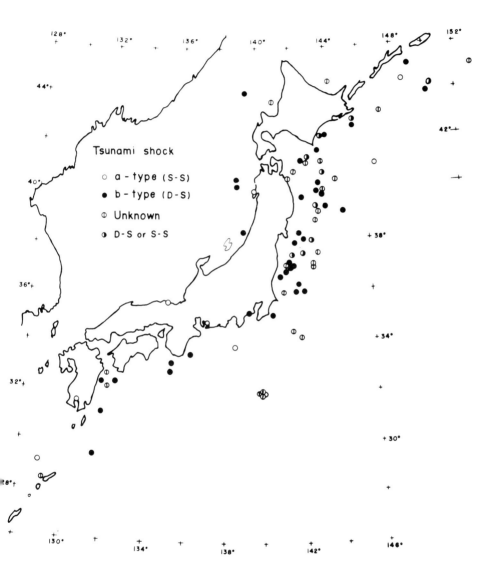

Fig. 5: Geographic Distribution of the Epicenters of Tsunamigenic Earthquakes Classified According to the Type of Faulting. S-S: Strike-slip fault type. D-S: Dip-slip fault type.

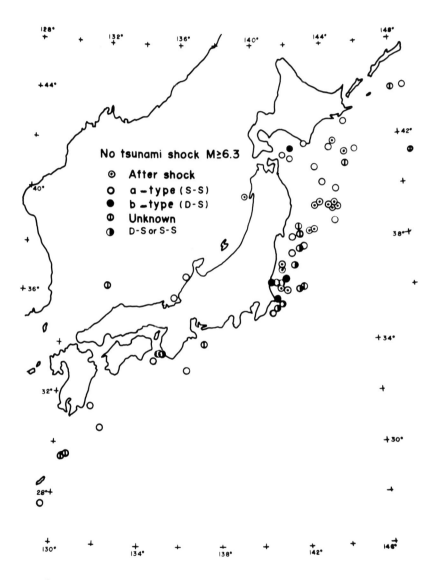

Fig. 6: Geographic Distribution of the Epicenters of Non-Tsunamigenic Earthquakes Classified According to the Type of Faulting.

TSUNAMI MAGNITUDE
AND TYPES OF FAULTING

The relationship between tsunami magnitude and the magnitude of a tsunamigenic earthquake was investigated in connection with the types of faulting as shown in Figure 7. From Fig. 7 it can be generally concluded that the greater the magnitude of an earthquake, the larger is the tsunami magnitude. Also, most large tsunamis occur in association with dip-slip type faulting. The tsunamigenic earthquakes marked by N in Fig. 7 are from the normal fault type, whereas those without N are from reverse fault type. Therefore, it may be concluded that most earthquakes accompanied by tsunamis in or near Japan are of the reverse fault type and that the relation of the tsunami magnitude to the reverse and the normal fault type is not clearly distinguishable.

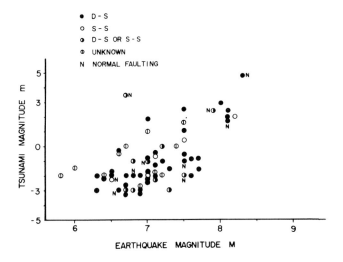

Fig. 7: Relationship Between Earthquake Magnitude and Tsunami Magnitude, Classified According to the Type of Faulting.

DISCUSSION

As is generally known, tsunamis accompany only those earthquakes whose epicenters are at shallow depths, less than 100 km., beneath the ocean or close to the shore; most tsunamis are generated by abrupt deformation of the sea bottom, either fault displacements or more general deformations caused by earthquakes, including deformations due to submarine volcanic explosions or landslides. Tsunami magnitude can be estimated from the generation mechanism of tsunamis, the dimensions of the area of tsunami origin, the speed and amount of displacement, and the water depth at the tsunami source. The dimensions and degree of crustal deformation of an ocean bottom is closely related to the earthquake magnitude, its focal depth, and the focal mechanism of the earthquake. Further, the area of tsunami origin is closely related to the area deformed by an earthquake and is approximately equal to the area of the aftershock activity.

Tsunamigenic earthquakes are always followed by a number of aftershocks. The generation of tsunamis accompanying earthquakes is closely correlated with dip-slip type faulting, i.e., the faulting of most tsunamigenic earthquakes is of the dip-slip type. The faulting of most non-tsunamigenic earthquakes having magnitudes greater than 6.3 is of strike-slip type. The process of tsunami generation is schematically illustrated in Figure 8.

Most aftershocks are generally not accompanied by tsunamis, but there are some aftershocks which are accompanied by tsunamis (such as 1938 Fukushima, 1952 Hokkaido, 1960 Sanriku, 1961 Ibaraki and Kushiro, 1963 Iturup, and the 1968 Aomori), although the tsunami magnitudes are usually comparatively small.

Abrupt deformation of a sea bottom is really caused by a main shock. By the change in stress or strain due to this deformation, aftershocks are caused. Consequently, we consider an aftershock not to partake in the production of deformation of the sea bottom. Since a tsunami results from sea-bottom deformation, which an aftershock does not produce, aftershocks are not accompanied by tsunamis. Therefore, if an earthquake which is considered to be an aftershock is accompanied by a tsunami, that earthquake is regarded as an independent shock and not as an aftershock of the main shock.

The crustal deformation due to a large earthquake may be classified into two types: the deformation may completely

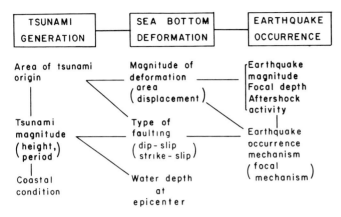

Fig. 8: Process of the generation of tsunamis related to the occurrence of earthquakes.

finish during the vibration of the main shock, or the deformation may continue for several hours or several months after the end of the vibration of the main shock. The former type of deformation is called the first kind of deformation and the latter type, the second kind of deformation. Any tsunami is attributed to the first kind of deformation. True aftershocks are generated by the release of the stress secondarily produced in the area of the crustal deformation caused by the main shock. Therefore, it is concluded that an aftershock does not produce new deformation of a sea bottom in the area deformed by the main shock and, consequently, an aftershock does not generally produce a tsunami. The conditions of tsunami generation are greatly dependent on the focal mechanism of earthquakes.

BIBLIOGRAPHY

Hirasawa, T. 1965. Source mechanism of the Niigata earthquake of June 16, 1964, as derived from body waves. *J. Phys. Earth.* Vol. 13, pp. 35-66.

Ichikawa, M. 1966. Mechanism of earthquakes in and near Japan, 1950-1962. *Papers in Meteorology and Geophysics.* 16, pp. 201-229.

Ichikawa, M. 1966. Statistical investigation of mechanism of earthquakes occurring in and near Japan and some related problems. *Kenkyu-Jiho*. 18, pp. 1-72.

Iida, K. 1967. Earthquakes and tsunamis (abstract), Symposium on Tsunami and Storm Surges, the Eleventh Pacific Science Congress, Tokyo 1966. *Proc. PSC Tsunami and Storm Surges Symposium*.

Iida, K. 1958. Magnitude and energy of earthquakes accompanied by tsunami, and tsunami energy. *J. Earth Sciences*. Nagoya Univ., 6, pp. 101-112.

Iida, K. 1963. Magnitude, energy, and generation mechanisms of tsunamis and a catalogue of earthquakes associated with tsunamis. *Proc. Tenth Pacific Science Congress Symposium*. IUGG Monograph 24, pp. 7-18.

Iida, K. 1965. Earthquake magnitude, earthquake fault, and source dimensions. *J. Earth Sciences*. Nagoya Univ., 13, pp. 115-132.

Japan Meteorological Agency 1958. Catalogue of major earthquakes in and near Japan (1926-1956). *Seismological Bull.* Supplementary volume, No. 1, pp. 1-91.

Japan Meteorological Agency 1966. Catalogue of major earthquakes in and near Japan (1957-1962). *Seismological Bull.* Supplementary volume, No. 2, pp. 1-49.

Japan Meteorological Agency *Seismological Bull.* 1963-1968.

Japan Meteorological Agency 1969. Report on the Tokachi-oki earthquake, 1968. *Technical Report*. No. 68.

Stauder, W. and Bollinger, D. A. 1964. The S-wave project for focal mechanism studies, earthquake of 1962, AF-AFOSR-62-458, Dept. Geophysics and Geophysical Eng., Inst. Technology, Saint Louis Univ.

Stauder, W. and Bollinger, G. A. 1965. The S-wave project for focal mechanism studies, earthquake of 1963, AF-AFOSR-62-458, Dept. Geophysics and Geophysical Eng., Inst. Technology, Saint Louis Univ.

Wickens, A. J. and Hodgson, J. H. 1967. Computer re-evaluation of earthquake mechanism solutions 1922-1962. *Publ. Dominion Obs. Ottawa*. 33.

2. Generation of the Tsunami Wave by the Earthquake

G. S. PODYAPOLSKY
Institute of Physics of the Earth
Moscow D-242, U.S.S.R.

ABSTRACT

The theoretical problem of generation of the gravity wave (tsunami) in the layer of elastic liquid (an ocean) occurring on the surface of elastic solid half-space (the crust) in the gravity field is studied with methods developed in the dynamic theory of elasticity. The source representing an earthquake focus is a discontinuity in the tangent component of the displacement on some element of area within the crust.

For conditions representative of the Earth's oceans, the solution of the problem differs very little from the joint solution of two more simple problems: the problem of generation of the displacement field by the given source in the solid elastic half-space with the free boundary (the bottom) considered quasi-static and the problem of the propagation of gravity wave in the layer of heavy incompressible liquid generated by the known (from the solution of the previous problem) motion of the solid bottom.

The theoretical dependence of the gravity wave parameters on the source parameters (depth and orientation) are obtained. In particular, a very rough estimation of the source energy passing into the gravity wave is given. In general, a portion of it corresponds to the estimation given by Iida from observation data.

The conclusion is made that the considered direct mechanism of the tsunami generated by an earthquake satisfies the main observed properties of this phenomenon and there is no need to include any additional hypothesis concerning significant influence of associated secondary phenomena (submarine landslides etc.).

INTRODUCTION

Development of the tsunami wave generated by a tectonic earthquake can be distinguished by three main steps: first, the transmission of the disturbance to the water mass produced by the tectonic dislocation along the fracture in the earth crust; second, the propagation of the gravity wave over a great distance across the deep ocean; and third, the transformation of the wave near the shore and other phenomena, called tsunami in the initial, narrower understanding of this term. The classic theory of tsunami as the long gravity wave in the layer of heavy noncompressible liquid on the rigid bottom concerns the second step. The most complex, and practically most important, third step is studied at the present time, more or less successfully, by means of model observations. But the first step has been studied theoretically very little because, for this step, the processes in the solid media beneath the bottom are important.

In this theoretical investigation, the model problem of a disturbance generated in an elastic liquid layer (the ocean) on an elastic solid half-space (the crust) in a uniform gravity field is studied. For the linear approximation satisfactory for the small deformations in the crust and for displacements small compared to the ocean depth H, the system of the equations of motion and boundary conditions for this problem is:

Equations of motion—

$$(c^2 \frac{\partial}{\partial x_j} - g_j) \frac{\partial u_k}{\partial x_k} = \frac{\partial^2 u_j}{\partial t^2}$$

for $0 < z < H$ (in the ocean)

$$[(a^2-b^2) \frac{\partial}{\partial x_j} - g_j] \frac{\partial u_k}{\partial x_k} + b^2 \frac{\partial^2 u_j}{\partial x_k \partial x_k} = \frac{\partial^2 u_j}{\partial t^2}$$

for $z > H$ (in the crust).

Boundary conditions—

(on the ocean surface)

$$\left[c^2 \frac{\partial u_k}{\partial x_k} - g u_3\right]_{z=+0} = 0$$

(on the bottom)

$$[u_3]_{z=H+0} = [u_3]_{z=H-0}$$

$$\left[\frac{\partial u_1}{\partial z} + \frac{\partial u_3}{\partial x}\right]_{z=H+0} = \left[\frac{\partial u_2}{\partial z} + \frac{\partial u_3}{\partial y}\right]_{z=H+0} = 0$$

$$m\left[(a^2-2b^2)\frac{\partial u_k}{\partial x_k} + 2b^2 \frac{\partial u_3}{\partial z} - g u_3\right]_{z=H+0} = \left[c^2 \frac{\partial u_k}{\partial x_k} - g u_3\right]_{z=H-0}$$

where: $x = x_1$, $y = x_2$, $z = x_3$ is the rectangular Descarte coordinate system with the axis z directed positively downward and the plane xy coincident with the ocean surface,

t is the time,
u_j are the components of the displacement relative to the initial state (but not non-stressed, because of the presence of gravity field),
g_j is the vector of gravity acceleration with the components: 0, 0, g,
a, b, c are the velocities of the longitudinal and transverse waves in the crust and of the sound in the ocean, respectively,
m is the relation of the density of the crust to that of the ocean.

This system of the equations of motion and boundary conditions was devised by Scholte (1934).

The disturbance source for modelling the earthquake focus is chosen as a discontinuity of the tangent displacement on a little square within the crust (the point source of Nabarro).

In the case of the extended source the displacement may be represented, in principle, by means of superposition of such point sources extended over all the focus area. This demands, however, detailed ideas about the process of breaking of the earthquake focus not yet known.

The solution of the auxiliary problem, the field of a Nabarro source in the infinite elastic solid, may be in the form of the triple superposition of the plane harmonic waves of different frequency ω, transmitted with the different horizontal phase velocity Θ^{-1} in the different azimuth directions ψ :

$$u_j^o (t,x,y,z) = \int_{-\infty}^{\infty} d\omega \int_{-\pi}^{\pi} d\psi \int_{0}^{\infty} N_j^o (\omega,\psi,\Theta,z) e^{i\omega[t-\Theta(x \cos \psi + y \sin \psi)]} d\Theta$$

The obvious expressions for the kernel $N_j^o (\omega,\psi,\Theta,z)$ obtained from the solution of the auxiliary problem are rather complicated and we shall not write them. They differ from known expressions of the displacement excited by the Nabarro source because the latter are obtained without taking into consideration the influence of gravity field. As in the case of non-gravital elasticity, the displacement of Nabarro source with arbitrary orientation may be represented in the form of the sum of four such sources of fixed orientation.

After the solution of the auxiliary problem, the solution of the principal problem may be constructed easily with the standard method: Expressions for the reflected waves in the crust and the interfering waves in the ocean are sought in the form of the analogous triple superposition of the plane harmonic waves, satisfying the corresponding equation of motion; the system of linear equations resulting from the substitution of these expressions into the boundary conditions yields the kernels defining the complex intensity of any wave. The components of the total displacement in the ocean are obtained after changing to the cylindrical coordinate system r, ϕ, z ($x = r \cos \phi$, $y = r \sin \phi$)

$$u_r = R \int_{0}^{\infty} e^{i\omega t} d\omega \int_{-\pi}^{\pi} \cos\psi \, d\psi \int_{0}^{\infty} \frac{n_r(\omega,\Theta,z) \, P(\omega,\Theta,\phi+\psi)}{\Delta(\omega,\Theta)} e^{-i\omega\Theta r\cos\psi}$$

$\cdot \, D(\omega,\Theta) d\Theta$

The expressions for the other components are distinguished only by the substitution of sin ψ for u_ϕ and $-i$ for u_z, instead of cos ψ under the second integral symbol, and for u_z by the substitution of the other function $\eta_z(\omega,\theta,z)$ instead of $\eta_r(\omega,\theta,z)$. The denominator $\Delta(\omega,\theta)$ is the determinant of the system of linear equations, and $\eta_r(\omega,\theta,z)$, $\eta_z(\omega,\theta,z)$, $P(\omega,\theta,\phi+\psi)$, $D(\omega,\theta)$ are the multipliers defined in the numerator on the basis of some physical ideas. Furthermore, $\eta_r(\omega,\theta,z)$ and $\eta_z(\omega,\theta,z)$ are the components of the total displacement of the harmonic interfering wave at any depth z under the ocean surface to the horizontal component at the surface; so $\eta_r(\omega,\theta,0) = 1$. $P(\omega,\theta,\phi+\psi)$ is the source function, including the dependence on the parameters of the source and on the azimuth ϕ. All these functions are very complex expressions and are not given here.

The methods of study and estimation developed in the dynamic theory of elasticity may be used for the integrals describing the exact solution. Different "waves", in the physical understanding of this term, are connected with singular points of the integrand in the integrals of such a type. The gravity wave in the ocean is connected with a pole of the integrand corresponding to the denominator $\Delta(\omega,\theta)$ being equal to zero. Study of the roots of the dispersion equation

$$\Delta(\omega,\theta) = 0,$$

giving the dependence of phase velocity θ^{-1} on the frequency ω, provides the following conclusions:

1. The dispersion equation has no roots within the first quarter of the complex plane θ; such roots should be connected with waves arising when propagating through the ocean.

2. The dispersion equation has an infinite set of positive real roots occurring when the gravity terms of the equation are regulated (the main and upper modes of Rayleigh waves in the liquid layer on the half-space of the non-gravity theory of elasticity), and another positive real root (the gravity root) connected primarily with the gravity terms. The gravity wave in the ocean is connected with this root. This conclusion has been pointed out by Scholte.

3. The dispersion equation has no positive real roots except the roots mentioned above in conclusion 2.

In Figs. 1 and 2 are shown the dispersion curves (versus $\Theta^{-1}(\omega)$, according to the dispersion equation) calculated for these values of the parameters: $a = 6$ km./sec., $b = a/\sqrt{3}$, $c = 1.45$ km./sec., $m = 3$, $H = 4$ km. The scale of ω in the horizontal axis is non-uniform (in proportion to $\sqrt[4]{\omega}$, and is marked by the values of the logarithm of the period $T = 2\pi\omega^{-1}$ in seconds. The dotted line shows the main mode of the Rayleigh wave calculated without taking into consideration the gravity field and the dispersion curve of the gravity wave according to the classic theory, described by this simple parameter form (parameter $\nu = H\omega\Theta$):

$$\Theta^{-1} = \sqrt{gH \frac{\tanh \nu}{\nu}} \; ; \; \omega = \sqrt{\frac{g}{H}} \, \nu \tanh \nu$$

Figure 2, differing from Fig. 1 only by the vertical scale, shows the dispersion curves for the gravity root at two ocean depths: 4 and 1 kilometers.

In Fig. 3, with the same horizontal scale, are shown the component factors $\eta_r(\omega,\Theta,z)$ and $\eta_z(\omega,\Theta,z)$ for the ocean surface ($z = 0$) and the bottom ($z = H$). The same values of the parameters and the same two ocean depths are assumed. The parameter Θ is defined as a function of ω according to the dispersion equation for the gravity root. The dotted line shows the position of these curves according to the classical theory:

$$\eta_r = \frac{\cosh[(1 - \frac{z}{H})\nu]}{\cosh \nu}, \quad \eta_z = \frac{\sinh[(1 - \frac{z}{H})\nu]}{\cosh \nu}$$

The deviation is apparent only for $\eta_r(\omega,\Theta,H)$; for $\eta_z(\omega,\Theta,H)$ the difference is not visible. $\eta_z(\omega,\Theta,0)$, according to the classical theory, is zero. It has been pointed out that $\eta_r(\omega,\Theta,0) = 1$ by the definition of component factors.

All the figures show that the essential difference between the classical conclusions and the real conditions of the earth-ocean occurs only in the region of very low periods ($T \sim 10^4$ sec.). For the ordinary tsunami wave ($T \sim 10^3$ sec.) it is permissible to use the "classical" simplification. For the functions η_r, η_z, and Δ, this classical simplification

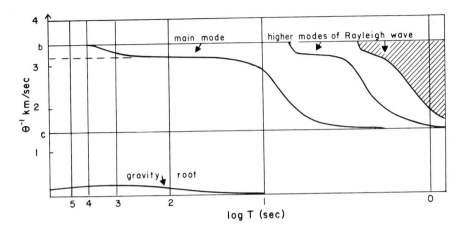

Fig. 1: Phase Velocity (km./sec.) versus the Period (seconds).

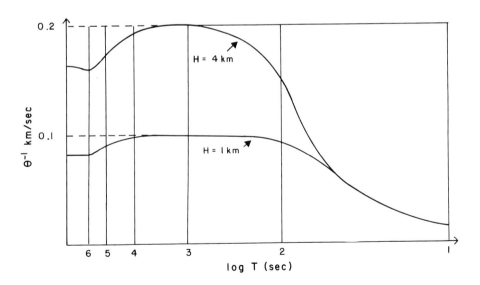

Fig. 2: Phase Velocity (km./sec.) versus the Period (seconds) for Two Ocean Depths: H = 4 km. and H = 1 km.

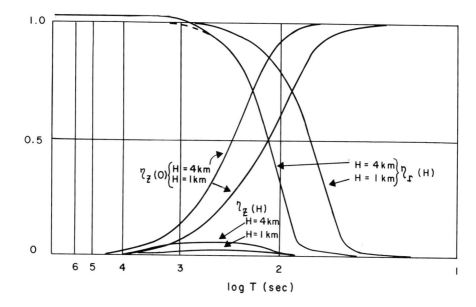

Fig. 3: Value of Component Factors $\eta_r(\omega,\theta,z)$ and $\eta_z(\omega,\theta,z)$ at Two Positions: on the Surface ($z = 0$) and at the Bottom of the Ocean ($z = H$) for Two Depths of the Ocean -- $H = 4$ km. and $H = 1$ km.

may be made by simply neglecting some terms being obviously small beyond the above-mentioned region of very low periods. But for the source function $P(\omega,\theta,\phi+\psi)$, such simplification is more difficult because $P(\omega,\theta,\phi+\psi)$ is the algebraic sum of two terms connected with gravity waves caused separately by the longitudinal and transverse waves propagating from the source in the crust. The result is the small residue from the compensation of these two large terms.

The exact analysis shows that this "classical" simplification of the solution of the initial Scholte's problem corresponds to the successive solution of the two simpler problems:

1. The problem about the displacement field of a Nabarro source in the elastic half-space with the free boundary in the quasi-static reduction (neglecting not only gravity but also dynamic terms in the equations of motion; the free boundary coincides with the ocean bottom).

2. The classic problem about the displacement in the layer of heavy noncompressible liquid excited by the known (from the solution of the previous problem) motion of the solid bottom. Apparently such a reduction to two simpler problems is permissible in the case of the ocean of variable depth.

The reduction from the exact solution of the Scholte's problem to simpler formulas describing the gravity wave contains the next approximations and transformations:

    1. The inside integral over θ is estimated by the residue in the gravity root.

    2. The simple integral over ψ is estimated by the method of stationary phase. This estimation is analogous to the substitution of the Bessel functions in the integrals, describing the solution of problems with the cylindrical symmetry by their approximations for large values of their argument. It is satisfactory under the assumption that the epicentral distance z is large in comparison with the length of the wave under consideration.

    3. The classical approximation of the integrand is used.

    4. The integral character of the Navarro source being the product of the average magnitude of the discontinuity of the displacement and its area is taken to depend on the time as:

$$F_\infty \, \varepsilon(t)$$

where $\varepsilon(t)$ is the step function and F_∞ is the ultimate value of the product. This supposition means that the characteristic time interval (period) of the gravity wave is long, relative to that of the source action. Fortunately this supposition is more or less true: in the opposite case, the unknown details of source action would be essential for the gravity waves.

    5. Integrating over the frequency ω changes to integrating over the above-mentioned parameter ν.

These steps give the next single integrals describing approximately the components of displacement in the gravity wave:

$$u_{r,z} \approx \frac{F_\infty}{(2\pi)^{3/2} \, H\sqrt{rH}} \, \text{Re} \left[\int_0^\infty \eta_{r,z}(\nu,z) \, P(\tfrac{h}{H}\nu,\phi) e^{i[f(\nu,t,r) \pm \tfrac{\pi}{4}]} \frac{\sqrt{\nu}}{\sin h \, \nu} \, d\nu \right]$$

where $\eta_{r,z}(\nu,z)$ are the component factors for the classical reduction, their simple dependences on the parameter ν being given above. The double sign: upper for r, the lower for z - component. The phase function:

$$f(\nu,t,r) = \sqrt{\tfrac{g}{H}} \, \nu \tan h \, \nu \, t - \tfrac{r}{H} \nu$$

The most interesting for this work is the source function, which is not obtained directly by the classical methods. It is:

$$P(\tfrac{h}{H}\nu,\phi) = [P_1(\phi) + \tfrac{h}{H} \nu \, P_2(\phi)] \, e^{-\tfrac{h}{H}\nu}$$

where h is the source depth below the ocean bottom, and $P_1(\phi)$ and $P_2(\phi)$ are the factors depending only on the azimuth ϕ and source orientation.

$$P_1(\phi) = \frac{ik^2}{1-k^2} \sin \zeta \sin \phi \, [\cos \zeta \cos \lambda \sin \phi - \sin \lambda \cos \phi]$$

$$P_2(\phi) = \cos \lambda \cos (2\zeta) \cos \phi + \sin \lambda \cos \zeta \sin \phi$$
$$+ \, i \sin \zeta \, [\cos \lambda \cos \zeta \, (1+\cos^2\phi) + \sin \lambda \sin \phi \cos \phi] \, ,$$

where ζ = the dip of the discontinuity plane,
λ = the angle in the discontinuity plane between the direction of the discontinuity of displacement and that of the greatest dip,
k = b/a.
The azimuth is measured from the dip direction.

The appearance of the second term inside the brackets in the formula $P(\tfrac{h}{H}\nu,\phi)$, obtained from the classical approximation from Scholte's solution, is connected with the above-mentioned compensation effect of the two large terms. Hence the additional decompensation raised by the little difference in the decreasing of these terms at the arising of source depth is essential. As the result of the solution of the two simplified problems, this structure at the source function is connected directly with the fundamental solution for the quasi-static elasticity. Because of the presence of this second term, the decreasing of the absolute value $P(\frac{h}{H}\nu,\phi)$ required by the presence of the exponent factor $e^{-\frac{h}{H}\nu}$ does not begin directly from the ocean bottom, but from some optimal source depth which corresponds approximately to $\frac{h}{H}\nu = 2\pi\frac{h}{L} = 1$ where L is the wave length. For the real tsunami wave, this optimal depth source may be some tens of kilometers. This effect is seen in Fig. 4, where the dependences of $P(\frac{h}{H}\nu,\phi)$ on $\frac{h}{H}\nu = 2\pi\frac{h}{L}$ for four standard sources at the azimuth, 0°, 45°, and 90° are shown (for any standard sources, symmetry relative to both dip and strike axes occurs).

Qualitative analysis of the approximate expressions obtained for the gravity wave in the ocean, strengthened by some calculation, shows that for the deep source (h > H)

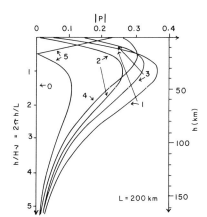

Source	ζ	λ	ϕ	N of curve
1	0°	0°	0° 45° 90°	1 2 0
2	45°	0°	0° 45° 90°	1 3 4
3	0°	90°	0° 45° 90°	0 2 1
4	90°	90°	0° 45° 90°	0 5 0

Fig. 4: Dependence of the Source Function $P(\frac{h}{H}\nu,\phi)$ on Depth ($\frac{h}{H}\nu$) for Four Source Types with Three Azimuths Each.

this expression describes a long single wave propagating almost without change in form with the velocity of $v = \sqrt{gH}$ and decreasing as $r^{-1/2}$ because of the cylindrical divergence. The maximum of the displacement at the ocean surface may be roughly estimated

$$u_r \sim k_r \frac{F_\infty}{H \sqrt{r(H+h)}},$$

$$u_z \sim k_z \frac{F_\infty}{(H+h) \sqrt{r(h+H)}}$$

where the factors k_r, k_z depend slightly on the orientation and azimuth, but in general being about $5 \cdot 10^{-2}$.

If the source is at a small depth below the bottom, the long wave is complicated with the superposition of the dispersive short-wave oscillations, decreasing very rapidly with deepening of the source. Figure 5 shows a series of the dependencies on the time or horizontal (continuous line) and vertical (dotted line) components of the displacement at the ocean surface, the source being directly at the bottom (h = 0), for H = 4 km. and different distances r. For 200, 400, and 600 km., there is shown also (low frequency solid line) horizontal component for the deep source (h = 20 km.). All the curves have been displaced along the time axis, so that the zero of the time really corresponds to the moment $t = r/\sqrt{gH}$ (the step of 100 km. by r corresponds to the shifting $\Delta t = 500$ sec. in time).

All calculations have been made for the point source or an element of the discontinuity surface. The really extended source exciting the tsunami wave would reach a large depth; so the long wave should be excited by all or a major part of the discontinuity surface, but the short wave oscillations only by the narrow stripe near the ocean bottom. Then the oscillation should be relatively small. Sometimes the superposition of two sufficiently different periods has been observed in reality. Then the short period has been treated as connected exactly with the seismic shock and the long-period as connected with accessory landslide phenomena. The natural explanation of this effect given above is worthy of attention.

The gravity wave is related to its source; hence the part of the seismic energy of the source given to the wave

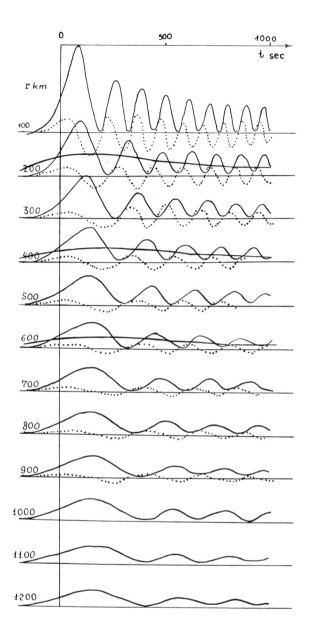

Fig. 5: Horizontal (continuous) and Vertical (dotted) Time Histories of Ocean Surface Displacement for 12 Distances from the Source: Source at the Bottom of the Ocean. At 200, 400, and 600 km. stations, the horizontal component for a source at greater depth (h = 20 km.) (solid line) is shown as a lower frequency half-cycle.

can be evaluated, in principle. This evaluation is difficult in practice. In contrast to the gravity wave energy, the energy of the source itself cannot be evaluated assuming instantaneous action of the source. Some additional ideas about the development of the source in time are necessary.

Assume that h>>H, that the discontinuity of the displacement propagates in the discontinuity plane symmetrically from a center with a velocity close to the velocity "b" of the transverse waves in the crust, that the ultimate area of the discontinuity is a circle of some radius R, and that the magnitude of the discontinuity decreases from the center to the periphery in accordance with some natural law. It is possible to derive a very rough formula (the error may be an order of magnitude or more):

$$\frac{\text{Energy of the gravity wave}}{\text{Energy of the source}} \simeq 5 \cdot 10^{-6} \left(\frac{R}{h}\right)^2 R \text{ (kilometers)}$$

This formula is in satisfactory agreement with the estimation obtained by Iida from observations.

BIBLIOGRAPHY

Scholte, J. G. 1934. Over het Verband tussen Zeegolven Microseismen. *Ned. Akad. Wetenshap.* Verslag, gewone Vergader, Afdeel, Natuurk., v. 52.

3. A Model Experiment on the Generation
of the Tsunami of March 28, 1964
in Alaska

William G. Van Dorn
Scripps Institution of Oceanography
University of California
San Diego, California

ABSTRACT

Analysis of the wave record for this tsunami, obtained from a special tsunami recording station at Wake Island, suggests that the tsunami originated from a line source in the form of a monopolar uplift of half-breadth about 20-30 km. Dispersive waves superimposed on the broad uplift are predicted by theory for such an uplift.

A laboratory experiment is described showing that the essential features of such a source can be represented by a one-dimensional model.

INTRODUCTION

In a previous paper (Van Dorn, 1964) the source mechanism of the tsunami of March 28, 1964 in Alaska was discussed in a qualitative fashion. On the basis of observed land-elevation changes and patterns of early water motion, the tsunami was construed to have resulted from a dipolar deformation of a large sector of the shallow coastal shelf bordering the Gulf of Alaska, resulting in immediate flow away from the uplifted pole (largely under water), both southeasterly toward the open Gulf, and northwesterly toward the depressed pole (largely above water). This latter flow was presumed

(Wilson, 1968) to have been arrested soon by land and reflected back out into the gulf as a second pulse about 1.8 hours later.

This paper attempts to show, by more detailed analysis of the wave record obtained at Wake Island and by a one-dimensional model experiment, that the above interpretation is essentially correct, with the exception that there appears to be no evidence of significant reflection from the coastline.

SOURCE CONDITIONS

The earlier geometrical reconstruction of the region of uplift (Fig. 1) was shown as comprising that area between the dipole axis A-A and the northerly end of the Aleutian Trench B-B, which borders the continental shelf. This figure shows

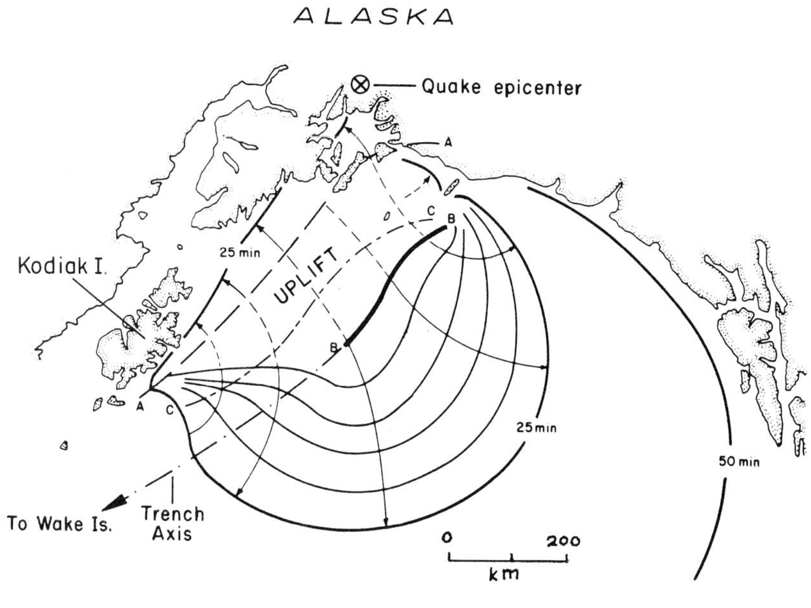

Fig. 1: Gulf of Alaska showing wavefront reconstruction of the region of uplift.

a diagrammatic progress of the reconstructed wave front radiating out into the Gulf of Alaska at 5-minute intervals. Within 25 minutes following the major quake, the wave front has filled the Gulf and by 50 minutes has traveled one-third of the way to Hawaii. The great-circle direction to Wake Island, where this tsunami was recorded at a special offshore station, is shown as nearly tangent to the trench axis, as well as to the dipole hinge-axis. Other reconstructions have been given (Wilson, 1968; Pararas-Carayannis, 1965), but they do not differ enough from that given here to affect the substance of the following. The significant feature of the motion described is that, although the uplifted region was rather long and narrow and apparently radiated its principal energy in a southeasterly direction normal to the dipole axis, the wave front became rapidly circular in deep water. Tide-gauge records from Eniwetok and Truk atolls (Spaeth, 1965), both of which lie nearly in the same direction as Wake Island, substantiate the Wake record in indicating that the first motion of the tsunami was a rise in sea level, thus ensuring that the wave motion reaching Wake came from the positive (uplifted) sector of the dipole source as a result of strong refraction in the source region. Because most of the uplifted region was under water, while a large fraction of the depressed sector was on land, the net effect of the dipolar ground motion was to add water to the ocean. Thus the oceanic effect was essentially a monopolar uplift.

ANALYSIS OF THE WAKE RECORD

The pertinent section of the Wake record (Fig. 2) consists of a sudden increase in elevation above the local tide datum by about 12 centimeters, followed by several cycles of decreasing amplitude and duration before the record finally returns to the tidal datum after 70-80 minutes. The theory of shallow water wave generation (Kajiura, 1963) provides a means of distinguishing between one- and two-dimensional sources comprising a uniform uplift of the sea floor along a line, or over a rectangular area, respectively. When viewed at a great distance, the leading wave from a line source will be the highest wave of the train, whereas with a radial source it will be some succeeding wave, depending upon the distance. Moreover, the crest of the leading wave will be retarded further behind the wave front, traveling at velocity $c = \sqrt{gh}$ where g = gravity, h = depth, for the line source than for the radial source, if they have the same width in the direction of interest and that width is large compared with the

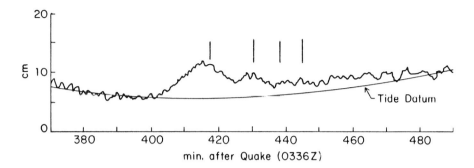

Fig. 2: Tide gauge recording at Wake Island for Alaska earthquake of 1964.

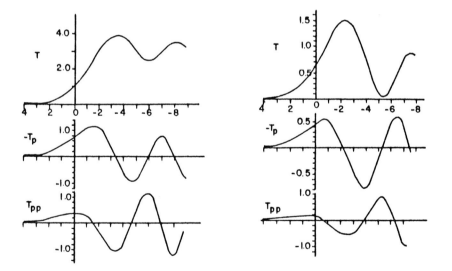

Fig. 3: Functions for wave-train characteristics predicted by shallow water theory. Line sources on left and two-dimensional sources on right.

depth of water at the source. These relative characteristics remain unchanged irrespective of variable water depth over the travel path if it is the same in both cases.

From Fig. 2 it is clear that the leading crest is the highest; thus it is appropriate to think of the Alaskan earthquake displacement as a long line source, despite the subsequent distortion of the wave pattern by refraction. Correspondingly, any calculations involving energy conservation or behavior should take refraction effects into account. The functions governing the wave train characteristics predicted by the shallow water theory are shown in Fig. 3 for line sources (left) and two-dimensional sources (right). The upper traces refer to wave systems produced by abrupt monopolar uplift; the center traces to dipolar motions with zero net displacement; and the bottom pair to impulsive sources initiated from rest. Clearly, the Wake record most closely resembles the upper left trace, the further quantitative correspondence with which can be demonstrated as follows.

The functions of Fig. 3 are plotted to a common abscissa scale in units of a parameter

$$p = (6/T)^{1/3} (T - R) \qquad (1)$$

where $R = r/h$ is dimensionless distance from the origin of the disturbance to a remote observation point, $T = t\sqrt{g/h}$ is dimensionless time from the instant of source occurrence, g = gravity, and h is the water depth (assumed constant) over the travel path between source and observation point. Thus p is a dimensionless time scale which increases like $T^{1/3}$ at great distances ($R \simeq T$). Although the position of the first positive maximum of each of these functions depends in a different way upon the dimensionless source parameter,

$$p_a = (6/T)^{1/3} (a/h),$$

representing the time (in units of p) for a long gravity wave to propagate across the half-breadth of a source of dimensionless width a/h, the respective spacings of subsequent crests depend also upon the nature of the source. Specifically, the functions $-T_p$ and T_{pp} are successive derivatives of

$$T(p) \equiv \int_p^\infty Ci_3 (p/3) \, dp$$

where

$$Ci_3(p/3) = 1/2\pi \int_0^\infty \cos(m^3 + pm)\,dm, \qquad (2)$$

is an Airy integral, and $m - m^3/6 \simeq \gamma$ approximates the wave frequency $\gamma = (m \tanh m)^{1/2}$ for small values of the wave number $m = 2\pi h/\lambda \ll 1$.

For the uniform uplift Hb of a line source of width 2a the surface elevation η at $R \gg 1$ is given by (Wilson and Tørum, 1964, p. 59)

$$\eta(p) = \frac{Hb}{2\pi}[T(p - p_a) - T(p + p_a)] \qquad (3)$$

and the source breadth can be determined from the arrival time of the leading wave $(p = p_1)$, where p_1 is the first maximum of the function $T(p)$. Equation 3 shows that the principal spectral contributions to the waveform at a distance result from delta functions representing the abrupt leading and trailing edges of the disturbance. If the source is very broad $(p_a > 4)$, $T(p + p_a) \to 0$, and only the leading edge contributes to the waveform; if it is narrow $(p_a \sim 0)$, the wave amplitude is smaller and the waveform is given by

$$\eta = \frac{Hb}{2\pi}(6/T)^{1/3}(-T_p). \qquad (4)$$

For extremely broad sources $(p_a \gg 4)$, $m^3/6 \ll 1$, and $\gamma \simeq m$, and the waveform approximates the seafloor deformation in shape, increasing only in length in proportion to $T^{1/3}$. If a broad source is anywhere discontinuous in uplift, it will have superimposed upon it a waveform of the type given by Equation 4.

From the general character of the observed ground motions in Alaska, the ratio a/h may be presumed to be of order 10^3, such that $p_a \gg 1$ for any possible values of $(6/T)^{1/3}$. Thus we can reasonably suppose that the waveform recorded at Wake conforms to the model of an extremely broad source with a local discontinuity. This view is supported by the large local uplift reported along the major uplift axis southwest of Montague Island (Plafker, 1965).

For the Alaskan tsunami, however, conditions are complicated by the fact that the source was in very shallow water (<200 m.) compared with the range of depths (4000-6000 m.) over the remainder of the travel path to Wake Island; nor is it clear which point along the source axis constituted the effective origin of the wave system. Nevertheless, a test of the dispersion predicted by Equation 4 can be obtained without knowing p_a by matching the observed wave arrival times to an equivalent source disturbance propagating in water of uniform effective depth h given by (Van Dorn, 1961)

$$h^{-1/2} = \frac{1}{n} \sum_{1}^{n} h_i^{-1/2}, \quad i = 1, 2, \ldots n,$$

where h_i represents local water depths assumed constant over small, equal increments of the least-time (refracted) travel path to Wake Island. The effective depth from Wake over the 5800 km. travel path to the center of the line B-B (Fig. 1) giving the deep-water boundary of the reconstructed source is 5400 meters.

If we now let T_i represent the arrival time of the ith crest from a localized uplift at a remote point $R = T_0 >> 1$, we can write, from Equations 1 and 4,

$$T_0 = T_1 + p_1 (T_1/6)^{1/3} = T_i + p_i (T_i/6)^{1/3} \qquad (5)$$

where the p_i correspond to consecutive positive maxima of $-T_p = Ci_3(p/3)$. For $R >> 1$, we can let $T_i = T_1(1 + \delta)$, $\delta << 1$, and Equation 5 becomes

$$\Delta T_i \equiv T_1 - T_i \doteq (p_1 - p_i)(T_1/6)^{1/3} \qquad (6)$$

$$- p_i(T_1/6)^{1/3}(\delta/3),$$

the last term of which can be neglected. Thus the crest intervals can be approximated knowing only the first crest arrival-time T_1. The pertinent data and results for the first four crests at Wake Island are given in Table 1. Computed crest arrivals are shown in Fig. 2 as vertical lines directly above the wave trace, showing that the approximate theory adequately predicts both the form and phase of the observed record.

Table 1: First Four Crest Intervals and Arrival Times, Wake Island, March 28, 1964.

Crest No.	p*	p_1-p_i	ΔT_i	ΔT_i (minutes)	Arrival Time 417 + ΔT_i (minutes)
1	-1.47				417
		5.70	32.2	12.6	
2	-6.97				430
		9.17	51.8	20.3	
3	-10.64				437
		12.28	69.5	27.2	
4	-13.75				444

*The p_i are taken as the positive maxima of $\pi/3^{1/3}$ $Ai(p/3^{1/3})$ = Ci_3, tabulated in Miller (1946), and $T = 417 \times 60\sqrt{g/h} = 1085$.

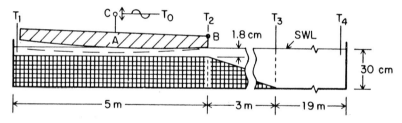

Fig. 4: Diagram showing setup of model experiment.

MODEL EXPERIMENT

As a further test of the one-dimensional source hypothesis, and to obtain an estimate of the source width, a model experiment was conducted in a wave channel constructed for studying the behavior of waves of very small amplitude. The experimental setup (Fig. 4) comprised a shallow-water section 5 m. long and 3.3 cm. deep at one end of the channel, terminating in a 1/10 slope to the full depth of 31.5 cm. over the remaining 19 m. channel length. On a scale of 1:17,000, this model represents a section of the continental shelf normal to the fault axis 90 km. in half-breadth and 565 m. deep, descending at the appropriate slope to the uniform depth of 5400 meters, corresponding to the effective depth computed for the travel path to Wake Island. In lieu of tilting the shallow shelf section to form the model tsunami, a long plunger A, hinged at B, and suspended from an automatic cycling mechanism C, was lowered into the water. In section, the plunger consisted of three wooden boards with spacers between them, with their lower edges slightly curved such that they just cleared the surface when fully raised, and were immersed full length to a maximum depth of 1 cm. when fully lowered. The total effective plunger width was only 16% of the channel width of 40 centimeters. Thus, when lowered rapidly, the plunger produced a positive half-wave 5 m. long with a maximum height of about 0.25 cm. at the end of the tank.

Surface elevations as a function of time were measured to an accuracy of 0.001 cm. by pressure transducers T_1-T_4, as shown, and the plunger motion was monitored by a separate displacement transducer T_0. Details of the tank and instrumentation have been given by Van Dorn (1966). The plunger and lowering mechanism are shown in Figure 5.

Fig. 5: Plunger and lowering mechanism for model experiment.

Of a number of experimental runs performed with various plunger speeds and initial conditions, the two which were considered to resemble the Wake record most closely are shown in Figures 6 and 7. In Fig. 6, the plunger (trace T_0) was started from its bottom-center position and cycled roughly sinusoidally to its initial position during an elapsed time of about 8 seconds, thus simulating an initial depression of the sea floor followed by an uplift of equal displacement. The other traces give the time-history of surface elevation at the transducer stations, including reflections from the far end of the channel. Trace T_2 shows that the tsunami at the edge of the continental shelf consisted of a leading trough, followed by a double pulse (A). The reflected train at the same station (B) exhibits a shallower trough and a dispersive train rather like the Wake record.

The result of a single, rapid plunger depression, corresponding to a simulated shelf uplift, is shown in Figure 7. In this case, the outgoing (A) and reflected (B) disturbances consist of single intumescences of roughly the same amplitude, but gradually developing a dispersive profile. The vertical lines over the crest of pulse B were computed by the same method as that for the Wake record, using the observed first-crest arrival time (34 sec.) and an effective depth $h = 21.3$ cm. to account for the increased travel time over the continental slope. Again, the agreement between the computed and observed arrivals supports the use of a one-dimensional dispersion model for the Alaskan source.

Making use of the known model source dimensions, we can now estimate the source half-breadth for Alaska. With reference to Fig. 7, let t_1 be the arrival time of the first crest at the shelf edge (T_2-A), and t_2 be its arrival time at any later station, say, T_4. The pulse duration at $t = t_1$ will be $2t_1$, since it consists of its direct and reflected image from the tank end at T_1. Letting Δt = pulse duration at T_4, we have, according to the theory for very long waves,

$$2t_1(t_2/t_1)^{1/3} = \Delta t. \tag{7}$$

But $t_1 = a/(g h_1)^{1/2}$, where a is the shelf half-breadth between T_1 and T_4 and Equation 7 can be written

$$a = (\Delta t/2)^{3/2} (g h_1/t_2)^{1/2} \tag{8}$$

Taking, from Fig. 7, $\Delta t = 24.5$ sec., $t_1 = 8.5$ sec., $t_2 = 21.5$ sec. and $h = 3.3$ cm., we obtain $a = 525$ cm., which is to be

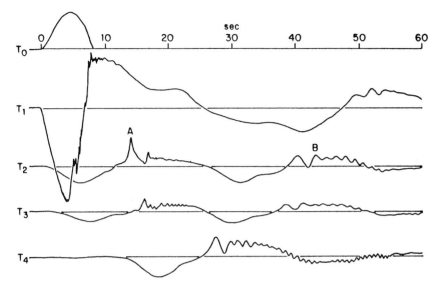

Fig. 6: Recording in model experiment for plunger started from bottom-center position and cycled for 8 seconds.

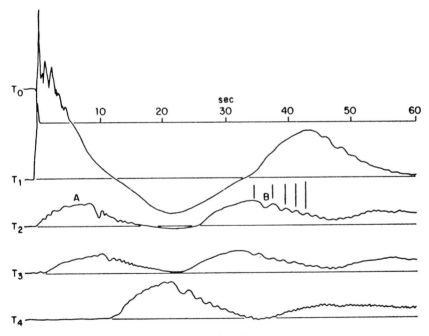

Fig. 7: Recording in model experiment for single, rapid plunger depression, corresponding to a simulated shelf uplift.

43

compared with the known shelf width (500 centimeters). This computation assumes that $p_a \sim 0$; so the first crest propagates at the velocity $c = \sqrt{gh}$. This can be shown to be a good approximation for all traces.

Similarly, for Alaska, from the Wake record, we take $\Delta t = 80$ min., $t_2 = 417$ min., and $h_1 = 100\text{-}200$ m., as reasonable mean depths for the continental shelf, giving

$$a = 23 \text{ km. } (h_1 = 100 \text{ m}),$$
$$= 33 \text{ km. } (h_1 = 200 \text{ m}).$$

These estimates are comparable to the width (~ 30 km.) between the smoothed 200 m. contour C-C and the presumed axis of maximum uplift A-A (Fig. 1), albeit only about half as wide as the zone extending to the trench axis B-B, previously defined by wave travel times (Van Dorn, 1964). The latter estimates appear to be more convincing as a result of the agreement between the above analysis and the model experiments.

ACKNOWLEDGEMENTS

This work was supported by the Office of Naval Research, under contracts Nonr 2216(16) and 2216(20).

BIBLIOGRAPHY

Kajiura, K. 1963. The leading wave of a tsunami. *Earthq. Res. Inst.*, 41:535-571.

Miller, J. C. P. 1946. *The Airy Integral*, British Assoc. Advancement of Science, University Press, Cambridge, 56 pp.

Pararas-Carayannis, G. 1965. Source mechanism study of the Alaska earthquake and tsunami of 27 March 1964, Part I, Water waves. *Tech. Rep. HIG-65-17*, Hawaii Institute of Geophysics. 28 pp.

Plafker, G. 1965. Tectonic deformation associated with the 1964 earthquake. *Science*, 148(3678):1675-1687.

Spaeth, M. G., Berkman, S. C. 1965. *The tsunami of March 28, 1964 as recorded at tide stations*. U. S. Dept. Comm. Coast & Geod. Surv., Rockville, Maryland.

Van Dorn, W. G. 1961. Some characteristics of surface gravity waves in the sea produced by nuclear explosions. *J. Geophys. Res.*, 66(11):3845-3862.

Van Dorn, W. G. 1964. Source mechanism of the tsunami of March 28, 1964 in Alaska. *Proc. 9th Conf. Coast. Engr., Am. Soc. Civ. Engr.* 166-190.

Van Dorn, W. G. 1966. Theoretical and experimental study of wave enhancement and runup on uniformly sloping impermeable beaches. Univ. of California, Scripps Inst. of Oceanography, *SIO 66-11*, 101 pp.

Wilson, B. W., Tørum, A. 1968. The tsunami of the Alaskan earthquake, 1964, engineering evaluation. *Tech. Mem. No. 25*, Coast. Engr. Res. Cen. 401 pp.

4. Relationship of Tsunami Generation and Earthquake Mechanism in the Northwestern Pacific

L. M. BALAKINA
Institute of Physics of the Earth
Moscow, U.S.S.R.

ABSTRACT

The features of the source mechanism for tsunamigenic earthquakes in the Northwest Pacific are studied by comparing the source mechanism of tsunamigenic and non-tsunamigenic earthquakes. The area studied included Kamchatka, the Kuril Islands, and Japan. Of the 47 earthquakes studied, 18 were tsunamigenic and 29 were non-tsunamigenic. For 14 of the tsunamigenic earthquakes, the fault-plane solution gave one possible fault plane to be predominantly dip-slip; for the other 4 tsunamigenic earthquakes, the accuracy of the solutions is not satisfactory for 3 of the 4, and for the fourth, the possible fault planes are inclined at 45° to the surface. For the 29 non-tsunamigenic earthquakes, the average value of shear horizontal motion is greater than those for the tsunamigenic earthquakes; however, in a few cases, solutions with steeply dipping fault planes occurred.

INTRODUCTION

The purpose of this work is to ascertain the features of source mechanism for tsunamigenic earthquakes in the Northwest Pacific. The source mechanisms of tsunamigenic and non-tsunamigenic earthquakes are compared.

The earthquake mechanism solutions obtained by Wickens and Hodgson (1967) and the author were used. Forty-seven tsunamigenic and non-tsunamigenic earthquakes of Kamchatka, the Kuril Islands, and Japan have been investigated for the period 1922-1962. The list of these earthquakes is given in Appendix 1.

There are several source mechanism solutions for some of the 47 considered earthquakes obtained by Wickens and Hodgson (1967) and by the author. In such cases, all solutions were analyzed.

In the northwestern part of the Pacific, 18 tsunamigenic shocks and 29 non-tsunamigenic shocks were investigated. Results of source mechanism determinations for these 47 earthquakes are given in Appendix 2. One of the possible fault planes for each of the 14 tsunamigenic earthquakes of this region dips steeply (dip angles are 65-90°). The motions in such fault planes are predominantly dip-slipping, normal or reverse ones; the shear horizontal displacements are practically absent. The second possible fault planes for these shocks are gently plunging.

We did not find such a feature of source mechanism in the case of 4 tsunamigenic earthquakes of the northwestern part of the Pacific, namely in the shocks of: 7.III.1927, near Honshu Island; 4.XI.1952, near Kamchatka; 26.II.1961, near Kyushu Island; 6.XI.1958, near South Kuril Islands. The accuracy of source mechanism solutions of the first three earthquakes is not satisfactory. The earthquake of 6.XI.1958 with magnitude $M > 8$ has a very reliable source mechanism solution. The shear horizontal motion for this earthquake is practically absent, but both possible fault planes are inclined less than 45° to the horizon.

For 29 non-tsunamigenic earthquakes which occurred in the same northwestern part of the Pacific, the outlined feature of source mechanism was not ascertained. The possible fault planes of these earthquakes have various dip angles; in a few cases, solutions with steeply dipping fault planes occurred. On the whole, for these 29 non-tsunamigenic earthquakes the value of shear horizontal motions are greater than those for the tsunamigenic earthquakes. As examples, the source mechanism solutions for tsunamigenic shock of 12.II.1961 and non-tsunamigenic shock of 7.V.1962 are given in Figures 1 and 2, respectively. The place of origin and magnitude values are approximately the same for both earthquakes.

The feature of source mechanism of tsunamigenic earthquakes in the northwestern part of the Pacific mentioned above

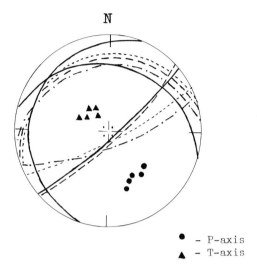

- P-axis
- T-axis

Fig. 1: Source Mechanism Solution for Tsunamigenic Shock of 12.II.1961.

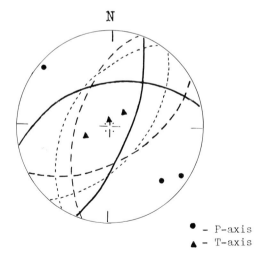

- P-axis
- T-axis

Fig. 2: Source Mechanism Solution for Non-Tsunamigenic Shock of 7.V.1962.

is apparently a significant one because the steeply-dipping fault planes show distinct mutual connections: the strike of these fault planes is similar to the strike of the main tectonic structures -- island arcs and oceanic troughs. Oceanic blocks for 14 tsunamigenic earthquakes mentioned before have undergone relative uplift on steep fault planes. Continental blocks, on the contrary, have undergone relative subsidence. The steep fault planes of 7 earthquakes do not dip under the continent, but under the ocean. This is at variance with the supposition that focal surfaces include only the faults dipping under the continents.

Tsunami generation is caused by the shocks having certain source mechanism. However, some other circumstances, as for instance earthquake magnitude and focal depth, may be important as well.

ACKNOWLEDGMENT

This work was suggested and initiated by S. L. Soloviev, who also provided the list of earthquakes for certain regions of the Circum-Pacific Belt.

BIBLIOGRAPHY

Wickens, A. J. and Hodgson, J. H. 1967. Computer re-evaluation of earthquake mechanism solutions. Publ. Dom. Obs., Ottawa, V. 33, N. 1, 560 p.

Appendix 1: List of Earthquakes Used for the Northwestern Part of the Pacific Ocean Basin.

No	Date	Hr.	Lat. N	Long. E	h	M	tsunami occurrence
1	2	3	4	5	6	7	8
1.	7.III.1927	09	35,6	135,1	10	7 1/2–7 3/4	tsunami
2.	25.XI.1930	19	35,1	139	0–5	7	
3.	11.IX.1935	14	42,7	145,1	60	7,1–7,6	
4.	30.VI.1936	15	50,5	160		7,4	
5.	23.V.1938	07	36,7	141,4	10	7,1	tsunami
6.	17.IV.1948	16	33,1	135,6	40	7,2	tsunami
7.	26.IV.1950	07	33,8	135,8	40	6,7	
8.	10.IX.1950	03	35,3	140,5	30–40	6,5	
9.	5.XI.1950	17	33,5	134,9		6,9	
10.	18.X.1951	08	41,4	142,1	40	6,5	
11.	6.XI.1951	16	47,8	154,2		7,2	
12.	4.III.1952	01	42,2	143,9	45	8,1	tsunami
13.	7.III.1952	07	36,4	136,2	20	6,8	
14.	9.III.1952	17	41,7	143,5	0–20	7	tsunami
15.	4.XI.1952	16	52,3	161		8 1/2	tsunami
16.	29.XI.1952	08	52,8	159,2		7 1/4	
17.	1.VII.1953	02	50,5	157,6	60	6	
18.	4.IX.1953	07	50,4	157		6 1/4	
19.	23.IX.1953	02	50	156		6 1/2	
20.	11.X.1953	13	49,5	156,4	90	7 1/2	
21.	25.XI.1953	17	34,1	141,9	40–60	7 1/2–8 1/4	tsunami
22.	26.XI.1953	08	34,3	141,6	30–40	6,6	
23.	23.XI.1955	06	50,4	157	60	7	
24.	29.IX.1956	23	35,5	140,2	60–70	6 3/4 – 7	
25.	11.X.1956	02	45,4	151,3	100	7 3/4	
26.	6.XI.1958	22	44,5	148,5		8 1/4	tsunami
27.	12.XI.1958	20	44,5	148,5		7 1/2	
28.	22.I.1959	05	37,5	142,3	20	6,8	tsunami
29.	4.V.1959	07	52,5	159,5		7 3/4	tsunami
30.	18.VI.1959	15	54	160		7	
31.	20.III.1960	17	39,8	143,5	20	7,5	tsunami
32.	29.VII.1960	17	40,2	142,6	30	6,7	tsunami
33.	10.I.1961	14	50	156,1	63	6 1/2	
34.	16.I.1961	07	36	142,3	40	6,8	tsunami
35.	19.I.1961	17	49,8	156,3	70	5 1/2	
36.	12.II.1961	21	43,8	147,8	50	7	tsunami
37.	23.II.1961	04	38,2	143,5		6,4	
38.	26.II.1961	18	31,6	131,5	40	7	tsunami
39.	23.IV.1961	09	44,6	150,4	20	6 3/4–7	
40.	18.VII.1961	14	29,6	131,8	20	6, 6–7	tsunami
41.	11.VIII.1961	15	42,8	145,4	60	7	tsunami
42.	19.VIII.1961	05	36	136,8		7	
43.	15.XI.1961	07	42,6	145,6	60	6,9	
44.	4.I.1962	04	33,6	135,2	40	6,4	
45.	12.IV.1962	00	38	142,8	40	6,8	tsunami
46.	23.IV.1962	05	42,2	143,9	60	7	
47.	7.V.1962	17	44	147		6 3/4	

Appendix 2: Source Mechanism Solutions for Earthquakes in the Northwestern Part of the Pacific Ocean Basin. The conventional signs are the same as in Wickens and Hodgson, 1967.

No	Date	Plane I				Plane II				P axis		B axis		T axis		author
		Az	dip	component strike	component dip	Az	dip	component strike	component dip	Az	PL	Az	PL	Az	PL	
1	2	3	4	5	6	7	8	9	10	11	12	13	14	15	16	17
1.	7.III.1927	256	63	0.79S	0.62T	146	56	0.54D	0.54T	290	4	196	44	24	45	W.H.
		261	50	0.69S	0.72T	137	57	0.63D	0.78T	110	4	203	32	14	58	W.H.
2.	25.XI.1930	87	32	0.95S	0.30N	192	81	0.51D	0.86N	341	45	108	30	212	29	W.H.
		281	86	1.00S	0.03N	111	88	1.00D	0.07N	145	44	302	86	55	2	W.H.
		100	84	0.88S	0.47T	7	62	0.99D	0.12T	140	15	21	62	237	24	W.H.
		101	88	0.99S	0.10T	10	84	1.00D	0.03T	146	5	26	84	236	5	W.H.
3.	11.IX.1935	4	69	0.68S	0.73T	253	47	0.87D	0.50T	34	13	293	39	138	48	W.H.
		66	42	0.14D	0.99T	235	49	0.12S	0.99T	240	4	150	5	4	84	W.H.
4.	30.VI.1936	322	66	0.36S	0.93T	186	32	0.63D	0.77T	337	18	241	19	108	63	W.H.
		331	40	0.09D	1.00T	144	50	0.08S	1.00T	148	5	57	3	295	84	W.H.
5.	23.V.1938	83	81	0.01S	1.00T	265	9	0.03D	1.00T	83	36	353	0	262	54	W.H.
6.	17.IV.1948	143	90	0.01S	1.00T	35	1	0.95D	0.32T	144	45	53	1	323	45	W.H.
		278	2	0.68S	0.74T	140	89	0.02D	1.00T	139	44	230	1	321	46	W.H.
7.	26.IV.1950	102	76	1.00S	0.08N	193	86	0.97D	0.24N	328	13	121	76	236	6	W.H.
		235	22	0.26S	0.97N	38	69	0.10D	0.99N	208	66	311	6	43	24	W.H.
8.	10.IX.1950	7	51	0.53S	0.85T	232	49	0.55D	0.83T	29	1	299	24	123	66	W.H.
		23	43	0.41S	0.91T	235	51	0.36D	0.93T	220	4	312	16	117	73	W.H.
		207	52	0.30S	0.95T	54	41	0.36D	0.93T	219	6	128	14	330	75	W.H.
9.	5.XI.1950	172	63	0.23S	0.97N	324	30	0.41D	0.91T	18	69	256	12	162	17	W.H.
		195	80	0.11S	0.94N	342	12	0.55D	0.84N	23	54	284	6	190	35	W.H.
		157	32	0.34S	0.94N	314	60	0.20D	0.98N	107	72	230	10	323	15	W.H.
		157	35	0.42S	0.91N	307	59	0.28D	0.96N	89	71	226	14	319	12	W.H.
10.	18.X.1951	268	17	0.24S	0.97T	103	73	0.07D	1.00T	99	28	192	4	289	61	W.H.
		112	45	0.13S	0.99T	302	45	0.13D	0.99T	117	0	27	5	209	85	W.H.

52

1	2	3	4	5	6	7	8	9	10	11	12	13	14	15	16	17
11.	6.XI 1951	143	89	0.10S	1.00T	41	6	0.98D	0.21T	148	44	53	6	317	46	W.H.
		235	32	0.9OS	0.45T	122	76	0.49D	0.87T	99	25	204	29	335	50	W.H.
12.	4.III.1952	93	74	0.05S	1.00T	275	16	0.05D	1.00T	95	28	5	1	271	60	B
13.	7.III.1952	245	89	0.95S	0.31T	336	72	1.00D	0.02N	109	14	332	72	202	12	W.H.
		65	70	0.98S	0.17N	156	80	0.34D	0.34N	290	20	97	67	200	6	B
14.	9.III.1952	123	70	0.36S	0.93T	355	30	0.71D	0.71T	140	20	42	20	273	60	B
15.	4.XI. 1952	265	86	0.99S	0.13N	355	83	1.00D	0.07N	130	8	328	82	220	2	W.H.
		122	80	0.92S	0.39T	28	68	0.38D	0.19T	163	8	55	65	257	21	W.H.
		254	87	1.00S	0.01N	344	89	1.00D	0.05N	120	2	270	87	30	1	W.H.
16.	29.XI. 1952	315	85	0.96S	0.28T	47	74	1.00D	0.09N	180	15	29	73	272	8	B
17.	1.VII.1953	268	79	1.00S	0.02N	358	89	0.98D	0.19N	133	9	273	79	42	7	W.H.
		324	39	0.03S	1.00T	146	51	0.02D	1.00T	145	6	235	1	334	84	W.H.
		151	41	0.54S	0.84T	11	56	0.42D	0.91T	354	8	87	21	244	68	W.H.
		145	51	1.00S	1.00T	325	39	0	1.00T	145	6	55	0	324	84	W.H.
18.	4.IX. 1953	312	64	0.47S	0.88T	182	38	0.69D	0.73T	332	14	235	25	88	61	W.H.
		328	39	0	1.00T	148	51	0	1.00T	148	6	238	0	330	84	W.H.
		155	40	0	1.00T	335	50	0	1.00T	335	5	65	0	155	85	B
19.	23.IX. 1953	148	44	0.58S	0.81T	13	55	0.50D	0.87T	352	6	85	24	249	65	W.H.
		148	52	0.02D	1.00T	326	38	0.03S	1.00T	147	7	237	41	336	83	W.H.
		77	57	0.78D	0.63T	323	58	0.77S	0.64T	290	1	21	41	200	49	B
20.	11.X. 1953	45	60	0.64S	0.77T	285	50	0.77D	0.64T	72	7	335	35	172	53	B
21.	25.XI. 1953	243	82	0.34S	0.94N	352	21	0.92D	0.39N	85	48	330	20	225	35	B
22.	26.XI. 1953	188	70	0.82S	0.57N	292	58	0.93D	0.37N	57	37	253	50	155	10	B
23.	23.XI. 1955	121	35	0.66S	0.75T	348	45	0.41D	0.91T	330	16	67	22	207	62	W.H.
		156	49	0.02D	1.00T	334	41	0.03S	1.00T	155	4	245	1	349	86	W.H.
		135	58	0.32S	0.34T	347	36	0.45D	0.87T	148	10	55	15	272	71	B
24.	29.IX. 1956	58	64	0.32S	0.95T	276	32	0.55D	0.83T	72	17	337	17	204	66	W.H.
		74	56	0.05S	1.00T	259	34	1.00D	1.00T	76	17	346	2	245	79	W.H.
		71	56	0.22S	0.98T	279	27	0.44D	0.90T	80	20	346	12	227	67	B
		91	55	0.05D	1.00T	266	35	0.07S	1.00T	89	10	179	2	282	80	B
		73	53	0.26D	0.97T	278	38	0.36D	0.92T	84	7	350	10	218	76	E
25.	11.X. 1956	241	26	0.88S	0.47N	357	78	0.40D	0.92N	150	52	272	23	15	29	W.H.
		240	30	0.99S	0.09N	335	86	0.50D	0.87N	128	40	247	31	10	34	B
26.	6.XI. 1958	118	45	0.17S	0.98T	314	45	0.17D	0.98T	305	0	36	7	215	82	B
27.	12.XI. 1958	117	53	0.57S	0.82T	347	49	0.59D	0.81T	141	2	50	28	233	65	B
28.	22. I. 1959	134	87	0.17S	0.99T	27	10	0.96D	0.30T	143	41	44	10	304	47	W.H.

(continued)

Appendix 2 (continued)

No	Date	Plane I			Plane II component			P axis			B axis			T axis		author
		Az	dip	strike	dip	Az	dip	strike	dip	Az	PL	Az	PL	Az	PL	
1	2	3	4	5	6	7	8	9	10	11	12	13	14	15	16	17
29.	4.V.1959	143	7	0.08S	1.00N	318	83	0.01D	1.00N	137	52	228	1	319	38	W.H.
		135	15	0	1.00N	315	80	0	1.00N	125	50	250	0	310	35	B
		200	15	0.82S	0.57N	315	80	0.24D	0.97N	135	50	230	15	325	35	B
30.	18.VI.1959	350	82	1.00S	0.08T	259	85	0.99D	0.14T	215	2	319	81	124	9	W.H.
		0	77	0.97S	0.26T	267	75	0.97D	0.26T	44	2	310	70	133	20	B
31.	20.III.1960	282	78	0.19S	0.98N	59	16	0.67D	0.75N	116	56	10	11	273	32	W.H.
32.	29.VII.1960	103	72	0.09S	1.00T	299	19	0.27D	0.96T	107	26	14	5	274	63	B
		318	85	0.12S	0.99N	80	10	0.83D	0.56N	145	48	46	8	310	39	
33.	10.I.1961	127	50	0	1.00T	308	40	0	1.00T	128	4	38	0	340	85	B
34.	16.I.1961	199	3	0.86S	0.31N	320	89	0.04D	1.00N	137	46	230	2	322	44	W.H.
		137	86	0.02S	1.00T	340	2	0.39D	0.92T	137	43	47	1	316	47	W.H.
35.	19.I.1961	127	50	0	1.00T	308	40	0	1.00T	128	4	38	0	340	85	B
36.	12.II.1961	135	76	0.11S	0.99T	339	16	0.40D	0.92T	140	30	47	6	306	59	W.H.
		151	77	0.01S	1.00T	335	13	0.06D	1.00T	152	24	61	1	330	58	W.H.
		156	70	0.08S	1.00T	350	21	0.23D	0.77T	160	25	68	4	258	60	W.H.
		137	80	0.09D	1.00T	289	10	0.45D	0.89T	132	36	226	4	328	65	W.H.
		137	80	0.28S	0.96T	13	20	0.82D	0.57T	150	33	50	16	299	51	B
37.	23.II.1961	173	60	0.45S	0.89T	307	40	0.60D	0.80T	154	5	248	23	41	64	P
38.	26.II.1961	238	51	0.84S	0.54T	126	65	0.72D	0.70T	95	8	192	40	355	48	W.B.
		60	56	0.29D	0.95T	210	36	0.41S	0.91T	47	10	140	14	285	72	B
39.	23.IV.1961	5	50	0.54D	0.84T	140	50	0.53S	0.85T	163	0	72	24	250	67	B
40.	18.VII.1961	294	53	0.17S	0.96N	98	38	0.23D	0.97N	154	79	18	8	287	8	W.H.
		202	45	0.10S	1.00N	30	30	0.10S	1.00N	299	86	116	4	206	0	W.
		179	59	0.09S	1.00N	350	31	0.14D	0.99N	13	75	267	4	176	14	W.H.
		200	45	0.13D	0.99N	31	45	0.13S	0.93N	295	85	115	5	25	0	W.H.
		292	65	0.17S	0.98T	90	28	0.37D	0.93N	136	70	118	9	284	20	P
41.	11.VIII.1961	80	59	0.67S	0.74T	320	51	0.75D	0.67T	108	5	15	35	205	54	W.H.
		111	54	0.30S	0.95T	319	40	0.37D	0.93T	123	7	31	14	239	74	W.H.
		109	66	0.18S	0.98T	314	26	0.39D	0.92T	117	21	23	16	245	67	W.H.
		103	60	0.32S	0.95T	317	34	0.49D	0.87T	117	13	23	16	245	69	W.H.
		113	66	0.31S	0.95T	332	30	0.57D	0.82T	127	18	31	16	265	65	B

1	2	3	4	5	6	7	8	9	10	11	12	13	14	15	16	17
42.	19.VIII.1961	87	61	0.19S	0.98T	287	31	0.33D	0.95T	95	15	2	10	240	72	W.H.
		94	41	0.21D	0.98T	258	50	0.18S	0.98T	265	4	174	8	24	81	W.H.
		108	50	0.29D	0.96T	263	44	0.31S	0.95T	96	0	187	11	354	79	B
		71	40	0.60S	0.80T	294	57	0.44D	0.90T	276	9	11	21	163	69	B
43.	15.XI.1961	121	60	0.52S	0.86T	350	40	0.63D	0.78T	142	10	49	25	154	62	B
44.	4. I. 1962	207	44	0.41S	0.91N	353	51	0.37D	0.93N	110	74	279	17	10	3	B
45.	12.IV. 1962	253	10	0.86S	0.51T	132	85	0.15D	0.99T	124	39	221	9	322	50	W.H.
		282	22	0.72S	0.70T	149	75	0.28D	0.96T	137	28	235	15	350	57	W.H.
		351	33	0.78D	0.63T	115	70	0.42S	0.91T	135	21	35	24	264	58	B
46.	23.IV. 1962	254	19	1.00S	0.09N	349	88	0.32D	0.95N	151	44	260	19	6	40	W.H.
		273	27	0.95S	0.32T	166	82	0.43D	0.90T	145	32	252	25	18	47	W.H.
		298	30	0.99S	0.16T	201	86	0.48D	0.87T	177	35	288	30	48	42	B
47.	7.V. 1962	286	51	0.51S	0.86T	150	48	0.54D	0.84T	308	2	217	24	42	66	W.H.
		299	42	0.10S	0.99T	127	49	0.09D	1.00T	123	4	214	4	351	85	W.H.
		346	41	0.75D	0.66T	109	65	0.56S	0.83T	133	13	36	30	247	58	B

5. Features of Tsunamigenic Earthquakes

W. M. ADAMS and A. S. FURUMOTO
Hawaii Institute of Geophysics
Honolulu, Hawaii
Contribution No. 296

ABSTRACT

　　Some seismological features of tsunamigenic earthquakes are studied, and the state-of-the-art for determining focal mechanisms is reviewed. The objective is to provide a basis for optimizing research on possible improvements for the tsunami warning system. The fact that only a single strike-slip earthquake is known to have generated a large tsunami indicates the potential value of additional real-time seismic data. Water-level gauges exist which could provide quantitative data in the source region, as opposed to "off-scale" reports. The prospects of using a satellite-based communication network are obvious. Hence, there are several possible ways for improving the tsunami warning system, so a decision on the optimal sequence of development must be made. The utilities required by statistical decision theory have not yet been estimated.

INTRODUCTION

　　The pragmatic value of any tsunami research is the possibility of using the knowledge gained to improve a warning system. The purpose of this paper is to review the seismological aspects of tsunamigenic earthquakes to ascertain if any of the recent findings in seismological research -- which might be used to improve a system -- have not been fully utilized.

FAULTS AND EARTHQUAKES

In general, seismologists in the Western Hemisphere have remained adamant for more than fifty years that faulting is the cause of earthquakes. This is surprising -- in view of the notable lack of direct evidence. In *A History of British Earthquakes*, C. Davison (1924) displays commendable intellectual honesty when he explicitly states, "The methods of investigation which I have used in studying British earthquakes are based on the theory that earthquakes are the results of successive steps in the growth of faults. No recorded British earthquake has ever been accompanied by perceptible crust-displacements at the surface." (He does, however, in discussing the Colchester earthquake of 1884, mention that, "In Mersea Island, two small fissures, or rather cracks, were formed by the earthquake." Presumably this was discounted as not being "crust-displacement".)

The need for such intense faith was considerably diminished by the study of Tocher (1958) on magnitude versus surface rupture length. However, another anomaly appeared: Tocher showed that rupture was characteristic for California earthquakes of magnitude greater than 6.5.

Most interesting is the exponential relation that he found (Fig. 1) instead of a bounding function. That is, a given magnitude produces a specific rupture length, not simply all lengths less than some maximum. Tocher explained this by restricting the strain accumulation to a layer. This assumption provides a 2-D model. It is not, however, obviously compatible with the distribution of foci from the surface down to 700 kilometers. Furthermore, the length was linearly related to earthquake energy, not by a second power. Hence it can be argued that the conditions horizontally perpendicular to the fault are the same as vertically!

Much emphasis is given to the rupture versus magnitude studies because, as Davison (1921, p. 91) noted, "...it seems probable that the proportion of submarine earthquakes accompanied by sea-waves does not differ greatly from the proportion of inland earthquakes that are accompanied by the formation of fault-scarps."

EARTHQUAKES AND TSUNAMIS

As the area of interest for tsunami research is the ocean, the relationship between earthquake magnitude and tsunami

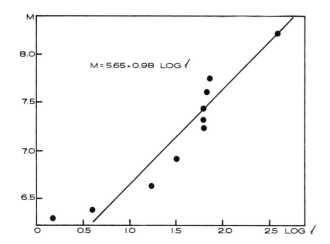

Fig. 1: Relationship Between Earthquake Magnitude M and Fault Length l (after Tocher, 1958).

magnitude must be considered. Iida (1963) presented such data and fit a straight line, obtaining m = 2.61M - 18.44.

The data and the best-fitted curve are given in Figure 2. From both the physical situation and the data, an upper bounding function would seem more useful. A possible upper bound was developed during another study (Adams, 1969).

The choice between these alternatives is not facilitated by the work of Cox (1968, p. 42) relating earthquake magnitude to maximum tsunami run-up in Hawaii and shown here as Figure 3. This might be interpreted as support for the hunch expressed by Davison (1921, p. 99) that "Seismic sea waves are associated as a rule only with earthquakes of great intensity." The graphs of both Iida (1963) and Cox (1968) show that for large earthquakes the earthquake-tsunami relationship is binary -- either a very large tsunami is generated or none at all. This is similar to the results of Hatori (1969), who finds an exponential relation, as shown in Fig. 4, not an upper bounding function.

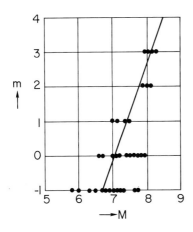

Fig. 2:
Relationship Between Tsunami Magnitude m and Earthquake Magnitude M (after Iida, 1963).

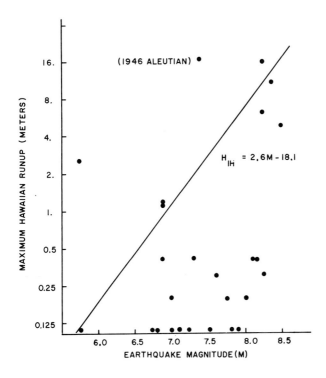

Fig. 3: Relationship Between Maximum Hawaiian Run-Up and Earthquake Magnitude (after Cox, 1968).

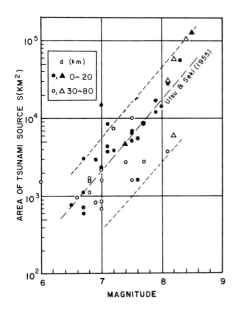

Fig. 4:
Relationship Between Area of Tsunami Source and Earthquake Magnitude (after Hatori, 1969). [o - Japan; Δ - other countries]

WATER DEPTH AND TSUNAMIS

According to the theory for tsunami generation provided by Keller (1963), 20-minute tsunami waves generated in a 12,500-foot deep ocean are effectively independent of the depth in the ocean at which the displacement is created. This is explained qualitatively in Adams (1969). A positive correlation of tsunami amplitude with water depth might occur for observations at a *fixed* depth, however. This can be explained theoretically as amplification during propagation by shoaling, according to Green's Law. This would only amount to a gain of about 1.2.

Iida (1963) has found a very strong dependence of the tsunami magnitude on the depth of water at the epicenter. This is an upper-bound relationship, as shown in Fig. 5, for tsunamis observed in Japan due to nearby earthquakes.

If the tsunamigenic earthquake is offshore from a coastline, as usually is the case, and the point-of-observation is

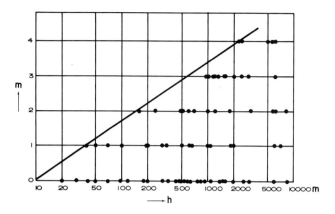

Fig. 5: Relationship Between Tsunami Magnitude M and Water Depth h (after Iida, 1963).

at a relatively great distance, then enhancement by reflection from the coastline can cause an amplification approaching 2. Since the depth of water usually increases monotonically with distance from shore, the reflection effect better explains high positive correlation of tsunami magnitude with water depth, but only for tele-tsunamis; it does not explain the findings of Iida.

FOCAL MECHANISM AND SEA-FLOOR SPREADING

Our understanding of tsunamigenic earthquakes would benefit substantially from a coherent world-model for earthquakes. Such a model involving sea-floor spreading has been compared with the results of focal mechanism studies by Isacks, et al. (1968). The data for selected areas were presented as evidence that the stress pattern is as shown in Figure 6. However, other areas are not so obliging, as noted by Ichikawa, (1969, p. 162) who considers the variation of inclination in the pressure axes for earthquakes near Japan: "...while pressure axes for earthquakes occurring in the crust lie horizontally, those for events in the uppermost part of the mantle dip gently toward the Pacific side and are nearly normal to the tendency of spatial distribution of earthquake foci. Furthermore, the pressure axes of deep earthquakes are approximately parallel to the tendency spatial distribution of earthquake foci." This is evidence against the proposed model. For despite the convenience of focal mechanism solutions in having two nodal planes and either of two types of force systems, as shown in Fig. 6, the deep and the intermediate earthquake mechanisms should at least be the same, relatively speaking. Some attempt was made by Isacks, et al. (1968) to explain such incompatibilities. The crustal plate was considered to separate under the effects of gravitational sinking. But the physics by which a relatively low density crustal section would "sink" downwards through a relatively high density mantle was not developed or referenced.

Alternate models can probably explain many of the phenomena, as noted by G. V. Keller (personal communication) who has proposed the model of a Poisson distribution of fissure eruptions from the mid-ocean ridges. The probability of a widespread flow is greater the longer the time interval.

Not all the difficulties lie in the model-building. A review of the irregularities in the earthquake mechanism solution was made by Hodgson and Adams (1958) and, more recently, by Stevens (1969) who finds only 38 earthquakes to have adequate data and unambiguous choice between force types 1 and 2. She says (p. 59), somewhat tautologically, "The most important

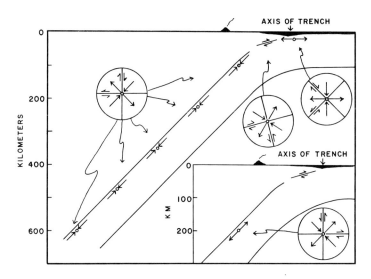

Fig. 6: Stress Patterns Consistent with Sea-Floor Spreading (after Isacks, et al., 1968).

task in earthquake mechanism studies at present is the application of seismic data to an accurate determination of the mechanism type in each seismic region," which agrees with the opinion of Miyamura (1969, p. 23), "Seismicity study to date has been based only on focal position and magnitude; focal mechanism should be the next important element."

Despite the present plight of focal mechanism studies, a choice between strike-slip and dip-slip can usually be made. A possible approach has been suggested by Furumoto (1969) by using ionospheric readings. Studies, including those by Iida (1969) and Watanabe (1969), substantiate the hypothesis that: Very few large tsunamis have been generated by dominantly strike-slip faults.

IMPROVEMENT OF THE PACIFIC TSUNAMI WARNING SYSTEM (PTWS)

There are four ways to change the PTWS. One may modify

1) the seismic data gathering,
2) the water data gathering,
3) the communication network,
4) the decision process

Plans for improving the communication network for the PTWS by the use of satellites were developed several years ago, and field tests by E.S.S.A. have already demonstrated the feasibility. Such a change in the communication network will be of greater value and should be instituted as soon as possible without waiting for the development of deep-sea buoy systems. The existing circum-Pacific seismic and tide stations can provide input data. Indeed, this approach of using existing sensors has the advantage of only requiring debugging of the communication network. If sensors and communication network are both modified, debugging is extraordinarily difficult.

The process of decision-making necessarily depends on the field data selected for sensing, and conversely.

If the utility of an early "NO GO" decision is high, then the capability of estimating rudimentary features of the focal mechanism from seismic data should be added to the system. This is technically feasible at present, but requires notable developments. This would involve modifying the seismic data gathering. Alternatively, high utility may be assigned to the accurate prediction of tsunami inundation, a task not presently within the capability of the PTWS. Estimating tsunami inundation accurately from current real-time seismic data has been shown to be impossible by Adams (1969). However, that study, in conjunction with an earlier one which presented a decision-making chart, makes clear that prediction of tsunami inundation could be sufficiently accurate if based on *quantitative* data of water-level fluctuation. Such quantitative data is desirable near the source area. Unfortunately, the water-level fluctuation in the source area for all tsunamis of interest to Hawaii will be beyond the range of most existing sensors. But fortunately, an instrument capable of recording and surviving large tsunami waves has been constructed by the Scientific Research Institute of Hydrometeorological Instrument Building (Shenderovich, 1961). The instrument is the GM23-II -- hydrostatic, battery-operated, 10-meter range, tsunami-switched, and fastened on a concrete base capable of withstanding the dynamic pressure of catastrophic tsunami waves. For coasts subject to waves greater than 10 meters, the recorders are located at about 8 meter intervals in height up the shore. Telemetry of the data would be desirable for purposes of the PTWS.

DISCUSSION

Just as a sequence of decisions must be made by the PTWS after a tsunamigenic earthquake, so must any improvement program make decisions concerning the sequence of research. These can be made in a variety of ways, from flipping-a-coin to statistical decision theory (Raiffa and Schlaifer, 1961). The latter approach of statistical decision theory has the advantage of making explicitly the assumptions and value judgments employed for decision-making. In applying this approach to improving a warning system for a natural hazard, there is a distinct possibility that too much time will be spent in gathering facts and improving estimates of utility or disutility. Hence hunch is still valuable for providing first approximations to the estimated utilities. Excessive delay can cost lives.

The time required for developments must not be ignored. The desired optimization is over a sequence of interrelated developments: such optimization comes within the scope of dynamic programming, as conceived and developed by Bellman (1961).

As a simple illustration of the temporal aspect in any improvement program, compare two possible alternative causes of action. Assume resources are fixed. Let one course of action be the concentration on improving one feature, such as the communication network, one measurable increment: let the other course of action be simultaneous improvement of all three features -- seismic sensors, water sensors, and communication network. The concentrated effort will increment the system capabilities sooner.

In a political environment where funding may be related to demonstrable success, the problem is more complex because the resources become variable. This aspect of the research improvement program is even analyzable, for example by the methods propounded by Forrester (1968) and Roberts (1964).

CONCLUSION

Our actual understanding of the physics of tsunamigenic earthquakes has increased only slightly over the past forty years, but methods of analyzing focal mechanisms have been developed which need but to be applied extensively. Already, qualitative findings can guide decisions for improving the PTWS.

The greatest immediate need is estimations of the utilities associated with improvements of the various features of the system.

BIBLIOGRAPHY

Adams, William Mansfield 1967. Analysis of a tsunami warning system as a decision-making process. *Proceedings of Symposium on Tsunami and Storm Surges*. The Eleventh Pacific Science Congress, Tokyo, 1966, pp. 1-22.

Adams, William Mansfield 1969. Prediction of tsunami inundation from current real-time seismic data. Hawaii Institute of Geophysics, *HIG-69-9*. 57 pages.

Davison, C. 1924. *A History of British Earthquakes*. Cambridge University Press, 416 pages.

Davison, C. 1921. A manual of seismology. Cambridge University Press, 256 pages.

Forrester, J. W. 1968. *Principles of Systems*. Cambridge, Mass.

Bellman, R. 1961. *Adaptive Control Processes; a Guided Tour*. Princeton University Press, 255 pages.

Cox, Doak C. 1968. Performance of the seismic sea wave warning system, 1948-1967. Hawaii Institute of Geophysics, *HIG-68-2*. 69 pages + 10 pages of appendices.

Furumoto, A. S. 1969. Ionospheric recording of Rayleigh waves for estimating source mechanisms. *Proceedings of International Symposium on Tsunami and Tsunami Research* jointly sponsored by Int'l. Union of Geod. and Geoph. and East-West Center, Univ. of Hawaii.

Hatori, Tokutaro 1969. Dimensions and geographic distribution of tsunami sources near Japan. *Bull. Earthquakes Research Institute*. Vol. 47, pp. 185-214.

Hodgson, J. H., Adams, W. M. 1958. A study of the inconsistent observations in the fault-plane project. *Bull. Seism. Soc. Am.* Vol. 48, No. 1, pp. 17-31.

Ichikawa, Masaji 1969. Mechanism of earthquakes in and near Japan and related problems. *The Earth's Crust and Upper Mantle*. Edited by P. J. Hart, Geophysical Monograph 13 Amer. Geoph. Union, Washington, D. C., pp. 160-165.

Iida, K. 1969. Generation of tsunamis, focal mechanism of earthquakes, and their sources. *Proceedings of International Symposium on Tsunami and Tsunami Research* jointly sponsored by Int'l. Union of Geod. and Geoph. and East-West Center, Univ. of Hawaii.

Iida, K. 1963. Magnitude, energy, and generation mechanisms of tsunamis and a catalogue of earthquakes associated with tsunamis. *Proceedings of the Tsunami Meetings Associated with the Tenth Pacific Science Congress*. Monograph 24, Int'l. Union of Geod. and Geoph., edited by D. C. Cox and E. McAfee, pp. 7-18.

Isacks, Brian; Oliver, Jack; and Sykes, Lynn R. 1968. Seismology and the new global tectonics. *Journ. Geoph. Research*. Vol. 73, No. 18, pp. 5855-5899.

Keller, J. B. 1963. Tsunamis -- water waves produced by earthquakes. *Proceedings of the Tsunami Meetings Associated with the Tenth Pacific Science Congress*. Monograph 24, Int'l. Union of Geod. and Geoph., edited by D. C. Cox and E. McAfee, pp. 154-166.

Miyamura, S. 1969. Seismicity of the earth. *The Earth's Crust and Upper Mantle*. Edited by P. J. Hart, Geophysical Monograph 13, Amer. Geoph. Union, Washington, D. C., pp. 115-124.

Raiffa, H., Schlaifer, R. 1961. *Applied Statistical Decision Theory*. Harvard University, 356 pages.

Roberts, E. B. 1964. *The Dynamics of Research and Development*. Harper and Row, New York, 352 pages.

Shenderovich, I. M., 1961. Tsunami recorders built by the Scientific Research Institute of Hydrometeorological Instrument Building. *Bull. of the Council of Seismology*. Academy of Sciences, U.S.S.R., Problems of Tsunamis, Moscow, No. 9, pp. 67-73. Translation available as JPRS 31, 019.

Stevens, Anne E., 1969. Worldwide earthquake mechanism. *The Earth's Crust and Upper Mantle*. Edited by P. J. Hart, Geophysical Monograph 13, Amer. Geoph. Union, Washington, D. C., pp. 153-160.

Tocher, Don. 1958. Earthquake energy and ground breakage. *Bull. Seism. Soc. Am.* Vol. 48, pp. 147-153.

Watanabe, H. 1969. Statistical studies of tsunami sources and tsunamigenic earthquakes occurring in and near Japan. *Proceedings of International Symposium on Tsunami and Tsunami Research* jointly sponsored by Int'l. Union of Geod. and Geoph. and East-West Center, Univ. of Hawaii.

6. Dimensions and Geographic Distribution of Tsunami Sources near Japan

T. HATORI
Earthquake Research Institute
Tokyo, Japan

ABSTRACT

The source areas of many small tsunamis which were generated in the vicinity of Japan during the period from 1959 to 1962 and those of the Tokachi-oki and the Hiuganada tsunamis of 1968 are estimated from inverse refraction diagrams. Adding those new data, the distribution of tsunami source areas during the last 76 years (1893-1968) is shown on a bathymetric chart. The geographic characteristic and the dimension of the tsunami sources are investigated. The features of tsunami sources for northeastern Japan are different from those of southwestern Japan. In northeastern Japan, the sources of small tsunamis are mostly located in the sea shallower than the depth of 2000 m. and the sources of large tsunamis lie on the steep continental slope near the trench. On the other hand, in southwestern Japan, the sources of large tsunamis lie near the coast. The major axes of source areas are parallel to the island arcs. The dimension of the source area of a tsunami generated by a deep earthquake is considerably smaller than that of a shallow earthquake. Moreover, if the tsunamis generated outside Japan are included, the dimension of the tsunami source seems to be different for different seismic regions.

INTRODUCTION

Large tsunamis which occurred in the sea adjacent to Japan have been investigated and the source areas of tsunami were estimated. Geographic distribution of the estimated source areas of tsunami was shown in a previous paper (Hatori, 1966). The source areas of many small tsunamis, however, were omitted from this map.

During the period from 1959 to 1962, thirteen small tsunamis were generated by earthquakes having a magnitude of the order of 6.8 in the vicinity of Japan. Two tsunamis were also generated by aftershocks of the 1968 Tokachi-oki earthquake. Making use of tide gauge records, source areas of these tsunamis were estimated. Including the tsunami data of the 1968 Hiuganada (Kajiura, et al., 1968), the 1968 Tokachi-oki earthquakes (Kajiura, et al., 1968) and others (Hatori, 1969), all the estimated source areas of tsunami during the 76-year period from 1893 to 1968 are shown on a bathymetric chart to indicate the geographic distribution of the source areas in relation to dimensions of these areas. Iida (1958) indicated a close relationship between the linear dimension of tsunami source and earthquake magnitude. Adding data of small tsunamis, tsunamis generated in other countries and land earthquakes accompanied by rupture, this relationship is re-examined with respect to the regional difference and the dependence on focal depth of the main shock.

ESTIMATION OF TSUNAMI SOURCE AREA

The source areas of tsunamis were estimated by means of an inverse refraction diagram based on the arrival time of the tsunami front at several stations and wave fronts were drawn at two minute intervals. Tide-gauge records of tsunamis at 26 stations were provided courtesy of the Japan Meteorological Agency (JMA) and other authorities, in addition to tsunami records obtained at Miyagi-Enoshima Tsunami Observatory belonging to the Earthquake Research Institute.

Tsunamis During the Period 1959 to 1962

During these four years, thirteen small tsunamis were generated in the sea adjacent to Japan. As shown in Table 1, three epicenters of earthquakes are located off Hokkaido, 8 off Iwate-Ibaraki and two off Kyushu. Details of these tsunamis are reported in the previous paper of the same title (Hatori, 1969). The principal features of tsunamis are as follows: Double amplitude is of the order of 20-30 cm. and the period is about 10 minutes. The initial motion of the tsunami front is mostly an up-wave, although the tsunami on July 18, 1961, generated off the south of Kyushu, began with a downward motion at all stations.

Figure 1 shows the estimated tsunami sources and distribution of aftershocks. For most of the small tsunamis, the length of the major axis of source is about 40 km. and the source region corresponds to the area of aftershock activity. Among these tsunamis, the tsunami on March 21, 1960 is the largest. Magnitude of the earthquake is 7.5 and the length of the tsunami source is 120 kilometers. A small tsunami was also generated by the aftershock (M = 6.7) on March 23. Off Ibaraki prefecture, two small tsunamis were generated by the main shock (M = 6.8) and its aftershock (M = 6.5) on January 16, 1961. As shown in Fig. 1, the sources of tsunamis generated by the aftershocks are located at a corner of the source area of the main tsunami. It is seen that the source areas of small tsunamis in the northeastern part of Japan are mostly located in shallow sea inside the bottom contour of 2000 meters.

Tsunamis Generated by the 1968 Tokachi-oki
Earthquake and Aftershocks

After the 1968 Tokachi-oki tsunami (tsunami magnitude, m = 2) was generated by the main shock (M = 7.9) at 9 h. 49 m. (JST) on May 16, 1968, two tsunamis were again generated by the aftershocks at 19 h. 39 m. (JST) on the same day and at 22 h. 42 m. on June 12 having magnitudes 7.5 and 7.2 respectively. Figure 2 shows the estimated source areas of tsunamis generated by the main shock and two aftershocks.

As reported in a previous paper (Kajiura et al., 1968), the source area of the tsunami corresponding to the main shock extends 230 km. in a southeast-northwest direction along the bathymetric line. Judging from the initial motion of the records, the subsidence of the bottom seems to have occurred in the northwestern part of the tsunami source, and the upheaval in the southeastern part. The second tsunami was generated about ten hours after the occurrence of the main tsunami. Therefore, waves of the second tsunami were mixed with free oscillations of shelf water due to the main tsunami. Although the generating area cannot be determined accurately, it seems to be located at the northern part of the main source area.

For the tsunami on June 12, the maximum double amplitude of 156 cm. with short period was observed at Shimanokoshi. The estimated source area lies to the south of the main source with the major axis of source directed north-south, extending

Table 1: List of Tsunamis In and Near Japan, 1893-1968.

Date (JST)	Epicenter Lat. N (°)	Epicenter Long. E (°)	Location	Depth (km)	M	m	L (km)	S x10^{3}km^2
1893 June 4	43.4	147.5	NE. Hokkaido		6.6	1		
1894 Mar. 22	42.5	145.1	"		7.9	2		
1896 Jan. 9	36	141	Ibaraki		7.3	0		
1896 June 15	39.6	144.2	Sanriku**		7.6	3		
1897 Feb. 20	38.1	141.5	Miyagi		7.8	0		
1897 Aug. 5	38.0	143.7	Sanriku		7.7	1		
1898 Apr. 23	39.5	143.6	"		7.8	-1		
1899 Nov. 25	31.9	131.4	Hiuganada		7.1?	0		
1901 June 24	28.3	129.3	Amami-oshima		7.9	0		
1901 Aug. 9	40.5	141.5	Hachinohe	Shallow	7.7	0		
1901 Aug. 10	40.5	141.5	"	"	7.8	0		
1909 Nov. 10	32.1	133.1	Hiuganada	"	7.4?	?		
1912 June 8	39.3	143.3	Iwate	"		0?		
1915 Nov. 1	38.9	143.1	Sanriku	"	7.5	1		
1923 June 2	36.0	141.4	Ibaraki	"	7.2	-1		
1923 Sept. 1	35.3	139.3	Kanto	"	7.9	2	170	12
1923 Sept. 2	35.1	140.4	Katsuura	"	7.4	0		
1927 Mar. 7	35.6	135.1	Tango**	10	7.5	0		
1927 Aug. 6	38.0	142.0	Miyagi	20	6.9	-1		
1927 Aug. 19	34.2	141.2	Boso			-1		
1928 May 27	40.0	143.2	Iwate	0-10	7.0		60	2.3
1931 Mar. 9	41.2	142.5	E. Aomori	0	7.6	-1	50	1.6?
1931 Nov. 2	32.2	132.1	Hiuganada**	20	6.6	-1		
1933 Mar. 3	39.1	144.7	Sanriku	0-20	8.3	3	500	57
1933 June 19	38.1	142.4	Miyagi	20	7.1	-1	70	3.8
1935 July 19	36.7	141.3	Ibaraki	0	6.5	-1		
1935 Oct. 13	40.0	143.6	Sanriku	40	7.2	-1	100	7.4
1935 Oct. 18	40.3	144.2	"	20-40	7.1	-1		
1936 Nov. 3	38.2	142.2	Miyagi	50-60	7.7	-1	80	2.8
1938 May 23	36.7	141.4	Ibaraki	10	7.1	-1	90	4.8
1938 June 10	25.3	125.2	Miyakojima	30-60	6.7	0		
1938 Nov. 5	37.1	141.7	Fukushima	20	7.7	0	120	8.8
1938 Nov. 5	37.2	141.7	"	15	7.6	0	100	5.5
1938 Nov. 6	37.5	141.8	"	0	7.5	0	100	7.0
1938 Nov. 7	37.0	141.7	"	0	7.1	0	120	8.5
1938 Nov. 14	37.0	141.5	"	60	6.0	-1	50	1.6
1938 Nov. 22	37.0	141.8	"	10	6.7	-1	60	3
1938 Nov. 30	37.0	141.8	"	5	7.0	-1		
1939 Mar. 20	32.3	131.7	Hiuganada	10	6.6	-1		
1939 May 1	40.0	139.8	Oga	0	7.0	-1	30	0.6
1940 Aug. 2	44.1	139.5	W. Hokkaido	0-20	7.0	2	170	9.4

Date (JST)	Earthquake					Tsunami*		
	Epicenter		Location	Depth	M	m	L	S
	Lat. N	Long. E		(km)			(km)	x10³km²
1941 Nov. 19	32.6	132.1	Hiuganada**	0-20	7.4	0		
1943 June 13	41.1	142.7	Sanriku	20	7.1	-1		
1944 Dec. 7	33.7	136.2	Tonankai	0-30	8.0	2	180	14
1945 Jan. 13	34.7	137.0	Mikawa	0	7.1	-1	25	0.4
1945 Feb. 10	40.9	142.1	Aomori	30	7.3	-1		
1946 Dec. 21	33.0	135.6	Nankaido	30	8.1	2	300	28
1947 Nov. 4	43.8	141.0	W. Hokkaido	0-30	7.0	0	60	1.9
1948 Apr. 18	33.1	135.6	Shionomisaki	40	7.2	-1		
1952 Mar. 4	42.2	143.9	Tokachi	45	8.1	2	90	3.7
1952 Mar. 10	41.7	143.5	SE. Hokkaido	0-20	7.0	-1		
1953 Nov. 26	34.3	141.8	Boso	40-60	7.5	1	160	10
1956 Mar. 6	44.3	144.1	NE. Hokkaido	0-20	5.8	-1		
1956 Aug. 13	33.8	138.8	Izu	40-60	6.5	-1		
1959 Jan. 22	37.6	142.4	Fukushima	30	6.8	-1	45	1.1
1959 Oct. 26	37.6	143.2	"	20	6.7	-1	35	0.7
1960 Mar. 21	39.8	143.5	Sanriku	20	7.5	-1	120	7
1960 Mar. 23	39.3	143.8	"	20	6.7	-1	45	1.1
1960 July 30	40.2	142.6	Iwate**	30	6.7	-1		
1961 Jan. 16	36.0	142.3	Ibaraki	40	6.8	-1	55	1.7
1961 Jan. 16	36.2	142.0	"	20	6.5	-1	35	0.8
1961 Feb. 27	31.6	131.8	Hiuganada	40	7.0	0	90	2.1
1961 July 18	29.6	131.8	S. Kyushu	60	6.6	-1	40	0.9
1961 Aug. 12	42.9	145.6	SE. Hokkaido	80	7(Pas)	-1	35	0.7
1961 Nov. 15	42.7	145.6	"	60	6.9	-1	40	0.8
1962 Apr. 12	38.0	142.8	Miyagi	40	6.8	-1	55	1.7
1962 Apr. 23	42.2	143.9	SE. Hokkaido	60	7.0	-1	40	0.9
1964 May 7	40.3	139.0	W. Aomori	0	6.9	-1	70	2.7
1964 June 16	38.4	139.2	Niigata	40	7.5	2	90	2.8
1968 Jan. 29	43.2	147.0	SE. Hokkaido	30	6.9	-1		
1968 Apr. 1	32.3	132.5	Hiuganada	30	7.5	1	60	1.6
1968 May 16	40.7	143.6	Sanriku	0	7.9	2	230	17
1968 May 16	41.4	142.9	"	40	7.5	0	100	5.1
1968 June 12	39.4	143.1	"	0	7.2	0	90	3.9

* m: Tsunami magnitude is defined by Imamura and Iida, classified by tsunami height as follows: m= -1; 0.5m or less. m= 0; 1m. m= 1; 2 m. m= 2; 4-6 m. m=3; 10-20 m. m= 4; 30 m or more.
L: Length of the major axis of source area.
S: Area of a tsunami source.
** Dimension of the source area is uncertain but the decided part of source is shown in Fig. 4.

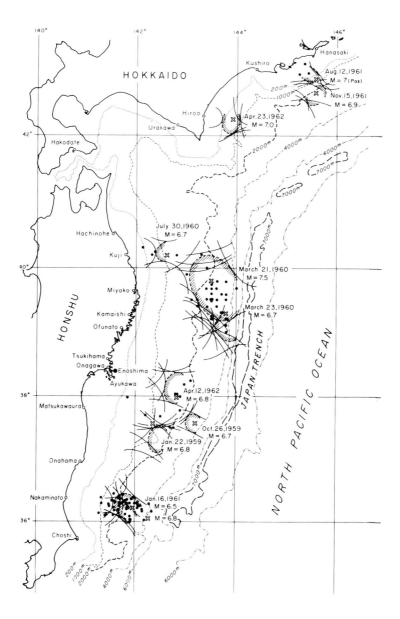

Fig. 1: Geographic Distribution of Estimated Source Areas of Tsunamis in Northeast Japan during the Period 1959 to 1962 and Aftershocks.

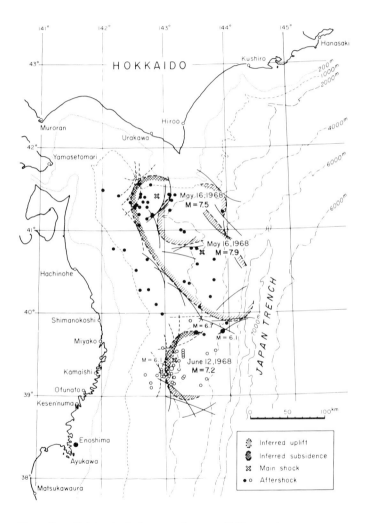

Fig. 2: Estimated Source Areas of Tsunamis Generated by the Tokachi-oki Earthquakes and Aftershocks.

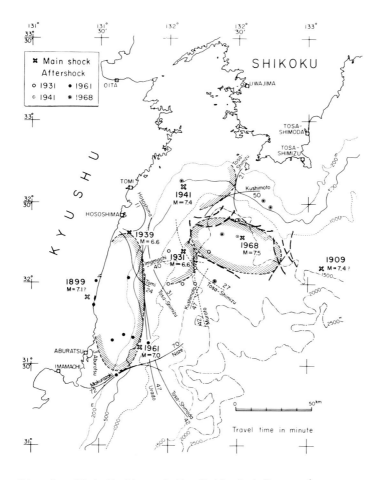

Fig. 3: Distribution of the Estimated Source Areas of the Hiuganada Tsunamis and Aftershocks.

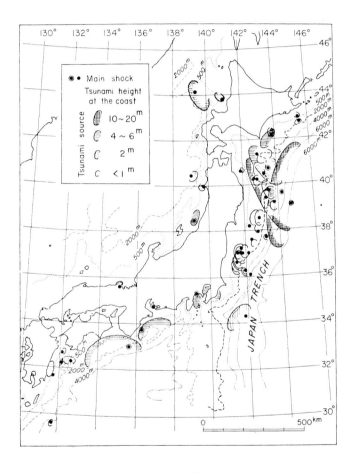

Fig 4: Geographic Distribution of the Estimated Source Areas of Tsunamis in the Sea Adjacent to Japan during the last 76 Years (1893-1968), Classified by the Tsunami Height at the Coast.

90 kilometers. Judging from the records, the initial motion of the tsunami is downwards to the north of Kamaishi and upwards in southern stations. Therefore, the motion of the bottom at the source area is considered exceedingly complex.

Making use of the seismological bulletin (JMA), the epicenters of the main shock and aftershocks are plotted in Fig. 2, where closed circles are those 24 hours after the main shock on May 16 and open circles are those 24 hours after the earthquake of June 12. The magnitude range of aftershocks is $4.4 < M < 5.9$, except the magnitude shown in Figure 2. The estimated source areas about coincide with the aftershock areas.

Tsunamis at the Hiuganada Region

The east coast of Kyushu (Hiuganada) is a region with a high level of seismic activity. Magnitudes of tsunamigenic earthquakes which occurred in this region were 7.5 or less. Several moderate tsunamis were observed at the tide stations in southwestern Japan. The estimated source areas of tsunamis generated since 1900 are shown in Figure 3. Double amplitudes of the 1961 tsunami, as well as that of the 1941 tsunami, are about 1 m. and that of the 1968 tsunami 2.4 m., the largest of all. The tsunami source of 1961 is the sea on the continental shelf, extending about 90 km. in an elongated shape parallel to the coast line directed north-south. In contrast to the source of 1968 tsunami, the major axis of the source is 60 km. in an east-west direction.

GEOGRAPHIC DISTRIBUTION OF TSUNAMI SOURCES

On the basis of the tsunami catalogue published by Iida, et al. (1967) and the seismological bulletin (JMA), the dimensions of the source areas of tsunamis generated in the vicinity of Japan during the period from 1893 to 1968 are shown in Table 1. Total numbers of estimated source areas are forty-five, including those newly estimated by the author. This is about two-thirds of the total tsunamis listed in the catalogue. The dimensions of source areas of tsunamis generated in other countries are given in Table 2. All the estimated source areas of tsunamis, classified by those magnitudes, and the epicenters of the related earthquakes are shown on a bathymetric chart (Figure 4). Generally speaking, this geographic distribution of the tsunami sources is similar to that of the estimated epicenters of historical earthquakes accompanied by tsunamis.

Table 2: World Data of the Tsunami Source.

No.	Date (GMT)	Earthquake					Tsunami		
		Epicenter		Location	Depth (km)	M	m	L (km)	S $\times 10^3 km^2$
		Lat.	Long.						
1	1952 Nov. 4	52.3N	161°E	Kamchatka	30-60	$8\frac{1}{4}$	3	600	60
2	1958 Nov. 6	44.2N	148.5E	Iturup	80	8.2	2	130	6
3	1960 May 22	41 S	73.5W	Chile		8.5	3	800	130
4	1963 Oct. 13	44.6N	149.5E	Iturup	60	$8-8\frac{1}{4}$	2	440	31
5	1963 Oct. 20	44.6N	150.5E	"	20-30	$7\frac{1}{4}-7\frac{1}{2}$	2	120	4.7
6	1964 Mar. 28	61.1N	147.7W	Alaska	20-50	8.4	3	700	105

1) S.A. Fedotov(1962), 2) S.A. Fedotov(1962), 3) T. Hatori(1968),
4) S.L. Solov'ev(1965), 5) S.L. Solov'ev(1965), 6) G. Pararas-Carayannis (1967).

As shown in Fig. 4, distribution of the source areas for northeast Japan is different from that for southwest Japan, the areas of small tsunami mostly lying below a bathymetric line of 2000 m. in the shallow sea and those of large tsunamis on the continental slope near the trench. In contrast, in southwest Japan, the areas of large tsunamis lie in and near the coast. In regions of Kyushu and the Japan Sea, the source areas of relatively small tsunamis exist near the coast. It is also seen that the major axes of source areas run almost parallel to the bottom contour. It is most remarkable that those axes change sharply in direction, following the island arc in the neighborhood of regions where two arcs meet, such as Hokkaido to Honshu and Shikoku to Kyushu.

DIMENSIONS OF TSUNAMI SOURCE

As is evident from Fig. 4, large tsunamis usually have large source areas. Figure 5 shows the relation between the length of the major axis of tsunami source and earthquake magnitude where circles are tsunamis which occurred in the vicinity of Japan, triangles are those in other countries, and crosses represent land earthquakes accompanied by crustal deformation (Okada, 1962; Dambara, 1966). Circles and triangles are divided according to the focal depth of the main shock. In cases of the earthquakes of November 14, 1938 (Fukushima, M = 6.0) and August 2, 1940 (W. Hokkaido, M = 7.0), the magnitudes of such earthquakes are abnormal, judging from tsunami

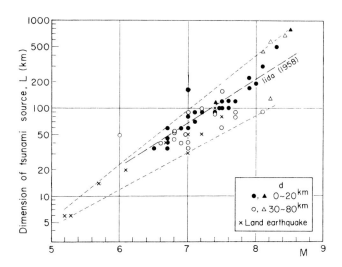

Fig. 5: Relation Between the Dimension of
Tsunami Source and Earthquake Magnitude.

magnitude and the source dimension. According to the United States Coast and Geodetic Survey, Richter magnitudes M_G of the earthquakes of November 14, 1938 and August 2, 1940 were 7.0 and 7.7, respectively. Dimension of the estimated source area for the tsunami on March 9, 1931 (E. Aomori, M = 7.6) is doubtful, judging from the tsunami records. In Fig. 5, the source dimension of tsunami with focal depth shallower than 20 km. (closed circle) is closely represented by the empirical equation given by Iida (1958):

$$\log L = 0.5 - 1.7.$$

It is seen, however, that those of deep earthquakes have a tendency to have shorter axes and those in other countries have longer axes. Mogi (1968) pointed out that the dimension of the aftershock area is considerably longer in Aleutian-Alaska than in Japan.

The estimated source areas of tsunamis are plotted against earthquake magnitude in Figure 6. This relation is markedly scattered. On the other hand, the relation between the after-

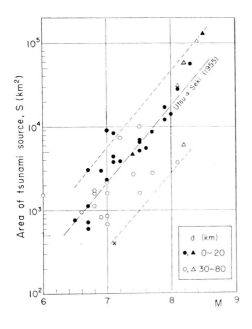

Fig. 6:
Relation Between the Area of Tsunami Source and Earthquake Magnitude.

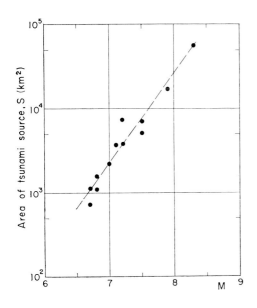

Fig. 7:
Relation Between the Area of Tsunami Source and Earthquake Magnitude, for Tsunamis which were generated off the Sanriku District.

shock areas of submarine earthquake and magnitude was given by Utsu and Seki (1955) as

$$\log A = 0.93\, M - 3.18.$$

In Fig. 6, the tsunami source areas due to shallow earthquakes have a close relation with the aftershock areas. However, for relatively deep (d = 30-80 km.) and large earthquakes, this relation is not so close. The source areas due to deep earthquakes have a tendency to decrease.

As shown in Fig. 7, for the tsunamis generated off Sanriku district by earthquakes with focal depth shallower than 40 km. the relationship between tsunami source area and earthquake magnitude is very well represented by the formula:

$$\log S = 1.07\, M - 4.12,$$

where S is measured in square kilometers. Thus, the dimensions of source area may be estimated by earthquake magnitude. This suggests that the generation mechanism at the limited region is similar but the feature of the tsunami source seems to be different for different seismic regions.

CONCLUSION

With the addition of the newly estimated source areas of tsunamis, the former map of the geographic distribution of the tsunami sources is revised. In northeast Japan, the regional variation of the feature of tsunami source is noticed; small tsunamis are generated in shallower sea and large tsunamis in deeper sea, divided by a bottom line of 2000 meters. In contrast, the tsunami sources in southwest Japan and the Japan Sea lie near the coast. The major axes of tsunami sources are in the direction along the island arc. The dimension of tsunami source seems to vary for different seismic regions and is relatively small for deep earthquakes.

BIBLIOGRAPHY

Dambara, T. 1966. Vertical movements of earth's crust in relation to the Matsushiro Earthquake (in Japanese). *J. Geodetic Soc. Japan*, *12*, pp. 18-45.

Hatori, T. 1966. Vertical displacement in a tsunami source area and the topography of the sea bottom. *Bull. Earthq. Res. Inst.*, *44*, pp. 1449-1464.

Hatori, T. 1969. A study of the wave sources of the Hiuganada Tsunami. *Bull. Earthq. Res. Inst.*, *47*, pp. 55-63.

Hatori, T. 1969. Dimensions and geographic distribution of tsunami sources near Japan. *Bull. Earthq. Res. Inst.*, *47*, pp. 185-214.

Iida, K. 1958. Magnitude and energy of earthquakes accompanied by tsunami and tsunami energy. *J. Earth Sciences*, Nagoya Univ., 6, pp. 101-112.

Iida, K., Cox, D. C., and Pararas-Carayannis, G. 1967. *Preliminary Catalog of Tsunamis Occurring in the Pacific Ocean*. Hawaii Inst. Geoph., Hawaii Univ., Data report No. 5, HIG-67-10.

Kajiura, K., Aida, I., and Hatori, T. 1968. An investigation of the tsunami which accompanied the Hiuganada Earthquake of April 1, 1968 (in Japanese). *Bull. Earthq. Res. Inst.*, *46*, pp. 1149-1168.

Kajiura, K., Hatori, T., Aida, I., and Koyama, M. 1968. A survey of a tsunami accompanying the Tokachi-oki Earthquake of May, 1968 (in Japanese). *Bull. Earthq. Res. Inst.*, *46*, pp. 1369-1396.

Mogi, K. 1968. Development of aftershock areas of great earthquakes. *Bull. Earthq. Res. Inst.*, *46*, pp. 175-203.

Okada, A. 1962. Some investigations on the character of crustal deformation. *Bull. Earthq. Res. Inst.*, *40*, pp. 431-493.

Utsu, T. and Seki, A. 1955. A relation between the area of aftershock region and the energy of main-shock (in Japanese). *Zisin*, [II], *7*, pp. 233-240.

7. The Tsunami Accompanying the Tokachi-oki Earthquake, 1968

Z. SUZUKI
Tohoku University
Sendai, Japan

ABSTRACT

A big earthquake, named the Tokachi-oki Earthquake, 1968 took place off the coast of Aomori Prefecture, northeastern Japan on May 16, 1968. The Japan Meteorological Agency estimated the magnitude to be 7.8 and the location of hypocenter to be at 40.7° N., 143.6° E. and 20 km. deep. The distribution of wave heights of the tsunami accompanying the earthquake is investigated based on the records of tide gauges and the data obtained the field survey along the Sanriku coast. The observed heights are compared with those in cases of the Sanriku Tsunami, 1933; the Tokachi-oki Tsunami, 1952; and the Chile Tsunami, 1960.

The source area of this tsunami is estimated from the records of tide gauges at 14 stations. The shape of the area is approximately an ellipse, of which the longer axis lies in the direction of N.N.W.-S.S.E. and the average dimension is about 200 km., the value being compatible with the empirical law between magnitude of earthquake and areal dimension of source domain. Drawing the refraction diagram from this estimated source, the theoretical distribution of wave heights is calculated. The results fit, in general, the observed data which demonstrates rather uniform height in the northern part and higher wave in the southern part.

Using Green's law, the wave heights at various points in the source area are compiled from the initial motion of tide gauges at many places. The estimated value differs from one point to another and suggests that the tsunami generating motion of the sea bottom is considerably complicated.

The distribution of wave heights within various bays is considered in comparison with those in other cases. When the seiche period of the bay is rather long, the wave heights in the cases of Chile, 1960, and the present tsunamis increase with the distance from the mouth of the bay, while they decrease with increasing distance in the case of Sanriku Tsunami, 1933. Within the bays having short period of seiche, on the other hand, the heights increase with the distance in every case. The effect of anti-tsunami facilities is obviously seen in Ofunato Bay, where a breakwater was constructed after the Chile Tsunami.

The relation between the directions of wave approach and the mouth of a bay is also studied and the result is qualitatively the same as that in the cases of former tsunamis.

INTRODUCTION

A big earthquake, named the Tokachi-oki Earthquake, 1968 took place off the coast of Aomori Prefecture, northeastern Japan on May 16, 1968. The Japan Meteorological Agency estimated the magnitude of the earthquake to be 7.8 and the location of hypocenter to be at $40.7°$ N. and $143.6°$ E. and 20 km. deep. About twenty minutes after the occurrence of the earthquake the Pacific coasts of Hokkaido and northeastern Honshu were assaulted by a tsunami which caused considerable damages to house and fishing facilities at many places.

Three parties of Tohoku University personnel made a field survey of Honshu coast from Imabetsu, Aomori Prefecture to Hachinohe, Iwate Prefecture and from Kamaishi, Iwate Prefecture to Niiyama-hama, Miyagi Prefecture in cooperation with the parties of Hokkaido University, which covered the southern coast of Hokkaido, and of Tokyo University, which covered a part of Honshu coast where our parties did not survey.

The data measured in the field survey and the records of tide stations are analyzed from various points of view in this paper. The comparison of the present data with those in previous tsunamis is also made. As is well known, this part of Japan has frequently suffered damage by big tsunamis. The field survey by various agencies has been made in such cases as the Sanriku Tsunami, 1933, the Tokachi-oki Tsunami, 1952, and the Chile Tsunami, 1960. The data of previous tsunamis referred to in this paper are taken from the Report of Earthquake Research Institute, Tokyo University (1934) for the Sanriku Tsunami, the paper by Suzuki et al. (1953) for the Tokachi-oki Tsunami, 1952, and the Report by the Committee for Field Investigation of Chilean Tsunami of 1960 and a paper by Kato et al. (1961) for the Chile Tsunami.

MEASUREMENTS

The height of tsunami wave from the water level at the measuring time was observed by means of hand-level and measuring tape. The wave heights were estimated by such marks as trace on wall of building, discolored grass on shore, or weeds left on beach. When no such trace was found, we were obliged to estimate the height based on the information given by inhabitants.

The observed values are reduced to the heights above the mean sea level of Tokyo Bay (TP) using the tide table. The data thus obtained are listed in a table with the degree of reliability of observed value. The degree is expressed in three classes of A, B, and C, which represent the highest, medium, and low accuracy, respectively.

DISTRIBUTION OF WAVE HEIGHTS OUTSIDE OF BAY

As well known, the wave height inside of a bay is much affected by the size, shape, distribution of water depth and other factors of the bay. Therefore, only the values observed at the places open to ocean are considered. The wave heights outside of the bay along the coastline are shown in Figure 1. The values are generally large from Miyako to north of Hachinohe, the maximum being 5 meters, and they gradually decrease to the north up to Shiriyazaki. South of Miyako the heights decrease rapidly down to 1-2 meters and increase again along the east coast of Oshika Peninsula. In Hokkaido, high values are obtained in the eastern part of Erimosaki. The following procedure is adopted to explain the observed wave height distribution.

SOURCE AREA

The usual way of determining the tsunami generating source is used based on the tide gauge records at fourteen stations in

Hokkaido and Honshu. An imaginary wave-front of long wave from a station is drawn using the bathymetric charts. The time back from the onset of tide gauge record is assigned to the wave front. We can thus construct a group of imaginary wave fronts starting from all the stations, corresponding to the time of occurrence of earthquake. The envelope of these fronts gives the source area of the tsunami. The estimated source is seen in Figure 2. The area is approximately expressed by an ellipse which nearly coincides with the aftershock area. The long axis of the source ellipse is about 280 km. long in the direction of N.N.W.-S.S.E. and the short axis is about 150 km. The epicenter of the main shock is located close to the center of the source area.

The relation between earthquake magnitude and size of source area has been studied by several authors (for example, Hatori, 1966). According to these studies, the area S (km.^{2}) of the source area is approximately expressed by

$$\log S = 1.25M - 5.5.$$

In the present case, S is calculated to be 2.0×10^4 km.^{2} as the magnitude is 7.8. The areal size of the above obtained area is roughly 6×10^4 km.^{2}, compatible with the calculation.

Since the source area is determined, the ray path of the tsunami wave can be drawn starting from a point on the margin of the area and the energy flux incident on a unit length of coast can be estimated. The refraction diagram thus drawn is seen in Fig. 2, which well explains the feature of observed wave heights in general. The high wave heights north of Miyako and east of Erimosaki, for example, are easily interpreted by this diagram. Two exceptions are the heights at the east coast of Oshika Peninsula and north of Hachinohe. The former might be due to some effect of the existence of the peninsula and the latter may be due to the run-up phenomenon, because the beach in the northern part of Hachinohe is very flat in contrast to the southern part.

MOVEMENT OF SEA BOTTOM IN SOURCE AREA

The initial motion of the present tsunami has a systematic feature. The first motion of tide gauge records in east of Erimosaki and southern Sanriku show a big amplitude with upward sense, while those in west of Erimosaki and northern

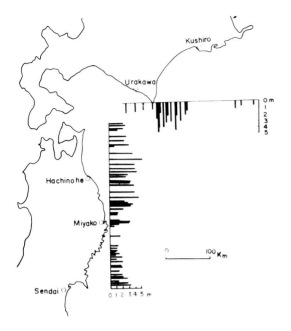

Fig. 1: Distribution of Wave Heights Outside of Bay.

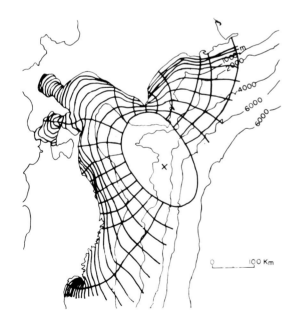

Fig. 2: Source Area and Refraction Diagram.

Sanriku are downward, although the amplitudes are comparatively small. According to the Green's law on the propagation of long wave, the wave height H is written as

$$H = A \cdot h^{-1/4} \cdot l^{-1/2},$$

where h and l are the depth of water and the distance between two neighboring ray paths, respectively, and A is a constant. Using this formula, the distribution of wave heights on the circumference of source area is calculated, based on the records at forty tide stations in Hokkaido and Sanriku coasts. The values calculated under the assumption that the water depth at the coast is 5 meters everywhere are shown in Figure 3. Interesting facts on the nature of sea bottom motion generating the tsunami may be learned from the figure. As seen in the figure, the distribution may be of a quadrant type, upward in northeast and southwest parts and downward in northwest quadrant. This distribution, however, is quite different from that of the first motion of the seismic wave, the latter being in push direction all over the region concerned.

Another interesting fact is that the change from downward to upward quadrants is very sharp. For example, a big amplitude with upward sign was observed at Hiroo station, while the record at Urakawa only 30 km. apart from Hiroo shows a weak downward motion.

These facts suggest that the movement of the ocean bottom in the source area is very complicated; an adequate model is not known at the moment.

WAVE HEIGHTS INSIDE OF BAY

In the Sanriku District, there are many large and small bays having natural periods of seiche range from 5-50 minutes. The observations of wave heights inside these bays, therefore, give good data for the study on different behavior of wave due to various conditions. To exhibit the behavior in a simple way, the median line of a bay is drawn from head to mouth of the bay and the distance between the head and the observation point is measured along this line. If the observed wave height is plotted against this distance, the amplitude variation inside of a bay is shown as in Figures 5, 6, and 7. The ratio of wave heights at mouth and head is taken as a measure of amplification factor of bay. The ratios for various bays are plotted in Fig. 4 against the natural period of seiche in abscissa. The ratio shows a large value at the period of about 10 minutes, as seen in the figure. We may take the record of the tide gauge at Enoshima, an island off Onagawa Bay, as the wave form outside of bay. This record gives a predominant period of about 10 minutes and, therefore, the large value of ratio mentioned above is reasonably thought to be due to resonance effect of bay. One exception in Fig. 4 is a point corresponding to 45 minutes, but this is the particular case of Miyako Bay: an explanation will be made later.

The comparison of the present tsunami with other cases of Sanriku (1933) and Chile (1960) Tsunamis is made. Three typical cases are given here in Figures 5, 6, and 7. The three tsunamis behaved quite similarly in Samenoura Bay, a small bay with the natural period of 16 minutes with a sharp change in water depth. As seen in Fig. 5, the amplitude increases monotonously from mouth to head and the amplification factor is approximately 2 in every case.

Miyako Bay, having a long period of seiche and a rather flat distribution of water depth, shows a different feature. The distribution of wave heights, as seen in Fig. 6, is similar for both the present Tokachi-oki and the Chile Tsunamis, the heights increasing monotonously from mouth to head. The wave heights in the case of Sanriku Tsunami, however, represent a flat feature for all the distances from head. Since the predominant periods of the present and the Sanriku Tsunamis are more or less the same and that of Chile Tsunami is very long, the difference cannot be attributed to the relation between predominant period of wave and seiche period. This might be due to the difference in the direction of approach of tsunami wave.

Another example is the case of Ofunato Bay seen in Figure 7. In the Sanriku Tsunami, the wave height decreases from mouth to head, the amplification factor being about 0.3, while the height is nearly the same in the case of Chile Tsunami. This difference is probably due to the relation between predominant and seiche periods, as stated in the papers studying the observations of Chile Tsunami. However, the flat distribution of wave heights in the present case is obviously much affected by an artificial construction. A new breakwater was constructed in Ofunato Bay after Chile Tsunami and the above fact implies that the breakwater works well in the present case.

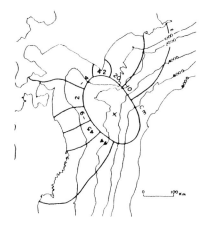

Fig. 3:
Distribution of
Wave Amplitude
Around Source Area.

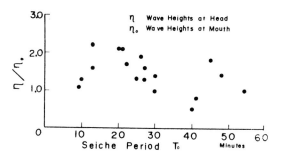

Fig. 4:
Amplitude Factor of
Bay Versus Seiche Period.

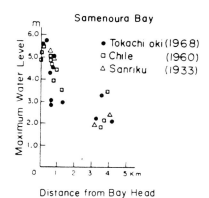

Fig. 5:
Wave Height Distributions
in Samenoura Bay.

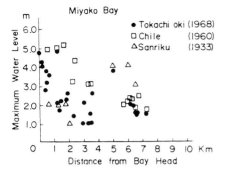

Fig. 6:
Wave Height Distributions in Miyako Bay.

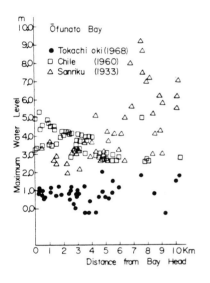

Fig. 7:
Wave Height Distributions in Ofunato Bay.

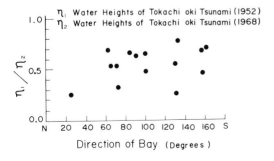

Fig. 8:
Ratio of Wave Heights in Two Tokachi-oki Tsunamis Versus Direction of Bay.

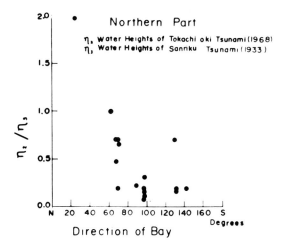

Fig. 9: Ratio of Wave Heights in the Present and Sanriku Tsunamis in Northern Coast of Sanriku District.

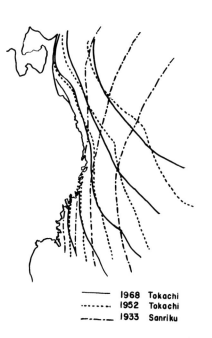

Fig. 10:
Wave Fronts of
Two Tokachi-oki
and Sanriku Tsunamis.

EFFECT OF DIRECTION OF BAY

As mentioned in the previous section, the wave heights inside of bay are sometimes affected by the direction of wave approach. This point is examined by comparing the present result with those in Sanriku and another Tokachi-oki (1952) Tsunamis. The ratio of wave heights observed at exactly the same point is taken to cancel the effect of bay. Hence the ratio is reasonably considered to be the energy ratio of two tsunamis coming to the mouth of bay.

The ratios between the present and another Tokachi-oki Tsunami are represented in Fig. 8 for various bays, the direction of bay being taken as abscissa. The direction of bay is expressed by the angle of the median line from north. As seen in the figure, the ratio is almost independent of the direction of bays, and the energy ratio between the two tsunamis is roughly 0.5.

In comparing the present Tsunami with the Sanriku Tsunami, the surveyed region is divided into two parts of northern and southern Sanriku at Hirota Bay. The ratio in northern Sanriku increases as the bay direction tends to north, as seen in Figure 9. The ratio in southern Sanriku, however, is not so dependent on the direction.

The wave fronts in the three cases are drawn to explain the above result. The fronts in both Tokachi-oki Tsunamis are parallel to each other, as is clear in Figure 10. The front of Sanriku Tsunami differs from that of Tokachi-oki Tsunami especially in the northern part of Sanriku coast. The direction of wave approach in the former case is nearly east, while the wave in the latter case came from northeast direction. This situation can explain the observation fairly well. A similar effect was reported in the case of former Tokachi-oki Tsunami by Suzuki et al. (1953).

As Miyako Bay faces the northeast, the present tsunami invaded almost directly into this bay, but Sanriku Tsunami came from the lateral direction. This may be the reason why the wave heights inside of Miyako Bay behaved in a particular way, as was stated in the previous section.

CONCLUSION

The tsunami accompanying the Tokachi-oki Earthquake on May 16, 1968 is studied, based on the observations in field survey and records of tide gauges. The conclusions obtained are summarized as follows:

1) The source area estimated from the tide gauge records is an ellipse with the long axis in the direction of N.N.W.-S.S.E.

2) The area is centered at the epicenter and roughly coincides with the aftershock region. The areal size is compatible with the law obtained in previous studies.

3) The wave height on the circumference of the source area shows a quadrant type distribution which is different from the mechanism based on seismic observation. The distribution suggests a complicated motion of the sea bottom in the source area.

4) The wave heights outside of the bay are well explained by the refraction diagram. Some exceptions may be due to the effect of run-up or some other reasons.

5) The behavior of wave inside of a bay is rather complicated. The relation between amplification factors of bays and seiche periods indicates the existence of a resonance effect. In some bays, a particular distribution of wave heights is seen in comparison with the cases of other tsunamis.

6) The ratio of wave heights in various tsunamis is affected by the direction of wave approach.

In conclusion, the present writer wishes to express his hearty thanks to the people in the surveyed region who gave us kind assistance in collecting data. The appreciation is also due to the members of Tokyo and Hokkaido Universities and other agencies who kindly permitted us to use their field data and marigrams.

BIBLIOGRAPHY

Aftershock Research Group of Tohoku University 1968. Observation of aftershocks of Tokachi-oki Earthquake, May 16, 1968. *Proc. Symposium for Prevention of Disasters*.

Hatori, T. 1966. Vertical displacement in a tsunami source area and the topography of the sea bottom. *Bull. Earthq. Res. Inst.*, Tokyo University., Vol. 44, p. 1449.

Kajiura, K., Hatori, T., Aida, I., and Oyama, M. 1968. On the field investigation of the Tokachi-oki Tsunami. Read at the *Meeting of JOTI*, Aug.

Kato, Y., Suzuki, Z., Nakamura, K., Takagi, A., Emura, K., Ito, M., and Ishida, H. 1961. The Chile Tsunami of 1960 observed along the Sanriku Coast of Japan. *Sci. Rep. Tohoku Univ.*, Ser. 5, Vol. 13, p. 107.

Papers and reports on the Tsunami of 1933 on the Sanriku Coast, Japan 1934. *Bull. Earthq. Res. Inst.*, Tokyo Univ., Suppl. 1.

Suzuki, Z., Noritomi, K., Ossaka, J., and Takagi, A. 1953. On the Tsunami in the Sanriku District accompanying the Tokachi Earthquake, March 4, 1952. *Sci. Rep. Tohoku Univ.*, Ser. 5, Vol. 4, p. 134.

The Committee for Field Investigation of the Chilean Tsunami of 1960. 1961. *The Chilean Tsunami of May 24, 1960*.

8. Statistical Studies of Tsunami Sources and Tsunamigenic Earthquakes Occurring in and near Japan

H. WATANABE
Japan Meteorological Agency
Tokyo, Japan

ABSTRACT

In statistical studies of tsunami sources and tsunamigenic earthquakes occurring in and near Japan, the relationships among tsunami source, aftershock, and earthquake mechanism are especially emphasized.

Assuming that earthquake mechanism is given by the nodal-plane solution, most tsunamigenic earthquakes are related to a fault of the dip-slip type; usually a reverse fault rather than normal fault.

In general, earthquakes of the dip-slip type produce larger tsunamis than those of other types, strike-slip and ambiguous type. On the other hand, there is no distinct difference between tsunami grade magnitude by reverse fault and normal fault. The foreshock and tsunamigenic mainshock both have the same fault type, but the fault type of the aftershock differs from that of the mainshock.

On the Pacific coast of northeast Japan, the areas of aftershock correspond to the tsunami sources except for the Tokachi-oki Tsunami of 1952. On the Pacific coast of southwest Japan and for the Tokachi-oki Tsunami, the areas of aftershock do not relate to tsunami sources. This relationship can be explained by considering earthquake mechanism and the tectonic line or belt. In the Japan Sea, areas of aftershock usually correspond to the tsunami source, but in the 1941 tsunami off west Hokkaido, the relationship between aftershock and tsunami source cannot be determined because the aftershock was not observed.

INTRODUCTION

Statistical studies of tsunami sources and tsunamigenic earthquakes occurring in and near Japan have already been made by Iida (1956, 1958, 1963a, 1963b) and Hatori (1966, 1969) Earthquake mechanism in Japan and its vicinity was investigated in detail by Honda (1962), Honda, Masatsuka, and Ichikawa (1967), and Ichikawa (1961, 1966a, 1966b, 1970). However, definitive results have not yet been obtained on the relationships between tsunami source, aftershock, and earthquake mechanism.

The purpose of this study is to determine, from a statistical point of view, the relationships between tsunami source, aftershock, and earthquake mechanism in and near Japan.

GEOGRAPHIC DISTRIBUTION OF TSUNAMI SOURCES AND EARTHQUAKES

The tables of tsunamis occurring in and near Japan were prepared by Watanabe (1968). Figure 1 shows the geographic distribution of tsunami sources from 1900 through 1968, classified by grade magnitude of tsunami (mt) and cause. The grade magnitude of tsunami is defined by Imamura and Iida. In most cases it is used in the same meaning as the tsunami magnitude but, defined exactly, they have different meanings. The data of the tsunamis used for these studies have been taken from Figure 1.

At the Japan Meteorological Agency, through Ichikawa's efforts (1970), we are now preparing a table of the computer redetermination of earthquake mechanism occurring in and near Japan. In the redetermination, we assumed the nodal-plane solution. Figure 2 shows the geographic distribution of these earthquakes whose magnitude (M) was six or more (M $\geq$ 6.0) and whose mechanism was redetermined. These occurred from 1926 through 1968 and are classified by the nature of faulting. From this figure we can see that most earthquakes on the Pacific Ocean side are produced by the dip-slip type of fault. On the other hand, most earthquakes in the island-arc side are produced by the strike-slip type fault. These facts are closely correlated with tsunami generation. To emphasize this fact, we show the examples of tsunamigenic earthquakes and the earthquakes with land damage in Figure 3. More than 70% of the tsunamigenic earthquakes are from dip-slip faults; about 60% of earthquakes causing land damage are from strike-slip faults.

TSUNAMI SOURCE, AFTERSHOCK, AND EARTHQUAKE MECHANISM

In this study we used the data of the Japan Meteorological Agency (JMA) and the United States Coast and Geodetic Survey (USCGS) on the aftershock, and the reports of Hatori (1969) and the author's recalculation on the tsunami source.

Now let us consider the Tokachi-oki earthquake and tsunami of 1968. Figure 4a shows the geographic and vertical distribution of tsunami source, aftershock, and earthquake mechanism from the origin time of the mainshock (M = 7.9), May 16, 00 h. 45 m. U.T. to 10 h. 38 m. U.T. The arrow signs in the figure show pressure direction and tension direction. Figure 4b shows the geographic and vertical distribution from the origin time of the first strong aftershock (M = 7.5), May 16, 10 h. 39 m. U.T. to June 12. The dotted line shows the area of the tsunami source of the mainshock and the solid line shows the area of the tsunami source of the aftershock. Figure 4c shows the geographic and vertical distribution from the origin time of the second strong aftershock (M = 7.2), June 12, 03 h. 42 m. U.T. to November 16. Figure 4d shows the sum of Figures 4a, 4b, and 4c. The area of the tsunami source of the first strong aftershock is contained within that of the mainshock; the earthquake mechanisms are almost the same. The area of the tsunami source of the second strong aftershock occurred in another area but still overlapped the mainshock area. The earthquake mechanism differs from that of the mainshock. In Fig. 5, we will explain the earthquake mechanism in more detail. Here, the tsunami source corresponds to approximately the area of the aftershock ($A_t \doteq A_a$).

Several interesting facts on the relationships between source, aftershock, and earthquake mechanism are shown in Figures 5 through 8. In these figures, the area of aftershock is shown by oblique lines and dotted points. In Fig. 5 are shown four examples -- the Tokachi-oki earthquake and tsunami pair of 1968, the Iturup pair of 1968, the Akita-oki pair of 1964, and the Niigata pair of 1964.

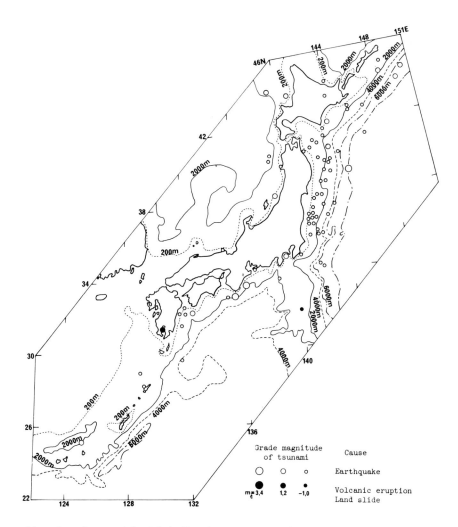

Fig. 1: Geographic Distribution of Tsunami Sources In and Near Japan from 1900 through 1968, Classified by Grade Magnitude of Tsunami (m_t) and Cause.

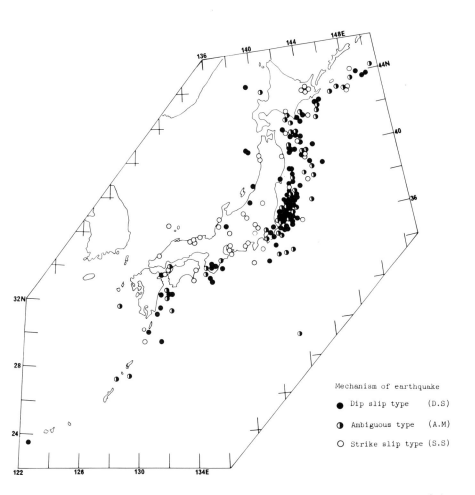

Fig. 2: Geographic Distribution of Earthquakes, Whose Magnitude (M) was 6 or more (M ≧ 6) and Whose Mechanism was Redetermined. These Occurred from 1926 through 1968 and are Classified by the Nature of Faulting.

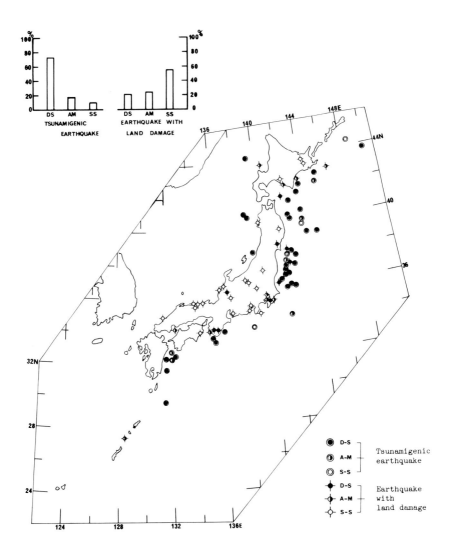

Fig. 3: Geographic Distribution of Tsunamigenic Earthquakes and Earthquakes with Land Damage from 1926 through 1968, as Classified by the Nature of Faulting. Figures on upper left side show Percentage of these Earthquakes.

Fig. 4: Geographic and Vertical Distributions of Tsunami Source, Aftershock, and Earthquake Mechanism, for the Tokachi-oki Tsunami of 1968.

a), b), c) ● $M \geq 6$
● $M < 6$
☼ Main shock

d) ● $M \geq 7$
● $7 > M \leq 6$
○ $M < 6$

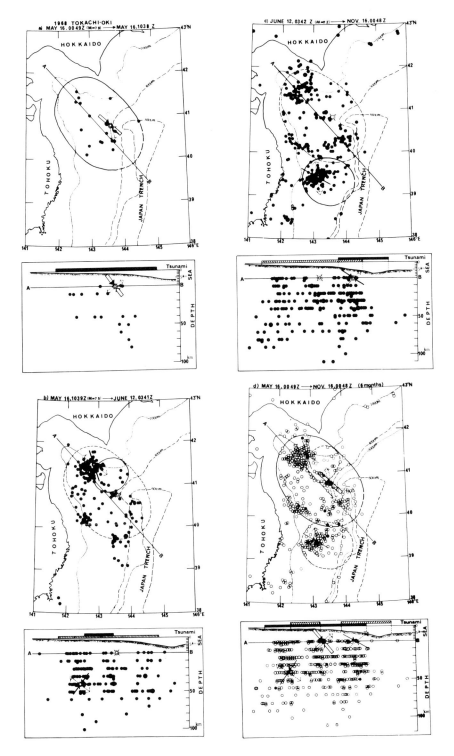

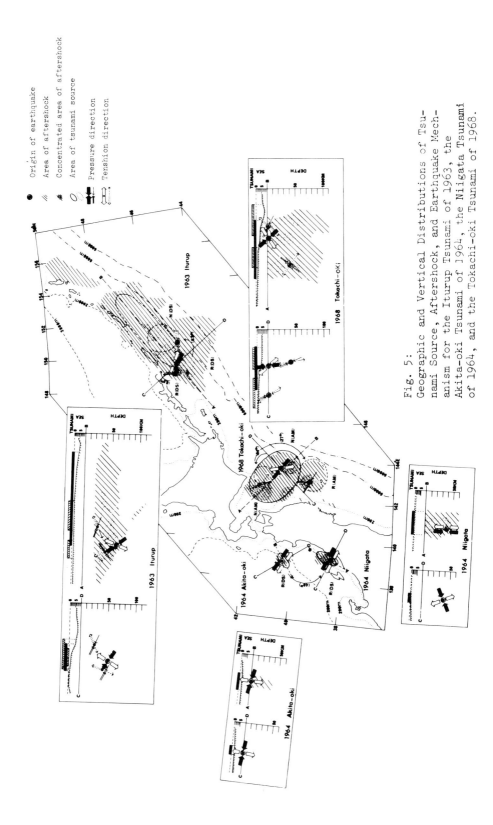

Fig. 5:
Geographic and Vertical Distributions of Tsunami Source, Aftershock, and Earthquake Mechanism for the Iturup Tsunami of 1963, the Akita-oki Tsunami of 1964, the Niigata Tsunami of 1964, and the Tokachi-oki Tsunami of 1968.

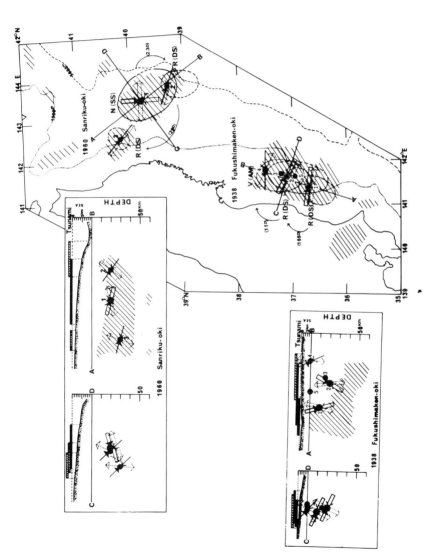

Fig. 6: Geographic and Vertical Distributions of Tsunami Source, Aftershock, and Earthquake Mechanism for the Fukushimaken-oki Tsunami of 1938 and the Sanriku Tsunami of 1960.

Some interesting facts on the Tokachi-oki earthquake and tsunami of 1968 have already been described. Now consider the earthquake mechanism.

The first strong aftershock occurred 10 hours after the origin time of the mainshock. Both are of the normal-fault type, but the mechanism of the second strong aftershock is the reverse-fault type and clearly differs from the mainshock. Often this can be seen on other examples and is pertinent to the generation of a tsunami by the aftershock. The maximum pressure exists in the direction of the long axis of the tsunami source, but the nodal plane is in the direction of the short axis and dips from the island-arc side to the Pacific-Ocean side. This differs from many other examples; it may be because this area is the curved zone of the Japan Trench.

In the Iturup earthquake and tsunami of 1968, a foreshock was accompanied by a tsunami. The foreshock and mainshock were due to reverse faults, but the aftershock was due to a normal fault. The area of aftershocks is larger than the area of the tsunami source.

The Akita-oki earthquake of 1964 may be considered the foreshock of the Niigata earthquake of 1964 because the epicenters are only one degree apart, the origin times are only one month apart, and the earthquake mechanisms are the same.

In Fig. 6, the Sanriku-oki earthquake and tsunami of 1960 is similar to the Tokachi-oki pair of 1968, except that the fault types are different. In 1938, eight small tsunamis occurred in the Fukushimaken-oki area. Five earthquake mechanisms have been redetermined; the mainshock and its tsunami occurred on Nov. 5, 08 h. 43 m. U.T. (M = 7.7). The mechanisms for the foreshock and for the mainshock were the same reverse-fault type. However, except for the last aftershock, the mechanisms of the aftershock were vertical normal faults. All these earthquakes have the same nodal plane -- along the long axis of the tsunami source parallel to the Japan Trench.

As shown in Fig. 7, the area of the aftershock of the Sanriku-oki earthquake of 1933 is smaller than the area of the tsunami source (Aa < At). The origin occurred in the Japan Trench and the nodal plane with normal faulting exists along the Japan Trench. In the Boso-oki earthquake and tsunami of 1953 the maximum pressure is almost parallel to the Japan Trench, although the other earthquakes are perpendicular to the Japan Trench. This may be because this area is in the curved zone of the Japan Trench, where it changes from a northeast to a northwest direction.

The Tonankai earthquake and tsunami of 1944 and the Hyuganada earthquakes and tsunamis of 1931, 1941, 1961, and 1968 have a special feature. The aftershocks are in a belt and are independent of the areas of tsunami source. These earthquake mechanisms are almost the same and the nodal planes are parallel to the trench. This is similar to most of those earthquakes which occurred in northern Japan.

As shown in Fig. 8, the earthquakes and tsunamis of Tokachi-oki in 1952 and of Tonankai in 1944 have the same shape as the areas of the aftershock, although the tsunami sources occurred in different areas. These facts will be discussed in the next section.

SOME STATISTICAL SUMMARIZATION

Figure 9 shows the relationships between tsunami grade magnitude (mt) and magnitude of earthquake (M), classified by the nature of faulting and the focal depth of earthquake (H). From this figure, we can see that the earthquake of dip-slip type clearly produces large tsunamis.

Figure 10 shows the same relationship as Fig. 9, classified by the fault type and the focal depth. There is no distinct relation between different fault types and tsunami grade magnitude. Almost all earthquakes accompanied by tsunami in and near Japan are of the reverse-fault type, but we must notice that the mainshock with normal faulting occurred in and near the Japan Trench.

Table 1 shows the fault type for the foreshock, mainshock, and aftershock. A summary of the results shows that the type of the foreshocks and mainshocks is of the same nature, but the fault type of the aftershock is different. It is interesting to speculate on the mechanism of the aftershock.

Figure 11 summarizes several geographic distributions of tsunami sources and earthquake mechanisms already shown. We do not have enough space to describe all of the examples; so we show a few typical ones only. In this figure, the tectonic belt which shows the abrupt change of Bouguer's gravity anomaly (or in other words, the belt along which the thickness of the earth's crust changes abruptly) is described. From this figure, several interesting facts can be noted:

1. In many cases in northern Japan, there is a close relationship between the area of aftershock and the tsunami source.

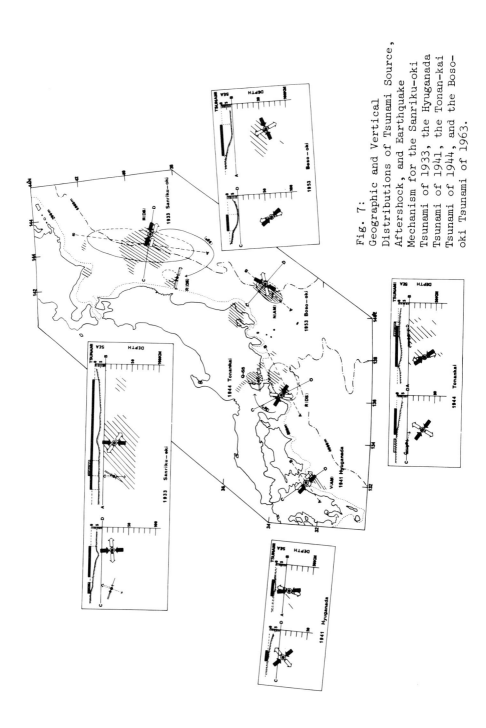

Fig. 7:
Geographic and Vertical Distributions of Tsunami Source, Aftershock, and Earthquake Mechanism for the Sanriku-oki Tsunami of 1933, the Hyuganada Tsunami of 1941, the Tonan-kai Tsunami of 1944, and the Boso-oki Tsunami of 1963.

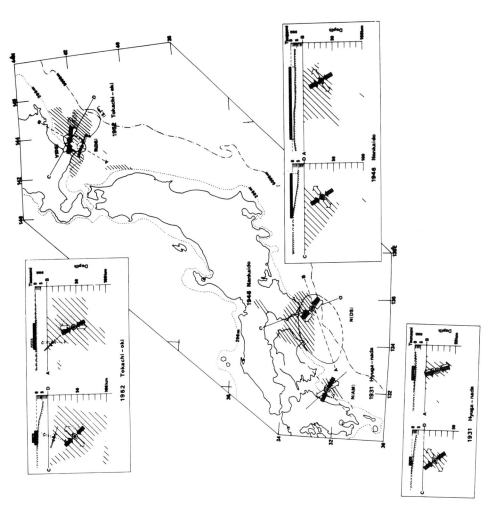

Fig. 8: Geographic and Vertical Distributions of Tsunami Sources, Aftershock, and Earthquake Mechanism for the Hyuganada Tsunami of 1931, The Nankaido Tsunami of 1946, and the Tokachi-oki Tsunami of 1952.

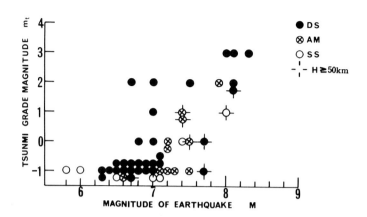

Fig. 9: Relationship between Tsunami Grade-Magnitude and Magnitude of Earthquake, Classified by the Fault Type and the Focal Depth of Earthquake. (H)

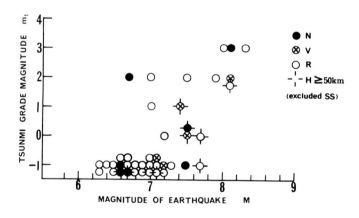

Fig. 10: Relationship between Tsunami Grade-Magnitude and Magnitude of Earthquake, Classified by the Fault Type and the Focal Depth of Earthquakes.

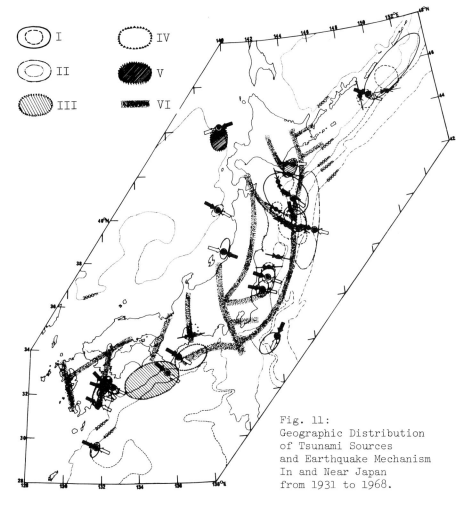

Fig. 11:
Geographic Distribution
of Tsunami Sources
and Earthquake Mechanism
In and Near Japan
from 1931 to 1968.

I: Area of tsunami source (A_t) corresponds to area of aftershock (A_a). $(A_t \leq A_a)$

II: Area of tsunami source is larger than area of aftershock.

III: Area of tsunami source does not correspond directly to area of tsunami source.

IV: Aftershocks are in a belt.

V: The relationship between aftershock and tsunami source is not known at all.

VI: Tectonic belt showing the abrupt change of Bouguer gravity anomaly.

Table 1: Fault Type for the Foreshock, Mainshock, and Aftershock.

Tsunami (Earthquake)	Foreshock	Mainshock	Aftershock		
1933 Sanriku-oki		N (DS)	R (DS)		
1938 Fukushimaken-oki	R (DS)	R (DS)	V (AM)	N (DS)	R (DS)*
1944 Tonankai		R (DS)	Q (SS)		
1952 Tokachi-oki		V (DS)	R (DS)		
1960 Sanriku-oki		N (SS)	R (DS)	R (DS)	
1961 Hyuganada		R (DS)	N (DS)		
1963 Iturup	R (DS)	R (DS)	N (DS)		
1964 Niigata	R (DS)	R (DS)			
1968 Tokachi-oki		N (AM)	N (AM)**	R (AM)	
	R	R	N, V		
		N, V	R		

* This shock occurred after more than 6 months from foreshock.
** This shock occurred after 0.4 days from mainshock.

114

2. In southern Japan, there is not a good relationship between the area of aftershock and the tsunami source. However, there is a relationship between the area of aftershock and the tectonic belt. For example, the area of the aftershock of the Tonankai earthquake of 1944 coincides well with the tectonic belt. Such an example is seen in the Tokachi-oki earthquake of 1952, although the earthquake occurred in northern Japan.

3. In the western Hokkaido earthquake and tsunami of 1940, which occurred in the Japan Sea, the relationship between the aftershock and the tsunami source was not known at all because the aftershock was not observed.

4. On many examples, the maximum pressure is perpendicular to the Japan Trench and the nodal line is parallel to the Trench. The earthquakes in and near the Japan Trench are of the normal-fault type, although their examples are few. But the earthquakes of the Japan Island arc side are almost all of the reverse-fault type.

CONCLUSIONS AND CONSIDERATIONS

Relationships between the tsunami source, aftershock, and earthquake mechanism have been investigated statistically. General conclusions and considerations are as follows:

In northern Japan, there are fairly good relationships between tsunami source, aftershock and earthquake mechanism. But in southern Japan, we cannot explain the relationship unless we consider the tectonic belt.

The mechanism of a strong aftershock accompanied by a tsunami differs from the mechanism of the mainshock. This is inconsistent with the conventional opinion that the mechanism of the aftershock must depend on the mechanism of mainshock. The crustal deformation accompanied by tsunami may be related with the mechanism of the aftershock. If so, as in the earthquake accompanied by land faulting, such a phenomenon must occur. But an aftershock accompanied by faulting was not observed in and near Japan.

The phenomenon of tsunami occurrence is clear evidence of the tectonic movement of the crust on the Pacific-Ocean side. It seems to support at least the idea of the new global tectonics presented by Isacks, Oliver, and Sykes (1968) that the Pacific-Ocean side in and near the Japan Trench shows the movement of the normal faulting with dip-slip type and the Japan island-arc side shows the movement of reverse faulting. Of course, the number of data used for statistical analysis are not enough to support a definite conclusion. In the future, the author will continue further research on these interesting and important phenomena.

ACKNOWLEDGMENT

The author wishes to thank Dr. M. Ichikawa for the use of his computer redetermination of earthquake mechanism.

BIBLIOGRAPHY

Hatori, T. 1966. Vertical displacement in a tsunami source area and the topography of the sea bottom. *Bull. Earthq. Res. Inst.*, *44*, pp. 1449-1464.

Hatori, T. 1969. Dimension and geographical distribution of tsunami source near Japan. *Bull. Earthq. Res. Inst.*, *47*, pp. 185-214.

Honda, H. 1962. Earthquake mechanism and seismic waves. *J. Phys. Earth*, *10*, No. 2, pp. 1-97.

Honda, H., Masatsuka, M., and Ichikawa, M. 1967. On the mechanism of earthquakes and stress proceeding them in Japan and its vicinity. *Geophys. Mag.*, *33*, pp. 271-279.

Ichikawa, M. 1961. On the mechanism of the earthquake in and near Japan during the period from 1950 to 1957. *Geophys. Mag.*, *30*, pp. 355-403.

Ichikawa, M. 1966a. Mechanism of earthquakes occurring in and near Japan, 1950-1962. *Pap. Met. Geophys.*, *16*, pp. 201-229.

Ichikawa, M. 1966b. Statistical investigation of mechanism of earthquake occurring in and near Japan and some related problem (in Japanese). *J. Met. Res.*, *18*, pp. 83-154.

Ichikawa, M. 1970. Computer redetermination of mechanism of earthquakes occurring in and near Japan, 1926-1968, and some statistical studies based on the solution. *Geophys. Mag.* (in preparation).

Iida, K. 1956. Earthquakes accompanied by tsunami occurring under the sea off the island of Japan. *J. Earth Sci.* Nagoya Univ., 4, pp. 1-43.

Iida, K. 1958. Magnitude of energy of earthquakes accompanied by tsunami, and tsunami energy. *J. Earth Sci.* Nagoya Univ., 6, pp. 101-112.

Iida, K. 1963a. Magnitude, energy, and generation mechanism of tsunamis. *I.U.G.G. Monograph No. 24*, pp. 7-18.

Iida, K. 1963b. A relation of earthquake energy to tsunami energy and the estimation of the vertical displacement in a tsunami source. *J. Earth Sci.* Nagoya Univ., 11, pp. 49-67.

Isacks, B., Oliver, J., and Sykes, L. R. 1968. Seismology and the new global tectonics. *J. Geophys. Res.*, 73, pp. 5855-5899.

Watanabe, H. 1968. Descriptive table of tsunamis in and near Japan (in Japanese). *Zisin, II, 21*, pp. 293-313.

9. Ionospheric Recordings of Rayleigh Waves for Estimating Source Mechanisms

A. S. FURUMOTO
Hawaii Institute of Geophysics
Honolulu, Hawaii
Contribution No. 297

ABSTRACT

Rayleigh waves from large earthquakes cause delayed oscillations in the ionosphere which can be detected and recorded by high-frequency Doppler sounding technique. The 5 MHz Doppler recording reflected the oscillations at 200 km. height and gave 100% coherence with Rayleigh waves of 20-30 second periods; the 10 MHz recording reflected the oscillations at 300 km. height and showed them to be Rayleigh waves of 90-170 second periods. The 10 MHz recording of Rayleigh waves was used to estimate the initial phase of the source of the Kuril earthquake of 11 August 1969 by Brune's simplified method. The initial motion seemed to be downward. Because of its rapidity, this approach to source mechanism estimation may prove useful for the Pacific Tsunami Warning System.

INTRODUCTION

When a large earthquake occurs in the Pacific area, the staff members of the Honolulu Observatory, E.S.S.A., are alerted to check for possible generation of a tsunami. Within a short time, sufficient data on P arrivals are sent by wireless from various seismograph stations to the Honolulu Observatory so that the staff members can deduce the following information:

(a) epicenter of the earthquake,
(b) origin time,

(c) magnitude, and
(d) depth of focus.

As these parameters are insufficient to determine whether a tsunami has been generated, and are certainly far less adequate to estimate the size of the tsunami if it has been generated, tidal stations near the epicenter are queried for tsunami information. The querying of tidal stations does consume time, and often the returning information is garbled. Nevertheless, until reports from tidal stations are available, there is little the staff members can do to reach a decision whether to continue the alert and escalate into a warning or to downgrade the alert.

While waiting for word from tidal stations, a very desirable piece of information is the source mechanism of the earthquake, for it has been shown statistically that tsunamis are associated with dip-slip type earthquakes rather than with strike-slip types (Iida, K., 1969). If the earthquake was found to be dip-slip type, the alert is continued and more tidal stations are queried. If the earthquake was found to be strike-slip type, the alert for most of the Pacific area can be called off except for regions close to the epicenter.

STATUS OF SOURCE MECHANISM STUDY

A short digressive discussion on the present status of source mechanism study is in order here. Many methods for source mechanism analysis have been developed, but very few are applicable in the short span of time available between the detection of the earthquake and the decision for tsunami warning. The methods will be discussed in the light of this criterion for quick application.

The first method to be developed for source mechanism analysis was the use of initial motions of the P wave. For this method, data from about forty stations evenly distributed throughout the world are necessary. Even then the solution may be ambiguous. This method is not feasible at the present state of the world communication network.

The S-wave method of analysis requires the knowledge of particle motion of S waves at various recording stations throughout the world. This, in turn, requires seismograms from matched horizontal seismometers at these stations. This method is also not presently feasible.

Surface waves have been used in source mechanism analysis. The method developed by Ben-Menahem (1961) requires digitizing of records and spectral analysis of the waves by high speed computers. The time required for this method is in the order of days at its most efficient performance. Furthermore, the seismograms are often not usable as Rayleigh waves and Love waves appear together.

The method of surface wave analysis developed by Brune (Brune, Nafe, and Oliver, 1960) is sufficiently simple in operation so that a solution can be obtained in the short time available during a tsunami alert. From a good recording of a dispersed train of surface waves, if the dispersive property of the earth between the epicenter and the seismic station is known, the initial phase of the earthquake can be derived. However, the operational difficulty in applying Brune's method at the time of a tsunami alert has been the unavailability of good seismograms. As tsunamigenic earthquakes are usually of large magnitude, greater than 7, the seismograms are usually off scale for surface waves. Low magnification seismometers have been set up at the University of Hawaii to overcome this difficulty, but so far the results have not been satisfactory. The interference of Rayleigh waves and Love waves render proper analysis quite difficult.

IONOSPHERIC RECORDING OF RAYLEIGH WAVES

Very recently, at the time of the Tokachi-oki, Japan earthquake of 16 May 1968, it was conclusively shown that Rayleigh waves from large earthquakes cause layers in the ionosphere to oscillate, and that these oscillations can be recorded clearly by monitoring the Doppler shifts of high-frequency radio waves (Yuen, Weaver, Suzuki, and Furumoto, 1969). As Rayleigh waves in the ionosphere have the advantage of being uncontaminated by Love waves, they are very useful in source mechanism estimates using Brune's method. Long-period waves of periods up to 200 seconds have been detected.

A short discussion of the phenomena of Rayleigh waves in the ionosphere will be given here for the benefit of seismologists. For a detailed description and background discussion, the paper by Yuen, et al. (1969) should be consulted.

In the ionosphere, electron density varies with altitude at any given time. The electron density also has diurnal and seasonal variations, but, during daylight hours over a short span of time of the order of tens of minutes, layering according to electron density is relatively stable and can be considered as stratified. A layer in the ionosphere with a given electron density is an effective reflector of radio signals of a certain frequency. To illustrate by an example, at about noon in early summer over the Hawaiian region, the layer at 200-km. altitude is a good reflector of 5-MHz radiowaves, while the layer at 300-km. altitude reflects the 10-MHz waves.

The reflecting property of the electron density layers can be used to detect disturbances in the ionosphere. Very conveniently, the WWVH time signal station on the island of Maui (Fig. 1) sends out continuous carrier waves at frequencies of 5 MHz, 10 MHz, and 15 MHz. These signals after being reflected by the ionosphere can be detected by receivers with dipole antennas. When the ionosphere is disturbed, the motion of the reflector causes a Doppler shift in the reflected carrier wave. The small shift in frequency, up to 2 Hz for the 5 MHz carrier wave, can easily be detected by comparing the reflected wave to a standard reference oscillator. For the purpose of monitoring ionospheric disturbances, the Radioscience Laboratory, Department of Electrical Engineering, University of Hawaii, has set up a Doppler-shift recording station on the University campus (Figure 1).

The instrumentation for a Doppler-shift recording station as described by Suzuki (1968) is given in the form of block diagrams in Figure 2. The diagram is self explanatory. Routinely, after a period of nine days, the data tape is scanned for unusual events. When an event is noticed, the section containing it is processed through a frequency spectrum analyzer. The end result is a plot of frequency shift versus real time, in the format commonly known as "sonogram".

When the data tape of the campus Doppler shift monitoring station was scanned routinely a few days after the Tokachi-oki earthquake, unusual oscillations were noticed at about the time of the earthquake. When seismograms from the low-magnification seismometers of station HIG, Hawaii Institute of Geophysics, were superimposed on the Doppler shift records, there was excellent visual coherence between the Rayleigh waves of 20 to 28 second periods and the 5 MHz signal (Figure 3). The 10-MHz Doppler record agreed with the envelope of the Rayleigh waves. Furthermore, when seismograms from the very long period instruments of station HON, E.S.S.A., were superimposed on the 10 MHz record, again there was excellent visual coherence during the interval when only Rayleigh waves were recorded by the seismometer (Yuen, et al., 1969). Calculations

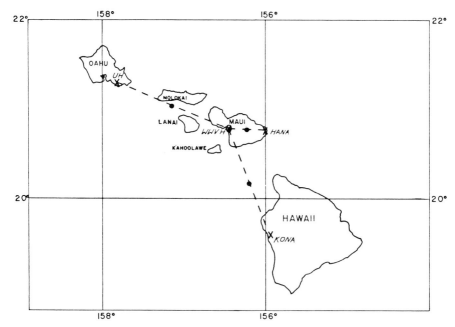

Fig. 1: Map of the Hawaiian Islands Showing Locations of WWVH Transmitter, Doppler Shift Monitoring Stations (UH, Kona, and Hana), and Midpoints Between Transmitter and Stations.

showed that the observed time lag between the recording of Rayleigh waves on the ground and the detection of Doppler shifts in the ionosphere agreed well with the time it would take for acoustic waves generated on the ground by Rayleigh waves to travel to the reflecting layers in the ionosphere. The discrepancy was within the allowable limits of error, as data on the acoustic velocity-height profile of the atmosphere-ionosphere are not very accurate.

These evidences showed conclusively that the vertical component of the ground motion of the Rayleigh waves generates acoustic waves in the atmosphere and that these acoustic waves travel upward into the ionosphere to cause the electron density layers to oscillate. As the Rayleigh waves have phase velocities of about 4 km./sec. and as the acoustic waves have a velocity of about 0.3 km./sec. in the atmosphere at ground level, the angle of refraction at the ground-air interface is about $5°$. Although the acoustic waves are almost an identical copy of Rayleigh waves recorded on vertical component seismometers, they are not true Rayleigh waves, as they do not have

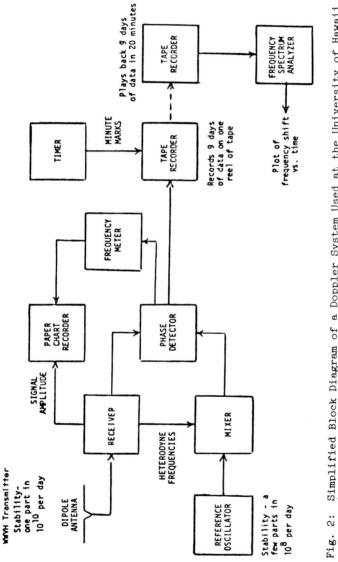

Fig. 2: Simplified Block Diagram of a Doppler System Used at the University of Hawaii (Suzuki, 1968).

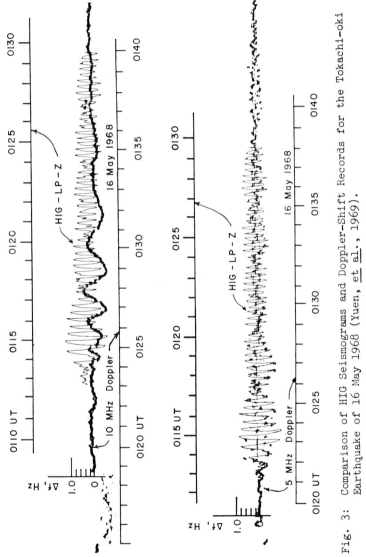

Fig. 3: Comparison of HIG Seismograms and Doppler-Shift Records for the Tokachi-oki Earthquake of 16 May 1968 (Yuen, et al., 1969).

the elliptical orbital motion characteristic of Rayleigh waves. We shall refer to them as Rayleigh-acoustic waves.

An extremely useful property, as well as a most interesting one, is that the ionosphere acts as a natural low-pass filter, as the Rayleigh-acoustic waves travel upward. Figure 4 shows the attenuation curves with respect to height for waves of different periods. The attenuation is presented as the ratio of the acoustic pressure variation at various heights ($\Delta p(h)$) to pressure variations at sea level ($\Delta p(0)$). At the 300-km. height, the attenuation for 30-sec. acoustic waves is at least an order of magnitude greater than that for 100-second waves. One added advantage of this type of low-pass filter is the absence of distortion by phase shifts.

The seismological implications of Doppler recording of Rayleigh-acoustic waves were so great that the Radioscience Laboratory established additional Doppler-shift monitoring stations at Hana on the island of Maui and at Kona on the island of Hawaii (Fig. 1), as soon after the Tokachi-oki earthquake as funds were available. On every station a second tape recorder was added so that there will always be a recorder in operation even while the tapes are being changed.

On 11 August 1969, at 21h 27m an earthquake of magnitude 7.8 occurred in the Kuril Islands region at $43.1°N.$, $147.7°E.$ (Terada, 1969). The epicentral distance from seismic station HIG was approximately 5540 kilometers. The Rayleigh waves travelling across the Pacific Ocean reached station HIG at 21h 50m (Figure 5). The accompanying Rayleigh-acoustic waves were detected at 21h 59m by the 5 MHz Doppler recorder and at 22h 01m by the 10 MHz recorder (Figure 5). The Rayleigh-acoustic waves were also detected by the Kona station on the island of Hawaii (see Figure 1). The Hana station was not fully operational at the time of the earthquake.

The Doppler-shift records from the campus station and Kona station are shown in Figure 6. Line B is the 10 MHz recording from the campus station; lines C and D are the 5 MHz and 10 MHz recordings respectively from the Kona station. Again the filtering effect by height of the ionosphere is noticed.

These records can be used for various seismic studies, such as power spectra of Rayleigh waves, phase-velocity dispersion, and source mechanism estimate. In this paper, source mechanism determination during a tsunami alert is studied.

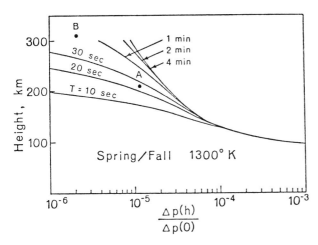

Fig. 4: Attenuation versus Height Curves Showing the Filtering Effect of the Ionosphere. Attenuation is Given as Pressure Variation Δp(h) at Height h to Pressure Variation Δp(0) at Sea Level (Yuen, et al., 1969).

ANALYSIS OF THE RAYLEIGH WAVE TRAIN BY BRUNE'S METHOD

As discussed in a previous section, the only rapid method for source mechanism analysis is Brune's method (Brune, et al. 1960). This method was applied to the 10-MHz record from the campus station, line B in Figure 6. The peaks in line B are designated by dots and the troughs by circles. Plots of peak numbers and trough numbers versus arrival times were made separately. The periods associated with the peaks and troughs were obtained in the conventional way by taking the tangents of the curves of the event number versus arrival times.

Of the peaks and troughs of line B in Fig. 6, only the events number 1, 2, 3, 4, and 5 were considered because the periods of these events were greater than 100 seconds. Long-period waves have the advantage of no regional variation in phase velocity. The values of phase velocity given by Ben-Menahem and Toksoz (1962) have been used.

In Table 1, the period, character, phase velocity, and the time lag of the events are listed. The time lags were measured from event 1.

Table 1: The Period, Character, Phase Velocity, and Time Lag of Rayleigh-Acoustic Waves from Kuril Earthquake, 11 August 1969.

Event	Period in Seconds	Character	Phase Velocity km./sec.	Time Lag Seconds
1	105	Trough	4.116	0
2	125	Peak	4.205	63
3	135	Trough	4.249	120
4	165	Peak	4.365	200
5	200	Trough	4.566	301

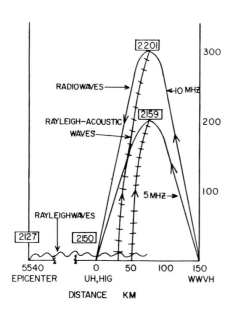

Fig. 5: Schematic Diagram of Events Associated with the Kuril Earthquake of 11 August 1969. The numbers in the rectangular are times.

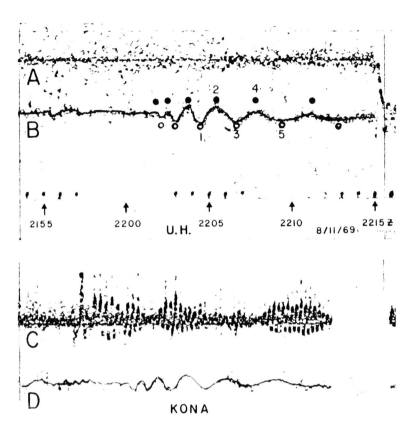

Fig. 6: Doppler Shift Records from the Kuril Earthquake. Line A is 5 MHz from Campus Station; Line B, 10 MHz, campus; Line C, 5 MHz, Kona; Line D, 10 MHz, Kona.

From the Doppler shift record, the arrival times of the Rayleigh-acoustic waves at the ionospheric reflector can be read off, but the time when the Rayleigh-acoustic waves were refracted into the atmosphere from the ground cannot be estimated accurately. The location where the refraction took place can be estimated readily. For the Kuril earthquake it was about 30 km. beyond station HIG down the travel path (Figure 5). The epicentral distance for the long-period Rayleigh-acoustic waves was, for practical purposes, 5570 kilometers.

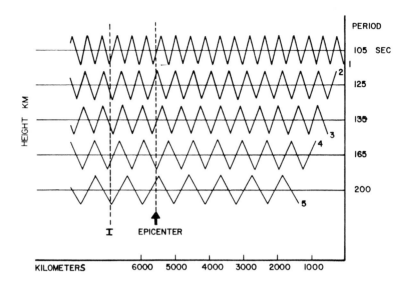

Fig. 7: Wave Trains from the Kuril Earthquake. Position I Indicates the Agreement of Phase from the Different Period Waves.

The analysis by Brune's method is shown in Figure 7. The zero distance is the point at which the Rayleigh-acoustic waves were refracted into the atmosphere. The arrival of the trough of event 1 of line B at the point of refraction was taken as reference time. At that time the peak of event 2 was a distance of 265 km. behind event 1, a distance equal to the product of phase velocity and time lag. The other events were also behind event 1, and their positions were found in a similar manner. Then continuous wave trains were constructed until a distance was found where the phases from all the trains were in agreement. Such a place has been designated as I in Fig. 7, and here all the wave trains have a trough.

The initial phase must be corrected by considering the phase shift of waves leaving a source, an advance of $\pi/4$ (Brune, Nafe, and Alsop, 1961). Hence the actual initial phase was $\pi/4$ before a trough. The initial motion was in a downward direction, at about $45°$ dip.

Tsunami generated by such motions may cause damage in the region near the epicenter but will be small at greater distances This proved to be the case, as the maximum tsunami amplitude at Nemuro, Hokkaido, was 130 cm., and at Urakawa, Hokkaido, 66 cm. (Iida, H., 1969); but the observations at Honolulu 5500 km. away showed amplitudes of about 5 centimeters.

DISCUSSION

From the point of view of a seismologist, the Doppler-shift recordings of Rayleigh-acoustic waves have expanded his field of study. The ionosphere is now within his realm of investigation; it is no longer merely an interesting, associated field. The Rayleigh-acoustic waves revealed the character of the train of long-period Rayleigh waves. Seldom have such waves been recorded, unalloyed by Love waves.

Fourier analysis of lines B and D in Fig. 6 will yield the amplitude spectra of Rayleigh waves from large earthquakes. By having an array of Doppler stations, phase-velocity dispersion of long-period waves can be measured. Up to the present, it has been assumed that there was no regional variation in phase velocity for mantle Rayleigh waves of periods greater than 100 seconds. This assumption was adopted for the calculations in this paper. The assumption can now be checked by measuring the dispersion across the array and comparing the results with global measurements that have been published (Ben-Menahem and Toksoz, 1962).

In source mechanism study, the feasibility of rapid analysis during a tsunami alert is a most welcome contribution of Rayleigh-acoustic wave recordings. Very-long-period seismometers have been in operation at station HON for several years, but the records from them have not been usable for source mechanism analysis. Usually the vertical component has been thrown off scale during large earthquakes. Strain seismometers are in operation at station KIP nearby, but in strain seismograms, the Rayleigh and Love waves are usually mixed. Theoretically, very-long-period vertical component seismometers can be useful in source mechanism estimates, but in practice, the only clear long period Rayleigh wave train records from large earthquakes were those from ionospheric recordings.

There are some difficulties involved in ionospheric recordings. The Rayleigh-acoustic waves must be recorded during daylight hours when the ionosphere is stable. The ionosphere is much disturbed near dawn and dusk. There are also some organizational difficulties too. During the Kuril earthquake the frequency spectrum analyzer was not in working order, and hence the records became available a few days later after the equipment was adjusted.

At the analysis end, once a record is available, results can be forthcoming rapidly. There is no need for a high speed computer; a desk computer such as the Olivetti Programma 101 will suffice. In fact, the calculations for this paper were done on such a desk computer.

The maintenance of Doppler shift stations is relatively low in cost. Furthermore, these stations are operated for ionospheric research and seismic results are by-products of the effort. A second tape recorder was installed at each station in order to insure uninterrupted recording, however, because of requests by seismologists.

ACKNOWLEDGMENTS

The Doppler shift monitoring stations are maintained by the Radioscience Laboratory, University of Hawaii, through National Science Foundation grants GA-1113 and GA-1594. Analysis was supported by funds from National Science Foundation grant GA-1084 and by State funds of the Hawaii Institute of Geophysics for seismology.

BIBLIOGRAPHY

Ben-Menahem, A. 1961. Radiation of seismic surface-waves from finite moving sources. *Bull. Seism. Soc. Am.* Vol. 51, No. 3, pp. 401-435.

Ben-Menahem, A. and Toksoz, M. N. 1962. Source-mechanism from spectra of long period seismic surface waves. 1. The Mongolian Earthquake of December 4, 1957. *Journ. Geophys. Res.* Vol. 67, No. 5, pp. 1943-1955.

Brune, J. N., Nafe, J. E., and Oliver, J. E. 1960. A simplified method for the analysis and synthesis of dispersed wave trains. *Journ. Geophys. Res.* Vol. 65, No. 1, pp. 287-304.

Brune, J. N., Nafe, J. E., and Alsop, L. E. 1961. The polar phase shift of surface waves on a sphere. *Bull. Seism. Soc. Am.* Vol. 51, No. 2, pp. 247-257.

Iida, K. 1969. Generation of tsunamis, mechanism of earthquakes, and their sources. *Proceedings of International Symposium on Tsunami and Tsunami Research* jointly sponsored by Int'l Union of Geod. and Geoph. and East-West Center, Univ. of Hawaii.

Iida, H. 1969. Kuril Islands earthquake. *Report No. 725.* Smithsonian Institution, Center for Short Lived Phenomena.

Suzuki, R. K. 1968. The ionospheric Doppler shift as a method of obtaining true height variations in the ionosphere. *Technical Report.* Radioscience Laboratory, University of Hawaii.

Terada, K. 1969. Kuril Islands earthquake. *Report No. 719.* Smithsonian Institution, Center for Short Lived Phenomena.

Yuen, P. C., Weaver, P. F., Suzuki, R. K., and Furumoto, A. S. 1969. Continuous, travelling coupling between seismic waves and the ionosphere evident in May 1968 Japan earthquake data. *Journ. Geophys. Res.* Vol. 74, No. 9, pp. 2256-2264.

10. Identification of Source Region from a Single Seismic Record

T. SOKOLOWSKI and G. MILLER
ESSA, Joint Tsunami Research Effort
University of Hawaii
Honolulu, Hawaii

ABSTRACT

In selected cases, there appears to be a similarity between seismic records of earthquakes which occur within a small source region. This similarity exists only between the records of a given seismograph station, not between the records of earthquakes recorded at different stations. If this similarity is consistent, it may be possible to identify in real time the particular fault system in which an earthquake occurs, assuming a sufficient amount of reference data exists. As the net ground motion along a given fault is typically consistent, it should be possible to improve the estimation of tsunami generation.

Forty seismograms representing six seismic regions were digitized and spectra computed for the duration between P and PP. The three seismic-recording stations were selected on the basis of having similar instrumentation and being at teleseismic distances from the epicenters. As a preliminary test of the hypothesis that it may be possible to identify the source region of an earthquake from a single seismogram, simple linear correlations were computed between an unknown spectrum and the average spectra of each of the groups. For some source regions this is a definitive test.

A method of "pattern recognition" which always worked within the chosen data set consisted of: (1) normalizing the log spectra and derivative log spectra, (2) forming averages by groups, (3) forming a permissible error envelope about these average spectra and spectral slopes, (4) testing the "unknown" spectrum for the number of values which fall within the envelopes of the various groups, (5) computing the variance of the residuals between the mean and test spectra, (6) scoring this testing as an integer value, best-through-worst score, (7) summing these integers by group, and, as a last step, calculating this sum as a percentage of possible score.

Further work is being done using automatic grouping techniques and so-called "adaptive pattern recognition" computer programs.

A much larger data sample will perhaps demonstrate the potential utility of this type of epicenter determination.

INTRODUCTION

In some instances, earthquakes occurring in different areas and recorded on the same seismograph system have different seismogram signatures. These signatures can be visually identified as originating from particular areas and the purpose of this study is an attempt to identify earthquake source areas using one recording station. Many factors can be attributed by these signature differences appearing on the seismogram. Among these factors the most predominant are the following: magnitude of the earthquake; response of the seismograph system; epicentral distance, hypocenter; local earth structure near the recording station; earth structure near the earthquake source; source and azimuth relative to the recording station.

These characteristic signature similarities appearing on the seismograms can be seen by considering one or more phases of a particular earthquake from the same area and recorded by the same seismic station. However, more obvious similarities and differences between earthquakes from the same area and different areas can be shown by considering the power spectra of the seismic records from the various areas (Alcock, 1969; Buchbinden, 1968). These spectral similarities or differences can be accomplished by considering the similarity of the above factors with respect to the earthquake source areas, the propagation path, the recording area, and instruments (Utsu, 1966). These factor similarities can be obtained by choosing the same type of seismograph system for recording the earthquakes, by using only those earthquakes that occur in a limited depth

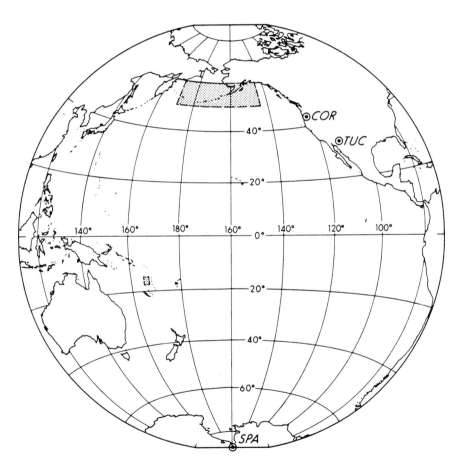

Fig. 1: Seismic Recording Stations Relative to Earthquake Source Areas (Stippled Areas).

range, considering only teleseismic distances between source and receiver, and considering earthquakes from various areas but recording them on the same seismic receiving station.

In this analysis only the phases from P through PP were considered. This involved only the first few minutes after the onset of the initial P motion.

The main region of this analysis was the Aleutian Islands chain, although seismograms from one particular area in the New Hebrides Islands region were processed and spectra computed. The earthquakes chosen from these regions were of normal depth and the magnitude range was approximately 5.0 - 6.0.

ANALYSIS

Only the Coast and Geodetic Survey World Wide seismograms were used in order that the type of recording instruments would be the same. The seismic recording stations chosen were South Pole (SPA), Corvallis, Oregon (COR), and Tucson, Arizona (TUC). SPA was only used for the area in the New Hebrides region. These stations were chosen such that the amplitude and distance to the earthquake areas were convenient for digitization. Figure 1 shows the location of the seismic stations and the earthquake regions. Figure 2 shows the Aleutian epicenters that were used in this study.

The distances from the epicenter to the stations were teleseismic. Seismograms from each of the stations for the various earthquakes were collected. Only the first part of the earthquake recording was used, including approximately the first few minutes of the recording after the initial onset of the P phase pulse. The phases included in this initial recording were the P through PP. The seismograms were enlarged approximately 3 times their original size. Records were then digitized on the University of Hawaii's Benson Lehner machine and processed so that the frequency intervals for all records were 0.051 cycles per second. The frequencies for this study range from approximately 0.2 to 2.0 cycles per second.

The signal-to-noise ratio for the records processed was high. The trace of the background noise was included from the seismogram for a few minutes immediately before the initial P phase pulse. This spectrum for the background noise showed that the effect of all frequencies was negligible; so further removal of the background from large signal-to-noise ratio records was neglected.

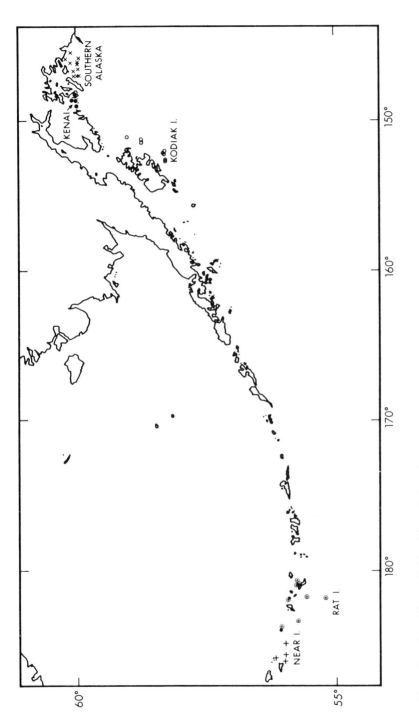

Fig. 2: Location of Aleutian Epicenters.

The frequency response curves of the World Wide Standard Seismograph System are supposed to be the same and we assumed this equality. Therefore, removal of the response curve from each of the earthquake spectra was not necessary for this study.

A rectangular data window was applied during the spectrum computation. The spectra were smoothed so that the spectral values at various frequencies were distinguishable; a minimum of irregularities in the spectral curve were eliminated. Also all spectra had their frequency intervals adjusted to the same value and the same number of spectral values. The number of spectral estimates was 33, which resulted in 48 degrees of freedom. The frequency interval for all the spectra is 0.051 cycles per second. The spectra for the various earthquakes for each area were plotted together to inspect the similarities and differences among the spectra of the same area and those of different areas. In this way the spectra from various earthquakes and areas could then be compared with respect to individual frequencies or to groups of frequencies.

DATA

A list of the earthquakes used in this study is shown in Table 1. The earthquakes are listed in chronological order; the various parameters given for each earthquake are taken from the *Seismological Bulletin of the Coast and Geodetic Survey*. The magnitude computations used are those of the Coast and Geodetic Survey and Pasadena (Cal Tech).

AREA SPECTRA

Approximately 40 log power spectra were computed for 6 different earthquake source areas. The maximum deviation between earthquakes in a particular area is approximately 2 degrees in both longitude and latitude. These areas are: New Hebrides, Southern Alaska, Kenai Island, Kodiak Island, Rat Island, and Near Island. The earthquakes used are those given in Table 1. Three representative earthquake recordings from different areas and recordings at TUC are shown in Figure 3. These areas, in particular, are the Kodiak Island, Near Island and Southern Alaska. The point to be made here is that in the time domain the earthquake recordings for these areas seem to be quite similar. Figure 4 shows plots of the log of the power spectra for the earthquake areas whose traces are shown in Figure 3. Each area in Fig. 4 has two

Table 1: Input Data Used in This Study.

DATE	ORIGIN TIME	LAT.	LONG.	REGION	MAG.	DEPTH
Mar 28, 64	09-52-55.7	59.7N	146.6W	Southern Alaska	5.5 6.2(PAS)	30 km.
Mar 29, 64	04-12-15.7	60.2N	145.5W	" "	5.3	15
Mar 29, 64	16-40-57.9	59.7N	147.0W	" "	5.6 5.8(PAS)	15
Mar 29, 64	16-53-26.6	60.3N	146.1W	" "	5.2	15
Mar 30, 64	07-09-34.0	59.9N	145.7W	" "	5.6 6.2(PAS)	15
Mar 30, 64	12-05-43.5	60.1N	147.0W	" "	5.0	25
Apr 4, 64	04-54-01.7	60.1N	146.7W	" "	5.6	40
Apr 8, 64	19-33-19.0	59.6N	147.0W	" "	5.1	15
Apr 8, 64	19-50-16.8	60.4N	145.9W	" "	5.3	10
Apr 9, 64	13-06-15.2	59.6N	146.1W	" "	5.1	15
Jul 11, 64	20-25-40.3	59.7N	146.2W	" "	5.6	40
Mar 29, 64	01-09-36.4	59.8N	149.2W	Kenai	5.5	20
Mar 29, 64	03-07-19.5	59.7N	148.8W	"	5.0	30
Mar 29, 64	10-08-02.4	60.9N	148.6W	"	5.3	20
Apr 10, 64	19-05-52.6	59.7N	148.2W	"	5.2	15
Mar 28, 64	08-33-47.0	58.1N	151.1W	Kodiak	5.6	25
Mar 28, 64	09-01-00.5	56.5N	152.0W	"	6.0 6.2(PAS)	20
Mar 28, 64	23-46-22.0	57.5N	151.1W	"	5.2 5.0(PAS)	33
Mar 29, 64	01-29-33.7	57.5N	151.3W	"	5.6	20
Mar 29, 64	09-06-44.8	56.6N	152.2W	"	4.8	15
Mar 30, 64	13-03-34.9	56.5N	152.7W	"	5.3	20
Mar 30, 64	16-09-28.4	56.6N	152.1W	"	5.5	25
Apr 4, 64	08-40-29.8	56.5N	152.6W	"	5.3	15
Feb 5, 65	22-15-59.5	51.5N	176.7E	Rat	5.6	25
Feb 6, 65	04-02-52.7	52.1N	175.7E	"	5.9	35
Feb 6, 65	06-48-29.7	51.8N	178.1E	"	5.0	40
Feb 7, 65	09-25-51.1	51.4N	179.1E	"	5.3	30
Feb 15, 65	01-25-08.8	51.4N	179.4E	"	5.8	42
Feb 18, 65	23-13-36.3	51.4N	179.1E	"	5.4	28
Feb 19, 65	18-52-42.1	51.1N	178.4E	"	5.6	35
Feb 5, 65	06-39-49.6	51.8N	175.1E	Near	5.7 6.5(PAS)	25
Feb 5, 65	09-32-09.3	52.3N	174.3E	"	5.9 6.5(PAS)	41
Feb 5, 65	20-47-13.3	51.9N	174.6E	"	5.7	35
Feb 6, 65	08-46-51.2	51.9N	174.0E	"	6.0	30
Aug 11, 65	07-18-45.3	15.6S	167.1E	New Hebrides	5.0	32
Aug 11, 65	20-13-59.1	15.7S	167.0E	"	6.0	51
Aug 13, 65	17-56-26.7	16.6S	167.6E	"	5.4	33
Aug 29, 65	18-31-24.6	15.7S	167.6E	"	5.1	14
Aug 31, 65	16-36-35.8	15.5S	166.8E	"	5.6	33

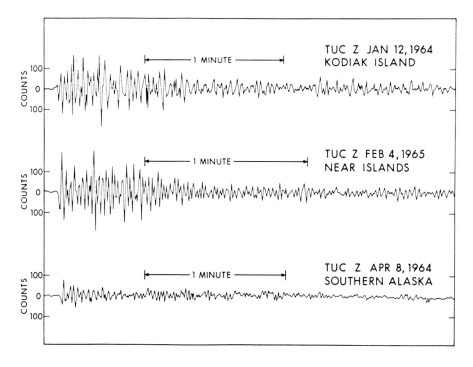

Fig. 3: Sample of Earthquakes Recorded at Tucson, Arizona, from Three Source Areas.

representative type spectra and each spectrum can be compared to the spectrum from its own area and to the spectra of the other areas. Visual inspection shows that each pair of spectra from a particular area has similar trace irregularities and the similarities to spectra from other areas are much less.

Spectra for an area located in the New Hebrides Island region were computed. This was the only area located outside the Aleutian Island chain. Logarithm plots of four spectra are shown in Figure 5. The earthquakes analyzed here were recorded by seismic station SPA.

The identification of an earthquake area by considering similarities or differences of spectra can be achieved by comparing the spectra obtained from earthquakes recorded by the same station. Spectra resulting from one particular area but

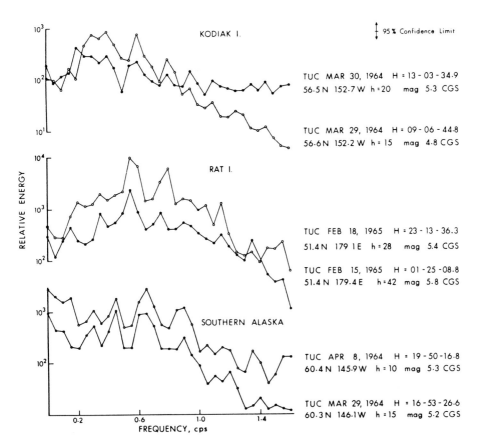

Fig. 4: Spectra from Seismograms from Three Different Source Areas but Recorded at the Same Station.

recorded at different distant stations need not be similar. In general, these spectra will show similarity only when obtained from the same seismic recorder.

Figure 6 shows a logarithm plot of the spectra from the southern Alaska area obtained from the recordings of the seismic stations TUC and COR. The spectra designated by a particular recording station show a high degree of similarity among themselves but are different when compared to spectra designated by another recording station. In Fig. 6 the spectra resulting from the TUC records shows the maximum spectral peak

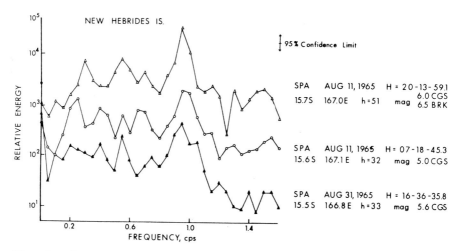

Fig. 5: Spectra Representative of the Particular Area in the New Hebrides Islands Using the South Pole Recording Station.

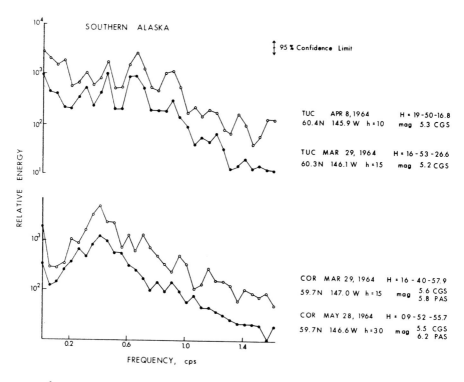

Fig. 6: Spectra from One Area Using Two Different Recording Sites.

at approximately 0.65 cps as compared to the maximum spectral peak at approximately 0.4 cps obtained from the COR records. Other spectral curve differences are also apparent.

Distinguishing the area spectra from each other was then attempted. A simple linear correlation technique was used but resulted in 70% accuracy of area identification. Another technique was applied which resulted in 100% accuracy for chosen spectra. In this method mean curve envelopes and maximum values of the unknown spectrum were compared to the same parameters of every known characteristic spectrum from all other areas. This comparison of parameters gave a coefficient which best correlated the unknown to a particular known area, thus identifying the unknown.

To achieve this end result, it was necessary first to obtain a characteristic spectrum (mean representative curve) of a particular area, the variation of spectral values about the mean curve, and the derivative and inverse of the mean curve. The mean spectrum was obtained by normalizing, on the maxima, all spectra from a particular area. The variation from this mean curve was obtained for each spectrum from a particular area. The variation from this mean curve was used in deriving the mean curve. The mean of these variation points for the total curve represented an envelope about the mean curve for a particular known area. Envelopes were then determined in the same manner for every other known area. Similarly, the derivative of the mean curve and its corresponding envelope were determined for each unknown. The maximum value for the mean curve and its inverse were also considered as characteristic of a particular area. Therefore, for each known area there are mean curve and derivative envelopes, the maximum value of the mean curve, and the inverse of the mean curve.

The same four parameters for an unknown curve were then determined. By comparing the normalized unknown curve to the known mean curve, an envelope is obtained that is characteristic of the unknown with respect to a particular known area. Unknown characteristic envelopes with respect to all known areas are obtained. In the same manner, the characteristic envelopes for the unknown derivative curves are obtained. The characterizing maximum values for the unknown are obtained by comparing the unknown curve maximum (mean and inverse of mean curves) to each of the known curve maxima.

The four comparison parameters are combined to yield a rank correlation coefficient given by

$$B_j = 1 - \frac{\Sigma C_i}{K} \qquad i = 1,4$$

where

B_j = rank correlation for a particular area,
ΣC_i = four characterizing parameters,
K = number of area times the number of parameters.

Each of the j coefficients is scanned for the highest value and identifies the unknown area. A computer is necessary to perform these calculations. An example of the output is given in Table 2.

Table 2: Sample Output of the Computer Pattern Recognition Program.

Comparison Areas	Mean Curve Envelopes	Mean Curve Derivative Envelopes	Maximum Mean Curve Envelopes	Maximum Mean Inverse Curve Value	Rank Coefficient
Southern Alaska	1.0850	3.1154	0.4899	0.4535	0.8333
Kenai	1.8680	5.5476	0.6833	0.6804	0.5000
Kodiak	2.2340	5.8390	0.7412	0.8107	0.1667
Rat Island	2.1624	6.2468	0.7311	0.8091	0.2083
Near Island	1.5904	4.9502	0.6692	0.5751	0.6667
New Hebrides	2.6943	6.1661	0.8591	0.6834	0.1250

Output: Southern Alaska.

Input: A spectrum from Southern Alaska treated as unknown.

ERRORS

The largest contributions to errors in the spectral estimates are due to: (1) human and instrumental errors during the digitizing process, and (2) gradual change of seismograph constants after calibration.

SUMMARY

It has been found that earthquakes recorded on the short period Z seismogram quite often yield spectra that are representative of a particular area and that in many instances short period vertical seismograms can be used to identify a particular area from which the earthquake originated. This identification is achieved by considering the energy distribution curves as a function of frequency for various earthquake sources. The energy distribution curves are different for different earthquake areas. An unknown seismogram can then be identified at a given station by comparison with other spectra received at the same station from the various areas. Although these spectra from the same area are similar, they are not exactly the same. Some spectral peaks vary considerably.

Identification of the source area can be made operational by use of the computer. Only four parameters were used to correlate a curve to a particular area although additional parameters may be included.

BIBLIOGRAPHY

Alcock, E. D. 1969. The influence of geologic environment on seismic response, *B.S.S.A.*, 59 [1], pp. 245-268.

Buchbinden, G. G. 1968. Amplitude spectra of *PcP* and *P* phases, *B.S.S.A.*, 58 [6], pp. 1797-1819.

Utsu, T. 1966. Variations in spectra of *P* waves recorded at Canadian arctic seismograph station, *Canadian Journal of Earth Sciences*, 3.

11. Recurrence of Tsunamis in the Pacific

S. L. SOLOVIEV
Soviet Tsunami Commission
Moscow B-296, U.S.S.R.

ABSTRACT

The catalogue of tsunamis in the Pacific has been supplemented and made more precise. Intensity of tsunamis according to modified Imamura-Iida's scale as well as coordinates and magnitude M of tsunamigenic earthquakes were estimated for many events.

The energy of some recent tsunamis has been estimated, and approximate correlations between the energy and other tsunami parameters known from usual observations have been found.

A catalogue of earthquakes in the Pacific with $M \geq 6$ and registered instrumentally in the 20th century is compiled. For all earthquakes, the magnitude M has been redetermined uniformly according to observations of Soviet and some European seismic stations. Completeness of the catalogue for the years 1928-1967 has been achieved.

The Pacific seismic belt has been divided into seismic zones having relatively homogeneous tectonic structure and seismic activity. Recurrence of earthquakes in different seismic zones has been calculated for different depths. Completeness of collection of macroseismic data in different seismic zones has also been estimated.

On the bases of all these data, approximate intensity-frequency and energy-frequency relations for tsunamis in different zones have been found; the probability of occurrence of a tsunami with given intensity after an earthquake with given magnitude at a given point of the Pacific has been re-estimated, as well as the probability of a tsunami to cross the Pacific. Small-scale tsunami zoning of the Pacific has been made.

Some peculiarities of the seismic process for tsunamigenic earthquakes have been investigated. In cooperation with L. M. Balakina, some peculiarities of the focal mechanism for tsunamigenic earthquakes in Kamchatka, Kuril Islands, and Japan have been noted.

INTRODUCTION

This paper is a progress report on the work started by the author together with Ch.N. Go in 1965. The aim of the work has been to furnish the Soviet tsunami-warning service with maps and tables showing the probability for tsunami with given intensity at Soviet shores to be excited by earthquakes with given magnitude somewhere in the Pacific.

The work was begun with compiling the fullest possible catalogue of tsunamis in the Pacific. The international *Annotated Bibliography on Tsunamis* published in 1964 has been used as the main guide-book for this purpose. The first draft of the catalogue was prepared in manuscript form in 1967; statistical results extracted from it have been reported by the author at the General Assembly of the International Union of Geodesy and Geophysics in Switzerland. Unexpectedly, it has been found that the probability of tsunami generation, other conditions being equal, is higher in the zones almost safe from tsunamis, such as California and New Zealand, than in the zones frequently suffering from tsunamis, such as Japan and Kamchatka.

Because of this, an effort was made in 1967-1969 to check and to supplement the subject catalogue as well as to prepare a catalogue of strong earthquakes in the Pacific with homogeneous magnitude determinations. For this purpose, almost all literature on seismicity of the Pacific kept in the libraries of the U.S.S.R. has been searched. National macroseismic catalogues proved to be of great importance for such work. The excellent *Preliminary Catalog of Tsunamis in the Pacific* by K. Iida, D. Cox, and G. Pararas-Carayannis, published in 1967, has been used also.

The finished list of tsunamis in the Pacific has been printed in 1969 and embraces about 820 items. Although it has

330 items more than the mentioned reference catalog by Iida, et al., these additional items are often (more than 50%) doubtful cases of feeble tsunamis.

During the work, the problem of dynamical classification of tsunamis has naturally arisen. The known scale of tsunami magnitudes proposed by A. Imamura and developed by K. Iida has been taken as the basis for this purpose, but some amendments to the scale seemed to be desirable.

At first it seemed necessary to distinguish strictly maximal (h_{max}) and mean ($\bar{h}$) heights of tsunami inundation at the area of the strongest tsunami action or at any other given shore. For instance, the maximal measured height of inundation during the Kamchatka tsunami of 1952 was equal to 18 meters. This height was reached by the water only at one point. The mean height of inundation near the earthquake and tsunami source, which stretched to 800 km., was equal to only 7-8 meters. For estimations of energy mean heights of inundation are required, but in old descriptions of tsunamis usually only maximal heights of water in a few points are indicated.

It is obvious that the difference between mean and maximal tsunami heights may be explained by the real shore differing from ideal straight form. Usually, the longer the shore, the more probable is the existence of some peculiar forms of shore relief capable of magnifying tsunami height. This explains the existence of some empirical correlation between the difference of $\bar{h}$ and h_{max} and the intensity of the tsunami (Figure 1).

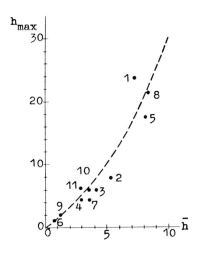

Fig. 1:
Correlation Between Maximal and Mean Heights, in Meters, of Tsunami Inundation. Numeration of tsunamis: 1-Sanriku, 3.III.1933; 2-Tonankaido, 7.XII.1944; 3-Nankaido, 21.XII.1946; 4-Tokachi-oki, 4.III.1952; 5-Kamchatka, 5.XI.1952; 6-Boso, 26.XI.1953; 7-Iturup, 7.XI.1958; 8-Chile, 22.V.1960; 9-Urup, 13.X.1963; 10-Alaska, 27.III.1964; 11-Niigata, 16.VI.1964.

Some discussion of terminology is expedient. If seismological terminology is applied to description of tsunamis, the grades of the Imamura-Iida scale must be designated as the intensity of the tsunami and not the magnitude of it. This is because the latter value must characterize dynamically the process in the source of the phenomenon and the first one must characterized it at some observational point -- the nearest possible point to the source included. The latter is the case in the Imamura-Iida scale. If the seismological terminology is not desired, then the term "magnitude" for grades of this scale is quite acceptable. In this work the grades of the Imamura-Iida scale will be denoted as intensity of tsunamis.

Analysis of the Imamura-Iida scale shows that the intensity i can be related fairly well to $\bar{h}$ by the formula

$$i = \log_2 \sqrt{2} \; \bar{h} \qquad (1)$$

$\bar{h}$ being connected with h_{max} by the correlation shown in Fig. 1 and in Table 1. No linear correlation between i and maximal height h_{max} can be found. The multiplier $\sqrt{2}$ explains the mean difference between maximal and mean heights for tsunamis of different intensities.

In practice Table 1 has been used to determine intensity of tsunamis. When quantitative data on tsunami heights have been absent, general qualitative descriptions of grades given in the original scale by Imamura-Iida and especially in the scale by N. N. Ambraseys have been used. These two scales correlate as shown in Table 1.

Accuracy of intensity determination naturally depends on the amount of available information and can vary from 1/2 to 1 1/2.

Epicenters and magnitudes of old tsunamigenic earthquakes included in the catalogue have been determined when it seemed possible. Epicenters have been determined from available macroseismic descriptions, taking into account contemporary status of knowledge of seismicity of given zones. Probable errors of the determined epicenters depend on the quality of macroseismic data and vary from 1/2 to 2 or 3 degrees.

The magnitude M of old earthquakes has been estimated also from macroseismic effect with the help of nomograms

Table 1: Correlation Between Mean and Maximal Heights of Tsunami Inundation and Grades of Scales by Imamura-Iida and Ambraseys.

Intensity of tsunami (Imamura-Iida)	Heights of inundation, in meters		Grades of the Ambraseys scale
	$\bar{h}$	h_{max}	
4.5	16	73.9	
4	11.3	40.3	6
3.5	8	22.9	
3	5.7	13.4	5
2.5	4	7.9	
2	2.8	4.8	4
1.5	2	3.1	
1	1.5	2.1	3
0.5	1	1.3	
0	0.7	0.9	2
-0.5	0.5	0.6	
-1	0.4	0.4	1
-1.5	0.25	0.3	
-2	0.2	0.2	1
-2.5	0.125	0.125	
-3	0.1	0.1	1
-3.5	0.06	0.06	
-4	0.04	0.04	1

proposed by N. N. Shebalin (1961, 1968). In some cases, empirical correlation between the duration of perceptible (without instruments) oscillations and the magnitude determined by the author, mainly on the basis of issues of *United States Earthquakes*, has been also used. This correlation is given in Table 2.

The mean error of magnitude determined from instrumental data is equal to ± 1/4, determined from surface effect is equal to ± 1/2, and determined from duration of earthquake is equal to ± 1.

Though accuracy of the determined epicenters and magnitudes is comparatively low, the author hopes the estimations made are of some value for further practical applications.

Table 2: Approximate Correlation Between Duration of Perceptible Seismic Oscillations and the Magnitude of Earthquake.

Duration of Oscillations	Magnitude of Earthquakes										
	$3\frac{1}{2}$	4	$4\frac{1}{2}$	5	$5\frac{1}{2}$	6	$6\frac{1}{2}$	7	$7\frac{1}{2}$	8	$8\frac{1}{2}$
Maximal	25	50	$1-1\frac{1}{2}$	2	$2\frac{1}{2}$	$3\frac{1}{2}$	$4\frac{1}{2}$	6	7	8	9
Mean	1	3	8	20	40	1	$1\frac{1}{2}$	2	$2\frac{1}{2}-3$	$3-3\frac{1}{2}$	4
Minimal	-	-	-	1	3	8	20	40	1	$1\frac{1}{2}$	2
	Seconds							Minutes			

For some tsunamis observed in many points and/or registered by mareographs, energy has also been estimated according to the usual formula (Takahashi, 1951)

$$E = \pi \rho g \; r \; \sqrt{gH} \; A^2 \; \tau \qquad (2)$$

where ρ = density of water,
g = acceleration of gravity,
r = radius of wave front, that is, approximate distance between the point of observation and the center of tsunami source,
H = depth of water at observational point (conditionally taken equal to 2.5 m. for all points),
A = mean amplitude, and
τ = duration of tsunami.

Analysis of data collected in the catalogue of tsunamis has led to the following correlation between total duration of oscillations of sea level, in hours, as the dependent variable and intensity of tsunami and magnitude of earthquake as the independent variables:

$$\lg \tau = -0.6 + 0.12 \; i + 0.24 \; m \qquad (3)$$

Equation 3 shows that the duration of tsunami depends on both the amount of water elevation in the tsunami source and the dimensions of the source.

It has been adopted conditionally that only 1/10 of total duration of sea-level oscillations must be used in calculations of energy if maximal amplitude of wave is taken instead of mean amplitude. A special nomogram has been built to facilitate calculations. In Fig. 2 the correlation obtained between log E and i is shown. So far it has been used only to estimate the value of i for tsunamis registered only by remote mareographs and not observed near the source.

In Fig. 3, correlation between maximal distance of tsunami registration of observation and intensity of tsunami is shown. In many cases, tsunamis with intensity as low as 0 can be registered by remote mareographs. Perceptibility of tsunamis occurring in America (especially in Chile and Peru) is higher than that of other tsunamis.

As mentioned earlier, an attempt was made to prepare (together with Ch.N. Go) a homogeneous catalogue of Pacific

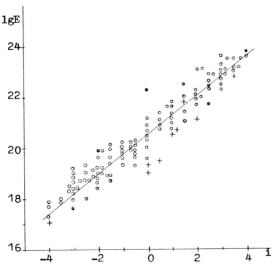

Fig. 2: Correlation Between the Energy of Tsunamis, in Ergs, and Their Intensities. + - Data for Japan Sea, Philippine Islands, Indonesia, U.S.A., Canada, Hawaiian Islands; o - Data for Other Zones.

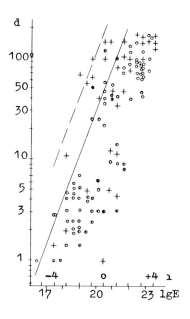

Fig. 3:
Correlation Between Maximal Distance of Tsunami Registration and Intensity of Tsunami.
+ is data for American tsunamis;
o is data for other tsunamis.

earthquakes with magnitude $M \geq 6$. This job is principally finished but some further refinement of the catalogue is still necessary. The catalogue has been compiled on the basis of the *Seismicity of the Earth* by B. Gutenberg and Ch. Richter, *International Seismological Summary, Bulletin mensuel du Bureau Central International de Seismologie, Seismological Bulletin of the U.S. Coast and Geodetic Survey, Seismological Bulletin of the net of Seismic stations of the U.S.S.R.*, and some other national bulletins.

For all earthquakes, magnitude M has been determined from displacements in surface seismic waves with the help of the author's nomogram (Soloviev, 1961) on the basis of observations of Soviet seismic stations and station Uppsala. In particular, all Soviet bulletins for 1928 and following years have been reviewed simultaneously with ISS. As a result, approximately 500 Pacific earthquakes with $M \geq 6$ for 1924-1948 and not in *Seismicity of the Earth* have been picked up. No more detailed and more reliable catalogue of earthquakes could be prepared by the authors. Nevertheless, the catalogue is not absolutely complete. Some gaps in data are evident, especially for the war years. Nevertheless, perhaps sufficiently reliable statistical conclusions can be obtained from the catalogue. Some zones have been outlined as shown in Table 3 under the assumption that seismicity, seismotectonics, and history of seismological and tsunami observations are homogeneous in each zone. For all these zones, magnitude-frequency relations have been found according to the usual formula

$$\lg N = a - b M \qquad (4a)$$

In the calculations, the time unit is equal to 100 years, the unit length of zone is equal to 1000 km., and values of M have been grouped in the following way: 6-6 1/4, 6 1/2-6 3/4, and so on.

Results of the calculations are given in Table 3. The mean value of the coefficient b equals 0.93. Individual values of b for the majority of zones differ insignificantly from the mean value. For Pacific ridges, Samoa-Tonga-Kermadec arc, and maybe for Taiwan, the values of b are significantly higher than the mean one. For some west and southwest zones (South Japan, Philippines, some parts of Indonesia, Fiji Islands) probably having block tectonics, values of b are lower than the mean value. To afford the quantitative comparison of seismic activity in different zones magnitude-frequency relations have been recalculated with the fixed value of b equal to 0.93. The frequency of earthquakes with M = 7 1/2 has been chosen as the measure of seismicity of zones. Only shallow-focus earthquakes ($\leq$70 km.) have been taken into account.

Values of N (7 1/2) show (see Table 3) that in the twentieth century the seismic process has been the most intense in the northwest corner of the Pacific (Kamchatka, Kurile Is., Hokkaido, east Honshu, Taiwan) and near New Britain (Figure 4). The intensity of seismic process has been somewhat lower at Aleutian Islands, Talaud and Sangihe Islands, South Solomon, Santa Cruz, and New Hebrides Islands, and Mexico. Other details are clear from Fig. 4 itself.

Intensity-frequency relations

$$\log n = \alpha - \beta i \qquad (4b)$$

have been determined for tsunamis in zones enumerated in Table 3. Other methods of statistical treatment, such as

Table 3: Some Data on Recurrence of Earthquakes and Tsunamis in the Pacific

Zone	Length km	a	b	N (7 1/2) original	N (7 1/2) by fixed b	N (7 1/2) adopted	T	Tectonics
Aleutian Is. & South-West Alaska	3100	8.43	0.92	11.0	9.8	10.4	-0.15	a
Kamchatka & Kurile Is.	2000	8.72	0.96	16.0	14.4	15.2	-0.2	a
Hokkaidō	330	8.90	0.82	17.0	14.5	15.7	-0.1	a
East Honshū	830	9.21	1.08	16.0	18.7	17.3	-0.4	a
South Japan	1000	6.80	0.80	6.3	5.0	5.6	-0.1	b?
Japan Sea	1600	7.07	0.86	2.4	1.8	2.1	+0.1	b?
Ryūkyū Is.	1200	7.68	0.92	4.8	4.8	4.8	-0.4	a
Taiwan	620	8.24	1.20	5.3	8.4	6.8	-0.6	b?
Luzon	1100	7.84	0.96	4.1	4.5	4.3	0.0	b
South-West Philippines	1100	5.27	0.60	5.3	2.1	3.7	+0.1	b
Philippine Deep	1500	7.05	0.80	7.5	5.0	6.2	0.0	a
Talaud & Sangihe Is.	720	5.52	0.60	14.7	5.5	10.6	-0.1	—
Banda Sea	1100	6.06	0.70	5.8	3.2	4.5	+0.2	b?
Sulawesi & Kalimantan	2100	7.86	1.00	1.1	1.4	1.2	+0.4	a?
Djawa & Lesser Sunda Is.	3000	8.63	1.10	0.9	1.1	1.0	-0.1	a?
Sumatera	2000	6.23	0.68	6.6	3.5	5.0	0.0	b?
Irian	700	5.75	0.68	6.5	3.3	4.9	+0.1	b?
New Britain & North Solomon Is.	1300	7.35	0.80	17.5	11.7	14.6	-0.3	a
South Solomon Is.	970	6.67	0.75	11.5	7.5	9.5	0.0	a
Santa Cruz & New Hebrides Is.	1200	8.04	0.90	10.5	11.5	11.0	-0.2	a
Fiji Is.	3100	7.00	0.84	1.6	1.1	1.3	0.0	b
Samoa, Tonga, Kermadec Is.	2900	10.70	1.30	3.3	9.3	6.3	-0.45	a
New Zealand, North Is.	550	7.58	0.94	6.0	6.0	6.0	-0.1	b
Chile & Peru	5500	7.67	0.85	4.5	3.3	3.9	+0.15	—
Central America & Mexico	3100	7.85	0.84	11.5	8.9	10.2	-0.35	—
California	1400	7.20	0.90	2.0	1.8	1.9	0.0	b
Canada	2000	7.94	1.00	1.4	2.0	1.7	0.0	b
Hawaiian Is.	225	7.00	0.92	5.6	5.6	5.6	+0.5	—

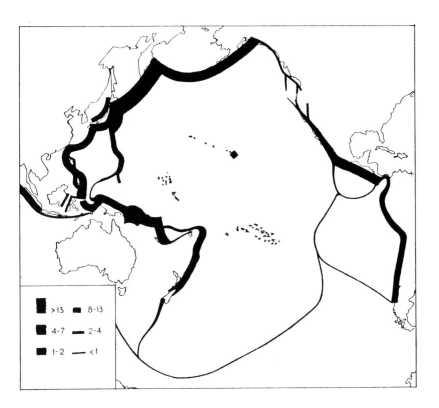

Fig. 4: Seismic Activity in the Pacific. Mean Frequencies of Earthquakes with M = 7 1/2 ± 1/4 during 100 Years in Zones with Length Equal to 1000 km. are indicated (for earthquakes with foci ≤ 70 kilometers).

Gumbel's statistics, are more suitable for investigation of tsunami recurrence, but relations 4 have been chosen at the first stage of the work due to their simplicity. Mean value of β for zones with most numerous data on tsunamis has been obtained equal to 0.31 (0.34 has been found in 1967). Then the intensity-frequency relations have been calculated for all zones with fixed value of β taken equal to 0.31.

Parameter $T = \log^{n(o)}/N$ (7 1/2), where $n(o)$ is the frequency of tsunamis with zero intensity, has been introduced as a measure of the mean ability of earthquakes in a given zone to generate tsunamis. The accuracy of the values of T has not been estimated so far. Naturally, it is not very high but in the author's opinion, the difference of T-values is significant (Table 3).

This difference can be partly due to some defects of the methods used. For example, the state of national seismic and sea-level services can influence the completeness of known data on tsunamis. To check this, the efficiency, e, of national macroseismic services has been estimated. For each event included in the catalogue of Pacific earthquakes, the number of settlements from which reports of given earthquakes were received have been taken into account, the number 10 being adopted when the actual number of reports exceeded 10. Then the efficiency, e, has been determined as the mean number of macroseismic reports from different places for one earthquake in a given zone (Table 3). According to Fig. 5a, the dependence of T on e is questionable, and, in any case, is not strong.

The proximity of tsunami sources to the shore can also influence the value of T, as seen from Fig. 2 and Fig. 5b, but this influence is not decisive.

Despite the defects of the data used and possible inaccuracy of calculation, the results obtained suggest that the difference in tsunami probability really exists. Perhaps it can be explained mainly by peculiarities of tectonics of zones. In zones of block tectonics with rigid superficial layers and superficial seismic activity, the probability of an earthquake exciting a tsunami is higher than in zones of so-called arc tectonics with relatively mild superficial layers and somewhat buried seismic activity (Figure 5c). This conclusion does not cancel the fact that such arc zones as Alaska-Aleutian Islands, Kamchatka-Kurile Islands, Hokkaido-Honshu, and others are sites of the most dangerous future tsunamis due to their uncomparably high level of seismic activity, maximal possible values of magnitude of earthquakes and some other peculiarities of these zones.

Analytically, the probability $P(i,M)$ of tsunami occurrences by a given earthquake can be expressed as the product of two probabilities:

$$P(i,M) = P_1(M) \cdot P_2(i,M) \qquad (5)$$

The first multiplier gives the general probability of tsunami occurrence independent of its intensity. The second

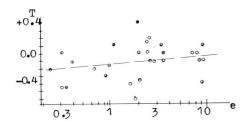

Fig. 5a:

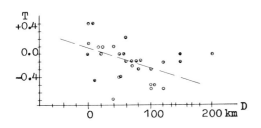

Fig. 5b:

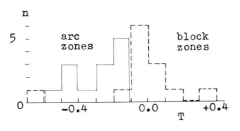

Fig. 5c:

Fig. 5: Dependence of Parameter T on Efficiency of Macroseismic Service e, Proximity of Tsunami Sources to the Shore D and Tectonics of Zones.

probability describes the distribution of tsunamis given the intensity.

The following results, corresponding formally to T = 0.15, were obtained in 1967 and have not been revised so far:

$$P_1(M) = \frac{1}{\sqrt{2\pi}} \int_{-\infty}^{2(M-7\ 3/4)} e^{-\frac{v^2}{2}} dv$$

$$P(i,M) = \frac{1}{\sqrt{2\pi}\ \sigma(M)} e^{-\frac{1}{2\sigma^2(M)}[i-\bar{i}(M)]^2}$$

$$\bar{i}(M) = -18.3 + 2.52\ M$$
$$\sigma(M) = 4.53 - 0.34\ M - 0.01\ M^2$$

Now only the earthquake magnitude method of tsunami prediction is used in the tsunami warning service. On the basis of the expressions given above, it can be shown that the efficiency of this method is inadmissibly low. Use of some new seismological parameters, such as mechanism of focus, can perhaps somewhat raise the efficiency of the method (see the paper by L. M. Balakina in this publication) but not in a decisive manner.

Elaboration of new methods of tsunami prediction and the registration of tsunamis in the open sea are necessary.

BIBLIOGRAPHY

Ambraseys, N. N. 1965. Data for the investigation of seismic sea-wave in Europe. *Un. Geodet. Geophys. Inst.* Monographie N 29.

Gutenberg, B. and Richter, C. 1954. *Seismicity of the Earth.* Princeton University Press, Princeton, New Jersey.

Iida, K., Cox, D., and Pararas-Carayannis, G. 1967. Preliminary catalogs of tsunamis in the Pacific. Hawaii Institute of Geophysics, University of Hawaii. *HIG-67-10 Data Report No. 5.* a200pp.

Iida, K. 1951. Earthquakes accompanied by tsunamis occurring under the sea off the islands of Japan. *J. Earth Sci.* Nagoya Univ. Vol. 4, N 1.

Shebalin, N. V. 1961. Intensity, magnitude, and focal depth of earthquakes. *Earthquakes in the U.S.S.R.*

Shebalin, N. V. 1968. Methods of application of data of engineering seismology for tsunami zoning. *Seismic Zoning of the U.S.S.R.*

Soloviev, S. L. 1961. Magnitude of earthquakes. *Earthquakes in the U.S.S.R.*

Takahashi, R. 1951. An estimate of future tsunami damage along the Pacific coast of Japan. *Bull. Earthq. Res. Inst.* Tokyo Univ., Vol. 29, N 1.

U. S. Coast and Geodetic Survey 1964. *Annotated Bibliography on Tsunamis.* Int'l. Union of Geod. and Geoph., Monograph No. 10, 233 pp.

12. The Tsunami in Alberni Inlet Caused by the Alaska Earthquake of March 1964

T. S. MURTY and LISE BOILARD
*Dept. of Energy, Mines, and Resources
Ottawa, Canada*

ABSTRACT

The tsunami in Alberni Inlet due to the Alaska earthquake of March, 1964, is studied theoretically. Some preliminary results are presented here. To evaluate the role of local resonance, the frequencies of natural oscillation of the Alberni Inlet and Trevor Channel have been calculated taking the topography into account. The propagation of the tsunami into the inlet has been calculated as an initial-value problem. Power spectral analyses of several tide-gauge records on the west coast of Canada during this period have been made to determine the energy contained in various frequency ranges. A new technique known as cepstrum analysis, originally developed by Tukey and colleagues, has also been attempted to isolate the influence of reflections of the trapped energy of the tsunami inside the inlet.

INTRODUCTION

The tsunami due to the Alaska earthquake of March 1964 caused severe damage on the west coast of Canada, especially at the head of Alberni Inlet (Wigen and White, 1964). Several aspects of this tsunami are being studied and some preliminary results are reported here. Figure 1 shows the locations of the tide gauges on the west coast of Canada which records have been used in this study. The insets show Barkley Sound which connects Alberni Inlet to the Pacific Ocean as well as the sections of Alberni Inlet that were used in the calculations. In these preliminary calculations the tsunami is assumed to propagate into Alberni Inlet through Trevor Channel and not through Junction Passage.

In Fig. 2 the vertical cross-sectional areas, surface widths, and average depths of the sections were plotted against the distance of the sections from the mouth of Trevor Channel. Since no analytical methods are capable of taking these irregular topographic features into account, numerical models are used. The length of Trevor Channel and Alberni Inlet together is roughly forty times the average surface width; hence the one-dimensional approximation is valid. Although, in principle, calculations could be made with a two-dimensional grid, in practice, it was almost impossible to adapt such a grid to this narrow system.

RESONANCE CHARACTERISTICS

For the interpretation of (1) tidal records, (2) spectra, (3) calculations on the dissipation of tsunami energy through construction of barriers, and (4) propagation of the tsunami into the inlet, a knowledge of the free oscillations of the inlet is essential. These were determined through a numerical integration of the channel equations. Let $\bar{M}(x,t)$ and $\bar{\eta}(x,t)$ represent the volume transport through vertical sections and the water level relative to the undisturbed level. The linearized channel equations, ignoring friction and, rotation are

$$\frac{\partial \bar{M}}{\partial t} = -gA \frac{\partial \bar{\eta}}{\partial x} \; ; \; \frac{\partial \bar{\eta}}{\partial t} = -\frac{1}{B} \frac{\partial \bar{M}}{\partial x} \tag{1}$$

where $A(x)$ is the area of the section, $B(x)$ is the surface width of the section, t is time, g is gravity, and the x axis is locally tangential to the main axis of the channel. For free oscillations, assume

$$\bar{M}(x,t) = M(x)\sin(\sigma t) \; ; \; \bar{\eta}(x,t) = \eta(x)\cos(\sigma t) \tag{2}$$

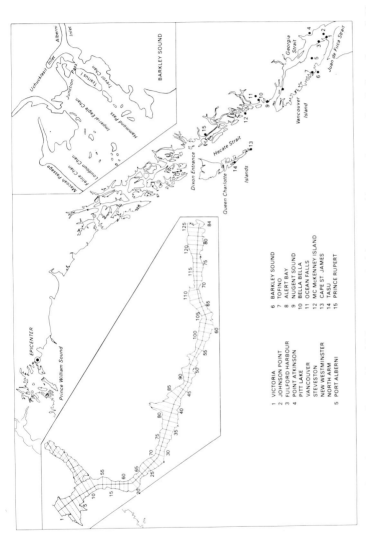

Fig. 1: The main diagram shows the tide-gauge stations on the west coast of Canada and the epicenter of the Alaska earthquake of March 1964 (on the scale of this diagram, the 6 stations listed under 4 could not be separated). The inset at right shows Barkley Sound which connects Alberni Inlet to the Pacific Ocean and the other shows the sections of Alberni Inlet used in the calculations. The numbers at the southern boundary refer to the section numbers when this was treated as a system by itself, while the numbers on the northern boundary were the section numbers when Trevor Channel was included.

1 VICTORIA
2 JOHNSON POINT
3 FULFORD HARBOUR
4 POINT ATKINSON
  PITT LAKE
  VANCOUVER
  STEVESTON
  NEW WESTMINSTER
  NORTH ARM
5 PORT ALBERNI
6 BARKLEY SOUND
7 TOFINO
8 ALERT BAY
9 NUGENT SOUND
10 BELLA BELLA
11 OCEAN FALLS
12 MC McKENNEY ISLAND
13 CAPE ST JAMES
14 TASU
15 PRINCE RUPERT

167

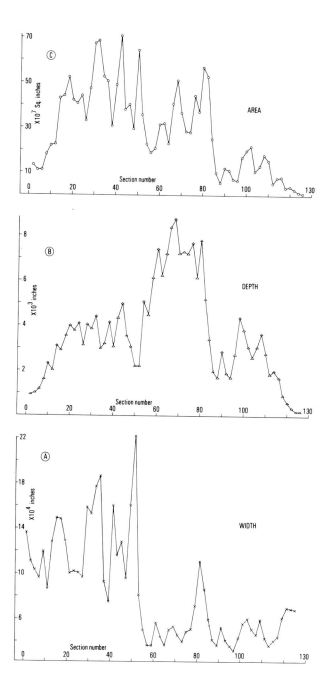

Fig. 2: A, B, C show the width, depth, and vertical sectional area averaged over each section for all the sections of Trevor Channel-Alberni Inlet-Uchucklesit Inlet.

where σ is the frequency of oscillation and $T(=2\pi/\sigma)$ is the period. Substitution of (2) into (1) gives two equations for the two unknowns M and η. For the Trevor Channel-Alberni Inlet system the finite difference forms of these two equations are

$$\eta_{i+1} = \eta_{i-1} - \sigma C_i M_i \quad \text{for } i = 2(2)124$$

$$M_{i+1} = M_{i-1} - \sigma E_i \eta_i \quad \text{for } i = 3(2)125 \tag{3}$$

where

$$C_i \equiv \frac{2 \cdot \Delta x}{g \cdot A_i} \quad \text{and} \quad E_i \equiv -2 \cdot B_i \Delta x$$

$2\Delta x$ is the distance between two successive even sections (at which M is calculated and A is prescribed), the same as the distance between two odd sections (where η is calculated and B is prescribed). The boundary conditions are M = 0 at the head of the inlet, η = 0 and M is arbitrary at the mouth. Using (3), the free periods T_1, T_2, T_3 of the first three longitudinal modes were determined.

Next, these periods were corrected for the effect of rotation. For the narrow Trevor Channel-Alberni Inlet system, the effect of rotation on the period is negligible; however, its effect on the structure of the oscillation is not because the oscillation is amphidromic (and not standing) with rotation. This effect was calculated by assuming (Mortimer, 1955) that the transverse slope is given by the geostrophic relation

$$fM = -gA \frac{\partial \eta^*}{\partial y}$$

where η^* is the additional change in level due to rotation and M is the volume transport calculated above. Under the assumption that the rotation effect was distributed equally between the north and south shores, the following expressions for the water level along the north and south boundaries of the inlet are obtained:

$$\eta_N^* = A(x)\cos[\sigma(t-t_N)] \; ; \; \eta_S^* = A(x)\cos[\sigma(t-t_S)]$$

$$A(x) = \sqrt{\eta^2 + \frac{f^2 B^2}{4g^2 A^2} M^2} \; ; \; t_N(x) = T-t_s \; ; \; t_s(x) = \frac{1}{\sigma}\tan^{-1}\left(\frac{fBM}{2gA\eta}\right)$$

Here t_N and t_S are the phases of high water along the north and south boundaries.

The effect of friction can be incorporated (Platzman and Rao, 1964) by adding the term $-FB/\rho$ to the right side of the first equation of (1), where F is an average value of the bottom stress over a section. Assuming a linear bottom stress $F = K\rho \bar{U} U^*$ where U^* is a constant velocity, and letting $\bar{M} = \text{Im}[-M(x)e^{i\sigma t}]$ and $\bar{\eta} = \text{Re}[\eta(x)e^{i\sigma t}]$, the following equations are obtained:

$$\sigma\eta = \frac{1}{B}\frac{dM}{dx}$$

$$\sigma M = -gA\frac{d\eta}{dx} + \frac{iKUB}{A}M \tag{4}$$

Here K is the skin-friction coefficient. Expanding η, M, and σ in powers of iKU and substituting these into (4), the frequency is given to second order by $\sigma = \sigma_0 - (KU)^2 \sigma_2$, where σ_0 is the inviscid frequency and σ_2 involves some integrals which can be evaluated using the results for the inviscid calculations.

The free periods were determined for Alberni Inlet and Trevor Channel-Alberni Inlet. There is a small inlet, Uchucklesit Inlet, which joins Alberni Inlet almost at right angles (Figure 1). It is awkward to take this into account in a one-dimensional calculation; however, this is included approximately by increasing the surface width of those sections of Alberni Inlet where Uchucklesit Inlet joins. Figure 3 shows the effect of rotation on the phase propagation of the fundamental mode. The inset shows the structures of the second and third modes without rotation.

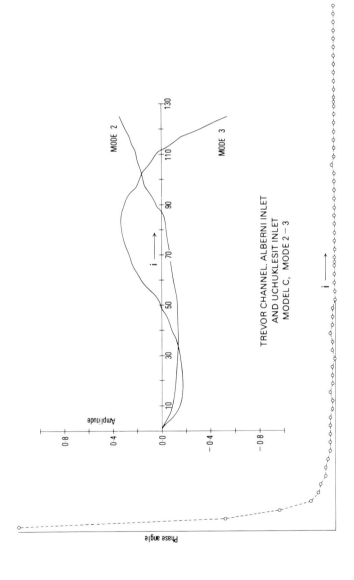

Fig. 3: The dashed line shows the phase of the water level on the north shore for Trevor Channel-Alberni Inlet-Uchucklesit Inlet with rotation. The phase at the south shore is a mirror reflection of this in the abcissa. The phase without rotation is zero and thus coincides with the abcissa. The index i denotes the section number. The inset shows the structures of the 2nd and 3rd modes without rotation.

Table 1 shows the topography periods for the first three modes. The periods calculated from the following Merian formula were also listed: $T_m = 4L/m\sqrt{gH}$, where T_m is the period of the mth mode, L is the length of the inlet and H is the average depth. As can be seen from the table, the Merian periods are badly off the topography periods for the first mode; this shows the necessity for the calculation of topography periods.

Table 2 shows the effect of friction on the free periods of the fundamental mode. As can be seen, friction does not change the period significantly although its effect is notice-

Table 1: Comparison of Merian Periods with Topography Periods.

System	Mode	Merian			Topography		
		Hrs	Min	Sec	Hrs	Min	Sec
Trevor Channel-Alberni Inlet	1	2	59	20	2	5	45
	2	0	59	49	1	0	38
	3	0	36	9	0	34	23
Alberni Inlet	1	2	1	15	1	26	2
	2	0	40	25	0	36	13
	3	0	24	15	0	24	14

Table 2: Effect of Friction on the Period of the First Mode.

(A) Alberni Inlet-Uchucklesit Inlet
(B) Alberni Inlet
(C) Trevor Channel-Alberni Inlet-Uchucklesit Inlet
(D) Trevor Channel-Alberni Inlet

System	Period								
	K = 0			K = 0.0017			K = 0.017		
	Hrs	Min	Sec	Hrs	Min	Sec	Hrs	Min	Sec
A	1	25	22	1	25	23	1	25	40
B	1	26	2	1	26	2	1	26	17
C	2	7	51	2	7	52	2	9	49
D	2	5	45	2	5	46	2	7	31

able. It can also be seen from the table that the addition of Uchucklesit Inlet does not necessarily increase the period of the system. When this small inlet is added, the period increased slightly, but when it is added to Alberni Inlet only the period decreased slightly. The reason for this difference in behavior may depend upon the fact that the period of a combined system depends upon how they are joined -- in series or in parallel.

PROPAGATION CHARACTERISTICS

The characteristics of the inlet with respect to tsunami propagation were determined through a calculation of the propagation of the tsunami into Alberni Inlet through Trevor Channel. This was treated as a one-dimensional, initial-value problem. Here the results of a simple model are presented: the vertical cross-sectional area A, surface width B, and depth H are only functions of the distance x from the mouth. To make these functions of the time t also requires a very large amount of planimetering work. The relevant equations of this problem are

$$\frac{\partial U}{\partial t} + U\frac{\partial U}{\partial x} = -g\frac{\partial \eta}{\partial x} - \frac{KU|U|}{H} \; ; \; \frac{\partial \eta}{\partial t} = \frac{-1}{B}\frac{\partial (AU)}{\partial x} \qquad (5)$$

where U is the velocity in the x direction (channel axis). Let i and n be indices representing x and t, respectively. Then the finite-difference forms of (5) are

$$U_{i+1}^{n+1} \left[1 + \Delta t \, |U_{i+1}^{n-1}| \frac{K}{H_{i+1}} - \frac{\Delta t}{4.\Delta x} (U_{i-1}^{n-1} - U_{i+3}^{n-1}) \right]$$

$$= U_{i+1}^{n-1} + g \frac{\Delta t}{\Delta x} (\eta_i^n - \eta_{i+2}^n) + \frac{\Delta t}{8.\Delta x} [(U_{i-1}^{n-1})^2 - (U_{i+3}^{n-1})^2]$$

and $\quad \eta_i^{n+1} = \eta_i^{n-1} + \frac{\Delta t}{\Delta x} \frac{1}{B_i} (A_{i-1} U_{i-1}^n - A_{i+1} U_{i+1}^n)$

As the boundary condition at the mouth of Trevor Channel, η is prescribed as a function of time through sinusoidal profiles with different amplitudes (one to two feet) and periods (15 to 90 minutes). At the head of Alberni Inlet, the Somass River discharges, on the average, 3910 cubic feet of water per second (Tully, 1949) and this is taken into account. Midway down the inlet, the Nahmint River discharges about 739 cubic feet of water per second, but this is ignored. The results of this crude model show the following features. The variation of water level at a given position in the inlet is roughly sinusoidal and does not offer any interesting features. However, the variation of the water level at a given time as a function of distance along the inlet is more interesting. Figure 4 shows this variation at three different times, assuming the water level was uniform at t = 0. To make this model realistic, at least two improvements are necessary: A, B, and H should be made functions of t also, and tide-gauge data at the mouth of Trevor Channel should be obtained.

SPECTRA OF THE TIDAL RECORDS

The tidal records for the 1964 tsunami were enlarged photographically and then digitized at intervals of 1.76 minutes. The power spectra of these were computed using the Fast Fourier Transform Technique. The tide-gauge records at North Arm, Vancouver, Pitt Lake, Point Atkinson, New Westminister, and Steveston showed negligible tsunami action where the astronomical tide completely dominated the power spectrum. Port Alberni, Tofino, and Tasu were selected as the most interesting stations for the following reasons. The maximum destruction in Canada occurred at Port Alberni and thus it is an important station. The Tofino gauge location has some features of an open ocean, though the tsunami is already slowed down on the continental shelf. Tasu is situated on Queen Charlotte Islands and is thus far away from the rest of the gauges.

Figure 5 shows the power spectra for Tasu, Tofino, and Port Alberni. Besides the tidal periods, the most prominent peak is at 61.6 minutes for Tasu. The source of this peak has not been determined satisfactorily. It could be a free oscillation on the continental shelf. For Tofino the two largest peaks are 87.3 and 31.8 minutes. The sources of these peaks are not certain. For Port Alberni the highest values of power occurred at 116.4 and 87.3 minutes, in addition to the tidal periods. In this case, the sources of these peaks are identified with some certainty because (Table 1) the

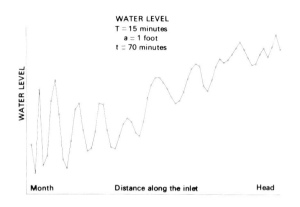

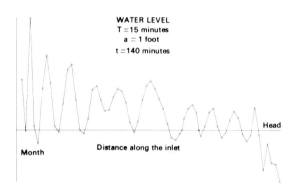

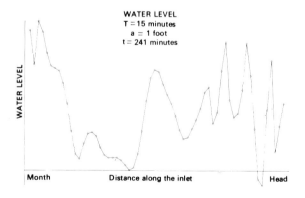

Fig. 4: The water level at the head of Alberni Inlet at times 70, 140, and 241 min. for a tsunami with 15-min. period and with an amplitude of one foot at the mouth of Trevor Channel. The largest and smallest amplitudes in these 3 cases are, respectively, 1.99, 0, 1.37, -0.64, 0.65, -0.07 feet.

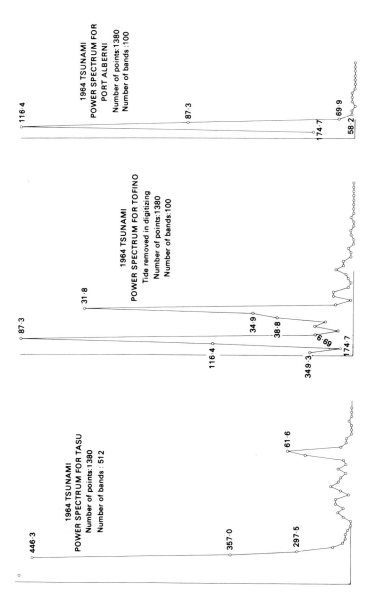

Fig. 5: Power spectra for Tasu, Tofino, and Port Alberni for the 1964 tsunami. The abcissa is the frequency and the ordinate is the variance normalized so that the total variance is 100. The numbers shown are the periods in minutes. For the Tasu and Port Alberni spectra, the tidal periods were not included in the plots because their inclusion would compress the nontidal peaks so drastically that the relative differences among them could not be seen.

periods of the fundamental mode for Trevor Channel-Alberni Inlet and Alberni Inlet are around 125 and 87 minutes respectively.

The cepstra of the tidal records are calculated using the concepts developed by Bogert, Healy, and Tukey (1962) with the aim of isolating any reflections that were present. The manner in which the cepstrum isolates the reflections can be seen from the following discussion. For a simple reflection, values of a time series $Y(t)$ were multiplied by a constant α, delayed by a time difference τ and added to the original series to give a new series $Z(t) = Y(t) + Y(t-\tau)\alpha$. If $\phi(f)$ is the power spectrum of $Y(t)$, then the power spectrum of $Z(t)$ is $\phi(f) [1 + 2\alpha \cos(2\pi f \tau) + \alpha^2]$ whose logarithm is approximately $\log \phi(f) + 2\alpha \cos(2\pi f \tau)$, ignoring terms of order α^2. Thus the effect of adding the reflection to the original series is obvious in terms of the log-power spectrum, to which is added a (nearly) cosinusoidal ripple whose parameters are simply related to the reflection parameters α and τ.

All the tide stations in the Strait of Georgia that showed little tsunami action have many peaks in the cepstra. Figure 6 shows the cepstra for Pitt Lake and Port Alberni. That for Pitt Lake could be just noise, because the logarithms were taken for quantities very close to zero. For Port Alberni there was one major peak at 6.4 minutes; surprisingly, the noise level was quite low. To reduce the noise, the low power part of the original spectrum was removed and the cepstrum computed. This procedure still showed the peak at around 6.5 minutes, but the amplitude had decreased considerably, as shown.

The cross-spectra (coherence and phase) were also computed between North Arm-Steveston, North Arm-New Westminister, Vancouver-Point Atkinson, Vancouver-Pitt Lake, Victoria-Fulford Harbour, Ocean Falls-Bella Bella, and Port Alberni-Tofino. As was expected, the tide stations in the Strait of Georgia showed strong coherence. Figure 7 shows the coherence and phase between Port Alberni-Tofino and Ocean Falls-Bella Bella. Surprisingly, the coherence is strong for these two cases in certain periods. For the 1960 Chilean earthquake tsunami, Loucks (1962) computed the cross-spectra between Barkley Sound and Tofino and found the coherence to be weak (0.2) in all periods and phase difference to be a smooth function of the period.

MISCELLANEOUS RESULTS

It has been stated (Van Dorn, 1968) that the tide-gauge records for a given station are similar for different tsunamis. That this may not be true in all cases can be seen from Figure 8. Comparison is made of five tidal records from the 1960 Chilean earthquake tsunami and the 1964 Alaska earthquake tsunami.

Stoneley (1963) considered the theory of propagation of tsunamis across the ocean and gave the formula for the time interval T between the arrival of the first crest and of the second crest at a tide station. This is generally called the period of the tsunami, i.e.,

$$T = 0.03058(\frac{x^2}{t})^{1/3} ,$$

where x is the distance between the source of tsunami and the tide gauge and t is the travel time (the time it takes for the first wave to arrive at the tide station). The multiplicative constant in the above expression holds when c.g.s. units are used. This formula was used to compute the period for several tide records. These are compared with observed periods in Table 3. The calculated periods were, on the average, an order of magnitude smaller than the observed periods. The calculations were repeated for three stations for the 1960 tsunami and, probably because of the much larger travel distances involved, the agreement was somewhat better. There appears to be no correlation between the period of the tsunami and the maximum rise or fall of the water level.

Yoshida (1959) stated that the portion of the tsunami containing the maximum energy (highest wave) will occur earlier for smaller travel distances and later for larger travel distances. In other words, the difference in time-of-arrival between the first and the highest wave increases with the travel distance. This hypothesis was examined for the 1964 tsunami through a comparison (Table 3) of the period of the highest wave (the time interval between the arrival of the first wave and the highest wave at the tide station) and the number of the wave that is highest. Some stations bear out Yoshida's hypothesis while others do not.

Yoshida also defined two types of tide stations, namely, Continental (along the continental edge of the ocean) and

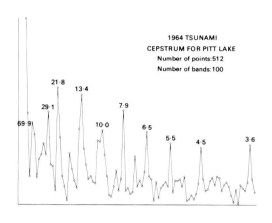
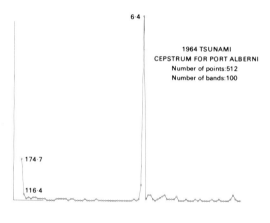
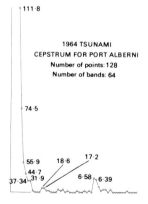

Fig. 6: Cepstra for Pitt Lake and Port Alberni (calculated in two different ways). Numbers shown are the periods in minutes.

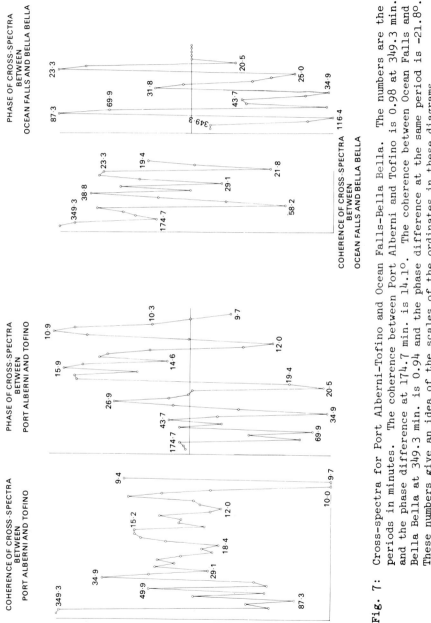

Fig. 7: Cross-spectra for Port Alberni-Tofino and Ocean Falls-Bella Bella. The numbers are the periods in minutes. The coherence between Port Alberni and Tofino is 0.98 at 349.3 min. and the phase difference at 174.7 min. is 14.1°. The coherence between Ocean Falls and Bella Bella at 349.3 min. is 0.94 and the phase difference at the same period is -21.8°. These numbers give an idea of the scales of the ordinates in these diagrams.

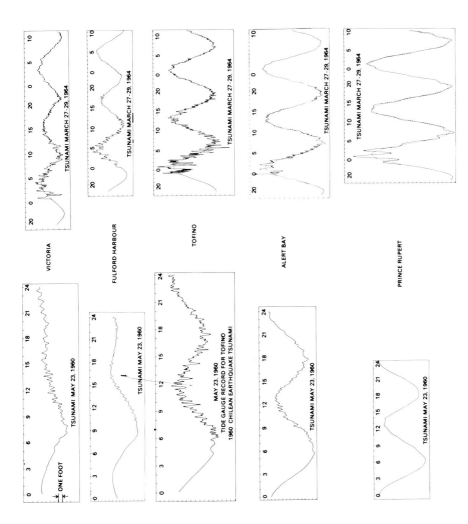

Fig. 8:
Comparison of Tidal Records for the 1960 Chilean Earthquake Tsunami and the 1964 Alaska Earthquake Tsunami.

Oceanic (or island) stations. For the two types, he plotted the period of the highest wave against the travel time and concluded that they show nearly a straight line with the slope and origin being different for each type. He found the slope to be 1/3 for Continental stations and 1/10 for Oceanic stations. Also, as the travel distance approaches zero, the first wave tends to be the highest for Oceanic stations, whereas the maximum wave will arrive about three hours later for the Continental stations. Table 3, column 8 shows the slope for the 1964 tsunami. Probably all these stations fall

Table 3: Some Relevant Miscellaneous Information About the 1964 and 1960 Tsunamis.

1 station
2 great circle distance
3 observed travel time
4 observed period
5 calculated period
6 maximum rise or fall
7 period of highest wave
8 ratio of period of highest wave to travel time
9 number of highest waves

1	2	3		4	5	6	7		8	9
	miles	hrs	min	min	min	ft	hrs	min		
1964 Tsunami										
Victoria	1285	4	26	50	7.1	- 4.8	0	16	0.06	9
Fulford Hr	1265	4	59	40	6.8	2.0	5	18	1.06	9
Vancouver	1247	5	44	120	6.4	0.6	1	45	0.31	
Pt Atkinson	1240	5	31	90	6.4	0.8	3	43	0.67	
Port Alberni	1201	4	24	87	6.8	>17	2	00	0.45	2
Tofino	1175	3	24	20	7.3	- 8.1	1	50	0.54	10
Alert Bay	1069	4	03	29	6.5	- 5.7	0	14	0.06	1
Ocean Falls	956	4	24	32	5.8	-12.5	0	25	0.09	3
Bella Bella	952	3	17	39	6.4	- 6.3	0	31	0.16	3
Tasu	815	1	57	70	6.9	- 6.3	0	19	0.16	16
Prince Rupert	783	3	16	92	5.6	8.9	1	20	0.41	2
1960 Tsunami										
Tofino	6820	17	10	60	13.7					
Cape St James	7112	17	10	28	14.1					
McKenney Is	7104	17	52	18	13.9					

under Yoshida's definition of Continental stations. As can be seen, there is large scatter ranging from 0.06 to 1.06. In regard to the number of the wave that is the highest, here again there appears to be no definite pattern against the travel distance.

Though a storm did not coexist with the 1964 tsunami, it may be useful to estimate, at least roughly, what might be the maximum contribution to the water level from an atmospheric pressure disturbance such as a travelling storm. To make this estimate, the theory of Rao (1969) for the effect of a travelling disturbance on a rectangular bay uniform in width and depth and open to the sea at one end is used. The maximum water level at the head of the bay is given by $\eta = \frac{2L}{gH} R$. For the Trevor Channel-Alberni Inlet, the length L is 292,922 ft. and the average depth H is 365 feet. For the wind stress R the formula ordinarily used in the literature is $R = 4 \cdot 10^{-6} \cdot W|W|$, where W is the wind speed (ft./sec.) along the length of the inlet. Assume a typical value for W as 75 ft./sec. (51 mph) and take $g = 32$ ft./sec.^{2}; then the maximum elevation of the water level at the head of Alberni Inlet is 1.1 feet. This estimate is rough and did not take into account the non-linear interactions between the tsunami and the storm surge.

The effect of a barrier of different sizes at different positions on the water level at the head of Alberni Inlet has been considered. Before embarking upon a complicated model with the topography taken into account, simple analytical solutions given by Fukuuchi and Ito (1966) for a bay assumed to be rectangular in shape and uniform in width and depth were examined. If B and B_1 are the widths of the bay and the imaginary outer sea, then the transmission coefficient at the mouth and the reflection coefficient at the head are given by

$$P = \frac{2B_1}{B} \Big/ (\frac{B_1}{B} + 1) \quad \text{and} \quad Q = (\frac{B_1}{B} - 1) \Big/ (\frac{B_1}{B} + 1)$$

Let T_o be the period of free oscillation of the bay, and let η be the water level at the head. If a is the amplitude and T the period of the tsunami at the mouth, the water level at the head is given by the following formulae:

At $t = 0$, $\eta = 0$. At $t = T_o$, $\eta = 2$ Pa sin $(\frac{2\pi T_o}{T})$

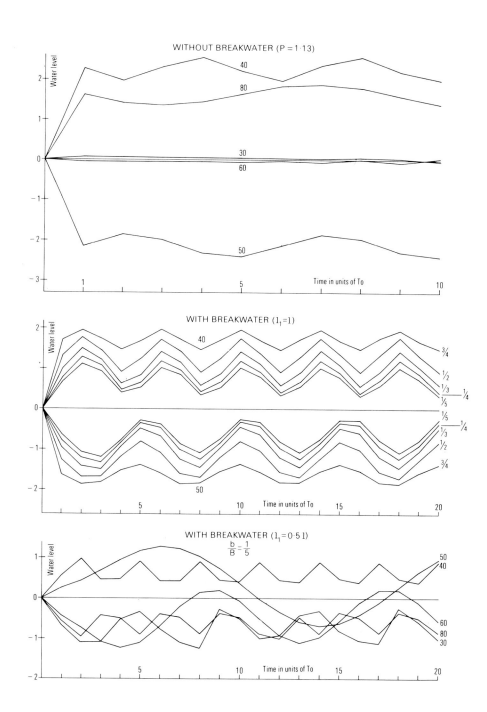

At $t = nT_o$ $(n=2,3,...)$

$$\eta = 2\, Pa \sin\left(\frac{2\pi T_o}{T}\right) + \sum_{m=2}^{n} 2\, P\, Q^{m-1}\, a \sin\left[\frac{2n\pi T_o}{T} - (m-1)\right.$$

$$\left. \cdot \left(\pi + \frac{2\pi T_o}{T}\right)\right]$$

The same equations could be used to calculate the water level after the barrier is constructed, provided the gap of the barrier is treated as the width of the imaginary outer sea.

Figure 9 (top) shows the water level at the head of Alberni Inlet as a function of time without a barrier and treating the mouth of Trevor Channel as the imaginary outer sea. The water level at the head is very sensitive to the tsunami period. Figure 9 (middle) shows the water level at the head for two different periods of the tsunami as a function of time for five different barrier gaps. Here b/B is the ratio of barrier gap to the average width of the inlet. The positive water levels are for T = 40 min. and the negative water levels are for T = 50 minutes. In this case, the barrier is located at the mouth of Trevor Channel. As is expected, increasing the barrier size (reducing the gap) decreases the water level at the head. Otherwise, qualitatively there is not much difference whether there is a barrier at the mouth of Trevor Channel. If the barrier is moved inside, then there are qualitative differences from the no-barrier case. As can be seen in Fig. 9 (bottom), the water

Fig. 9: The top diagram shows the water level at the head of Alberni Inlet for tsunamis with periods 30, 40, 50, 60, and 80 minutes. The ordinate is the amplitude in units of a (the amplitude of tsunami at the mouth of Trevor Channel). The middle diagram shows the water level at the head of Alberni Inlet with barrier (breakwater) at the mouth of Trevor Channel ($\ell_1=\ell$) for tsunami periods of 40 and 50 min. and for different ratios of barrier gaps to width of the inlet. The bottom diagram shows the water level at the head of Alberni Inlet with barrier midway in the system ($\ell_1=0.5\ell$) for tsunami periods 30, 40, 50, 60, and 80 min. and ratio of barrier gap to width as 1/5.

level at the head varies in time with a larger amplitude for the 30- and 60-minute periods than when there is no barrier, while the amplitude of variation reduces for the 40-, 50-, and 80-minute periods. To make this model realistic the topography has to be taken into account.

ACKNOWLEDGMENTS

We thank Mr. Fred G. Barber for enlightening discussions on several aspects of this work. We also had the benefit of occasional discussion with Dr. G. L. Pickard and Mr. S. O. Wigen. Mr. John Taylor and Dr. Curtis Collins clarified many doubts in the power spectral techniques. Mr. David Greenberg did part of the programming. Mr. Maurice Isabelle provided the illustrations, and Mr. G. C. Dohler provided a major part of the data.

BIBLIOGRAPHY

Bogert, Bruce P., Healy, M. J. R., and Tukey, John W. 1962. The quefrency alanysis of time series for echoes: cepstrum, pseudo-autocovariance, cross-cepstrum and saphecracking. *Proceedings of the Symposium on Time Series Analysis*. Edited by Murray Rosenblatt. John Wiley and Sons. Pp. 209-243.

Fukuuchi, Hiromasa and Ito, Yoshiyuki 1966. On the effect of breakwaters against tsunami. *Proceedings of Tenth Conference on Coastal Engineering*. Vol. II. Published by the American Society of Civil Engineers. Pp. 821-839.

Loucks, R. H. 1962. Investigation of the 1960 Chilean tsunami on the Pacific Coast of Canada. *M. S. Thesis*. Dept. of Physics. The University of British Columbia. Aug. 1962. 21 pages.

Mortimer, C. H. 1955. Some effects of earth's rotation on water movement in stratified lakes. *Verhandl. Intern. Ver. Theoret. Angew. Limnol.* 12:66-77.

Platzman, George W. and Rao, Desiraju B. 1964. The free oscillations of Lake Erie. *Studies in Oceanography* (Hidaka volume). Univ. of Washington Press. Edited by K. Yoshida. Pp. 359-382.

Rao, Desiraju B. 1969. Effect of travelling disturbances on a rectangular bay of uniform depth. *Archive Fur Meteorologie, Geophysik und Bio-climatology* (in press).

Stoneley, Robert 1963. The propagation of tsunamis. *Geophysical Jr. Roy. Astron. Soc.* 8(1):64-81.

Tully, John P. 1949. Oceanography and prediction of pulp mill pollution in Alberni Inlet. Fisheries Research Board of Canada. *Bulletin No. 83.* 169 pages.

Van Dorn, W. G. 1968. Tsunamis. *Contemporary physics.* 9(2): 145-164.

Wigen, S. O. and White, W. R. H. 1964. Tsunami of March 27-29, 1964. West Coast of Canada. *Dept. of Mines and Technical Surveys.* Aug. 1964, 6 pages.

Yoshida, Kozo 1959. A hypothesis on transmission of energy of tsunami waves. *Records of Oceanographic Works in Japan.* Feb. 1959, pp. 14-37.

TSUNAMI INSTRUMENTATION

13. Tide-Gauge Data Telemetry between the Tsunami Warning Center at Honolulu, Hawaii and Selected Stations in Canada

G. DOHLER
Dept. of Energy, Mines and Resources
Ottawa, Canada

ABSTRACT

The resolution on the international aspects of the Tsunami Warning System in the Pacific, as adopted during the fourth session of the Intergovernmental Oceanographic Commission, indicated the importance of providing timely warnings of the approach of tsunamis.

In 1966, Canada, realizing the importance of such timely warnings, installed equipment capable of reporting, unattended and automatically, water-level data of significance to the Tsunami Warning Center.

INTRODUCTION

For an effective and efficient warning system, interested parties must have direct communication links with tide gauge stations around the Pacific Ocean.

As the rental of telephone lines covering great distances for telemetry of continuous tidal observations is expensive, and since suitable radio channels are not easily available, a system providing instantaneous tidal data had to be developed which would:

1. Have available for comparison full predictions of high- and low-water levels and the corresponding times, or predictions at specific intervals for selected gauge stations.

2. Have access to a power source and a telephone system in the vicinity of the gauge.

3. Have access by telephone and/or telex to gauge values.

4. Store past water-level heights and the rising and falling tendency at the gauge station.

5. Incorporate a tele-announcer which would call the tsunami warning service automatically should a tsunami occur.

6. Have data available in a suitable format for easy interpretation by a small group of users.

TSUNAMI WARNING STATION LOCATIONS

Three locations were selected at which water-level data of significance to the Warning Center at Honolulu would be obtained. The stations are Tofino and Victoria, on Vancouver Island, and Langara Island at the northwesterly tip of the Queen Charlottes (Figure 1). The special equipment at these locations is operated in conjunction with equipment used for ordinary tide-gauging functions.

THE TELE-ANNOUNCING SYSTEM

The basic components of the system (Fig. 2) were already available through Hagenuk/Kiel, West Germany, with whose co-operation a number of special features were added.

The tele-announcing system (Fig. 3) consists of

(a) A data transmitter mounted over a stilling well; analog data are converted into electrical digits by means of a device which incorporates a chain pulley and float.

(b) A memory unit to store minimum or maximum water level information and ten successive water levels.

(c) An announcing unit which, when addressed, announces the existing water level, the rising or falling tendency, the last high or low water, or any of the other stored data.

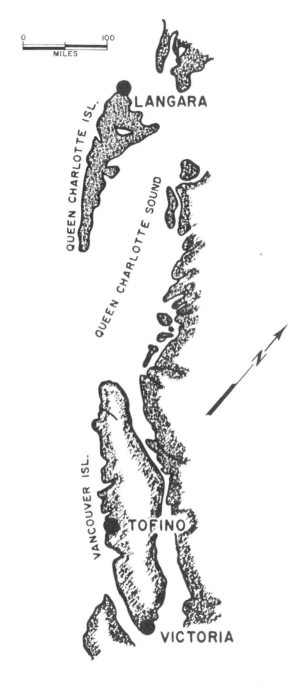

Fig. 1: Map of Queen Charlotte Sound Area.

Fig. 2: Components of Tele-Announcing System.

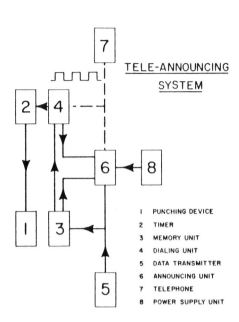

Fig. 3:
Block Diagram of
Tele-Announcing System.

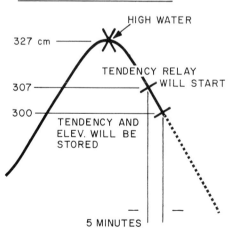

Fig. 4:
Sample Illustration of
the Tele-Announcing System
in Normal Tide Mode.

(d) A dialing unit which automatically calls a predetermined party in the event of an unusual water-level condition.

(e) A power supply to convert the 117-volt, 60-cycle line voltage into the voltages required by the operating parts of the unit.

(f) A punching device to collect, upon request, 24 hours of water level at one-minute intervals.

The tele-announcer takes the place of an ordinary telephone set and is addressed by dialing the number of the telephone lines. The ringing current activates a relay which closes the D.C. line and initiates the internal sequence of operation. Upon completion of this sequence, the unit disconnects itself and is ready for another telephone call.

In the event of an unusual water-level change, the dialing unit produces a series of pulses corresponding to a coded telephone number and switches on the tele-announcer. All necessary information is stored in a magnetic disk on a number of separate tracks which contain station identification, numerals, tendency, and end of announcement. The order in which the tracks are played back is controlled by a sequential switching arrangement. The sequence of numerals in the instantaneous water-level announcement is determined by the position of the digitizer in the data transmitter. The sequence of numerals of the stored data is determined by the memory relays in the memory unit.

At the Victoria station only, should an unusual water level occur, a Punch-Paper-Tape Recorder is activated to store water-level changes at one-minute intervals over a pre-set time period. This data is available to the authorities either immediately or shortly after the event. The punch-paper tape is coded in such a way that telex facilities can be used.

Modes of Operation

For a better appreciation of the tsunami warning gauges installed at Tofino and Victoria, it is imperative to understand the automated sequential operations performed by the instrument and the three basic modes of operation which consist of:

(a) The Normal Tidal Mode
(b) The Warning Mode
(c) The Sampling Mode

The Normal Tidal Mode (Fig. 4). The incident water level is available at any time from a mechanical-electrical digitizer connected to a float with chain and counterweight. Water-level changes corresponding to the existing value, which in turn is referred to a fixed and agreed upon datum plane, are transmitted in centimeters to the digitizer.

The instantaneous water-level announcement preceded by "automatic announcing service -- water level in centimeters" gives, in every mode of operation, precisely the level at the time of interrogation. The other two announcements during this normal operation are the tendency and the words "last high water" or "last low water", followed by the respective values thereof. The words "No tsunami registered" and "End of Announcement" conclude the cycle of this mode.

A tendency change will not activate the memory unit of the apparatus at the time the highest or lowest water elevation has been reached, nor will it be at the highest or lowest elevation stored in the memory unit unless the following conditions are met:

(a) The amplitude causing a reversal must be more than 20 centimeters for Tofino and 14 centimeters for Victoria.

(b) After this vertical change has been reached, the tendency must continue in the same direction for at least five minutes.

If these two conditions are fulfilled, the new tendency and the new water-level value are entered in the memory unit of the instrument. The tsunami-warning gauges at Victoria and Tofino are adjusted to the settings just mentioned.

It is important to understand this sequential-delay arrangement and, to illustrate how it operates, assume the tide is rising. Upon interrogation, the announcement is:

> Automatic Announcing Service
> Water Levels in Centimeters 306, 306
> Tendency Rising Last Low Water 213, 213
> No Tsunami Registered   End of Announcement

The water level is still rising and the point of high water will be reached at say 327 centimeters. The water must now fall 20 centimeters (14 centimeters at Victoria) before the tendency relay will start running for five minutes. After these five minutes, the new tendency will go into the tidal register and from now on the announcement will be ... Tendency Falling, Last High Water, 305 ... Observe that it will not announce 307 since this was the level at which the tendency delay started running and during the five minute time lapse, the water level has fallen further before the new high water was registered.

These delays are necessary to reduce the possibility of false messages. Any disturbances which are in excess of the aforementioned conditions, and which are different from normal tidal oscillation, will be regarded by the instrument as a tsunami. Human interpretation of these oscillations is necessary in arriving at a conclusion and in making decisions.

The Warning Mode (Fig. 5). For the stations Victoria and Tofino, the time difference between successive high and low waters has been established at 4.5 hours. If a change of tendency occurs within this period, an alarm signal will be triggered. For this reason, an adjustable timer has been employed which, after running for 4.5 hours, will disconnect itself from the circuitry. In other words, while this timer is switched on, the operation sequence of the instrument is changed from normal to warning, and the signal originated by a change of tendency will not, during that time, initiate storage of the instantaneous water level in the tidal register.

Instead, the level will be stored in the first of the 10 tsunami registers. Following this, the apparatus will dial a telephone number to deliver its warning message, ... Warning Tsunami, Warning Tsunami ... This message will be sent for about two minutes to a pre-assigned number. At the same time and by means of a set of slave contacts, other processes are initiated. At Victoria, for example, a punch-paper-tape recorder is switched on to operate for 24 hours, punching at one-minute intervals the course of the water level.

The Sampling Mode (Fig. 6). After the warning message has been delivered, the unit will change to the sampling sequence of operation. It will collect the water level nine more times at intervals of five minutes. This means that, some 7 to 8 minutes after a warning message has been initiated, all other values will enter the tsunami register at intervals of five minutes if the apparatus remains undisturbed by outside calls. If addressed before all the samples have been accumulated, the unit will recount to the caller the numerical value of the samples taken up to that particular time. For the remaining registers, however, only zeros will be announced.

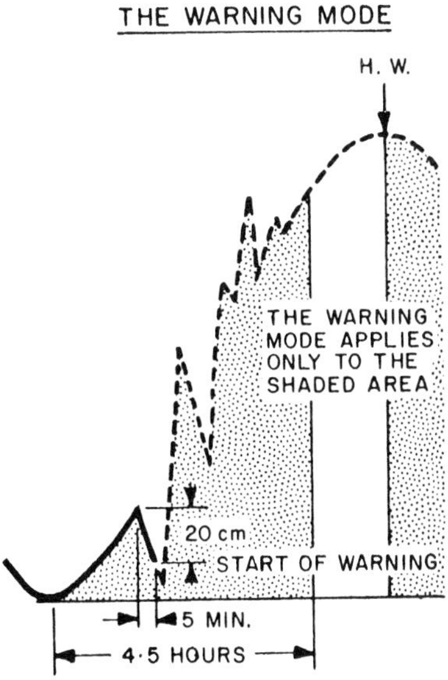

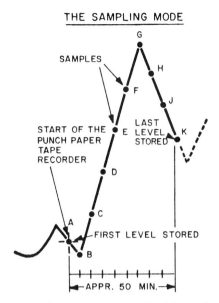

Fig. 5: Sample Illustration of the Tele-Announcing System in Warning Mode.

Fig. 6: Sample Illustration of the Tele-Announcing System in Sampling Mode.

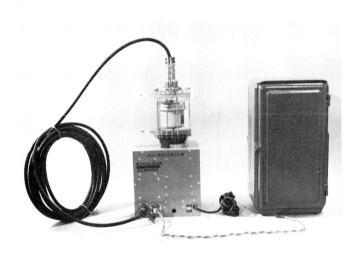

Fig. 7: Components of Pressure-Recording System: Sensor (center), Cable (left), and Recorder (right).

For example:

>Automatic Announcing Service
>Water Level in Centimeters
>250, 250
>Tendency Rising
>Tsunami Registered as follows:
>Sample a 259, Sample b 232
>Sample c 317, Sample d 294
>Sample e 322, Sample f to Sample k 000

Depending on how often the station was addressed during the sampling time, a complete announcement will be available some 50 minutes after the warning message was delivered. After this, all values will remain in the register until the timer (which in our case was set to 4.5 hours) has run out. It should be noted that, during this time, the station, if interrogated, will repeat the same information.

Should an oscillation, which does not meet the conditions described earlier, occur after the elapsed time of 4.5 hours and before a real tide change is reached, the instrument will start its three modes of operation in the manner previously described.

If the electrical supply to the station was interrupted during a tsunami sampling period, the instrument will be reset with the return of the power supply and will announce a normal tidal condition. Only further oscillations or a real tidal change after a power break will reset the station to announce in its proper sequence. The punch-paper-tape recorder will also stop collecting until the power is restored.

POTENTIOMETRIC PRESSURE TIDE GAUGE AND PULSE FREQUENCY SYSTEM

At the Tofino and Victoria stations, changes in water level are converted by means of a float and counterweight arrangement. At Langara Island, however, a potentiometric type absolute-pressure transducer is used.

The original pressure-recording system was developed by the National Research Council in 1957 at the request of the Tides and Water Levels Section for the purpose of obtaining tidal data in the Arctic, or in areas with long, low-lying foreshores. The system consists of the pressure head, the connecting three-conductor cable, and the recorder unit (Figure 7).

Principle of Operation (Figure 8)

The instrument determines the height of a column of water by measuring the hydrostatic pressure with a resistance potentiometer-type absolute-pressure transducer. Changes in pressure are converted to resistance changes in the transducer by means of a linkage system, which connects the sliding arm of the transducer potentiometer to a pressure-sensitive metal diaphragm. The position of the slider of the potentiometer, which is controlled by the total pressure applied to the transducer, is obtained by having this potentiometer form part of a servo-balanced resistance bridge circuit. The servo system in the shore recorder, sensing the unbalanced current in the bridge circuit caused by the movement of the transducer potentiometer slider, drives a precision 10-turn potentiometer in the correct direction to restore the bridge balance.

Atmospheric pressure variations which affect the total pressure measured by the submerged pressure transducer are compensated in the recorder unit. The punch-paper tape recorder can be operated either with a 110-volt A.C. or D.C. power supply.

Transducer. This element converts the physical change into an electrical potential. The pressure transducer is one side of a balancing bridge circuit with a ten-turn potentiometer, driven through reduction-gearing by a servomotor. The ideal transducer would have a voltage output exactly proportional to the pressure acting upon it. Practical transducers, however, are electrically defined in such terms as linearity, resolution, repeatability, hysteresis, and a few other parameters. In many respects, a practical transducer is obviously far from ideal and consequently, the transducer is the element in the operational chain which limits the accuracy of the records obtained.

Repeating Element. The electrical magnitude (voltage) is converted into a readout-configuration (punched tape). The unit is calibrated to make the readout actually represent the water level in feet and tenths. The practical readout here is a punch-paper-tape recorder which gives a four-digit, four-level punch configuration.

Mechanically coupled to the readout device is the repeating element, which consists of a ten-turn high-linearity potentiometer. The linearity and resolution of the

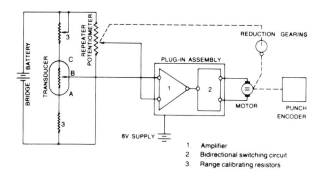

Fig. 8:
Schematic Diagram of the Pressure-Recording System.

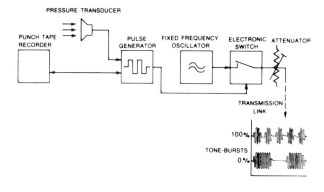

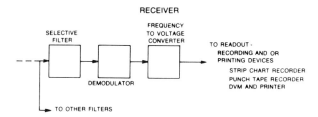

Fig. 9:
Block Diagram of the Pulse-Frequency System
Used for Data Transmission.

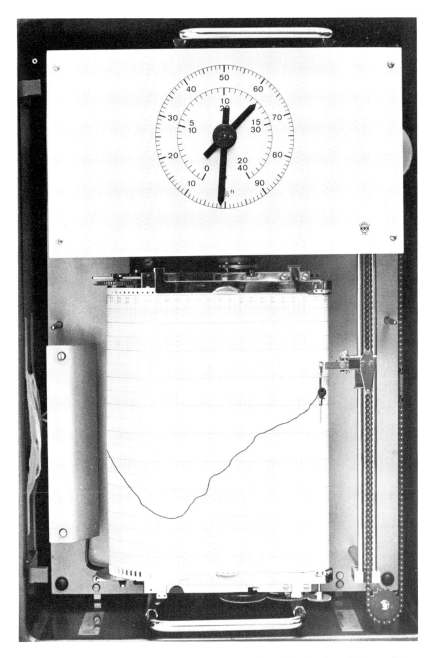

Fig. 10: Recorder at Central Station for Displaying Information Transmitted on Pulse-Frequency System.

repeater potentiometer are far superior to the corresponding parameters of the transducer.

Error-Sensing Amplifier. This portion of the circuitry compares the output voltage from the transducer and the output voltage from the repeater potentiometer. If there is a difference between these two voltages, the difference is amplified to the power level required to drive a small motor, which drives the repeater-potentiometer/punch encoder. The motor used is a permanent-magnet D.C. motor, and the direction of rotation is determined by the polarity of the difference in voltage.

As far as is known, the resolution of the best potentiometric transducer presently available is nominally given as 0.05 percent, a figure which has been accurately verified in practical tests. On this basis, a difference in voltage of approximately 2.5 mv. at a supply voltage to the transducer of 5 v. will produce, after amplification, enough power to operate the motor.

For example, a pressure transducer with a range of 0 to 100 foot-equivalent water pressure, at a resolution of 0.05 percent (one part in 2000), would indicate a theoretical resolution equal to 5/100 of a foot of water level.

Pulse-Frequency System (Figure 9)

Transmission of data is carried out by the pulse-frequency system. At the sampling stations, a potentiometric transducer which precedes the sound transmission, converts the resistance or voltage values into a pulse-frequency. This frequency continuously follows the changing measuring values in an envelope between 5 and 15 Hertz. The measuring value 0% equals 5 Hertz and value 100% equals 15 Hertz of the pulse frequency.

At the central or receiving stations, the modulated signal is transformed into D.C. pulses; a converter provides the necessary direct current to operate the recording or indicating equipment (Figure 10). Within the temperature range of -20° to +60° centigrade, the transmission error will not exceed 1.5%.

Structure. All functional units of the system are on plug-in printed circuit boards. A hermetically-sealed container protects the units from mechanical damage, and proper contact is ensured by indirect plug-in devices. The operation of each unit is described on the front plate, and the units are clamped to the cassette by a mechanical locking device. Uniformly wired cassettes are available to contain the various units, thus offering a choice of layouts.

Any transmitter may be used providing:

(a) it can be modulated with the corresponding audio channel,
(b) the power output is sufficient to cover the desired distance, and
(c) it can operate either continuously or from a time-dependent signal on a clear channel.

The receiving equipment should be capable of delivering a minimum of 3 v. R.M.S. undistorted audio output. Successful tests using Motorola equipment have covered distances of up to 100 miles in the 30 to 40 megacycle band. If the underwater unit and underwater cable prove capable of withstanding storms for the next year, the equipment will be installed at the Langara Island gauge to transmit data to Prince Rupert on a continuous basis.

CONCLUSION

The automatic announcing system has now operated reliably for over two years with very little maintenance. At the Tofino and Victoria stations, down time during the same period has been at a minimum. The possibility of modifying the equipment for interrogation with telex facilities throughout the normal tide cycle or whenever a tsunami has been recorded is under investigation. Evaluation of the potentiometer pressure tide gauge and pulse-frequency system will require a much longer period of operation.

It may be worthwhile to note that several attempts have been made in the past to have equipment at Langara Island, British Columbia, to measure the arrival of the "great waves" for the tsunami-warning service. This equipment installed by units of the Department of Energy, Mines and Resources is the first equipment actually to be operated at this location.

BIBLIOGRAPHY

Department of Energy, Mines and Resources 1967. *The Collected Papers and Reports of the Tides and Water Levels Section.* Volume I.

Hagenuk/Kiel, PA 791, 1968. Tonfrequenz-Fern Wirk System. TF 24.

National Research Council of Canada. *Proceedings of Hydrology Symposium No. 7.* Victoria, British Columbia.

14. Recent Advances in Tsunami Instrumentation in Japan

K. TERADA
*National Research Center for Disaster Prevention
Tokyo, Japan*

ABSTRACT

   In Japan, tsunamis are generally experienced along the seashore facing the Pacific Ocean. Furthermore, the earthquakes occurring in the Aleutian Islands and Chile sometimes cause tsunamis on Japanese coasts.

   Hitherto, the observations of such tsunami were done with ordinary tide-gauges installed at tidal stations of various coasts of Japan. Therefore, the record of tidal variation is available only at tidal stations which are generally located far from a meteorological observatory.

   Since 1955, several attempts have been made by the Japan Meteorological Agency (JMA) to find a way of remote reading of this tidal station data, and a specially designed tide gauge was successfully invented. In 1959, Japan was attacked by a severe typhoon, Vera, and more than 5,000 persons were killed, mainly by the storm surge.

   Therefore, the JMA developed the above remote-reading technique and utilized it for the warning of storm surges by obtaining instantaneous data of the coastal tidal stations by wireless signal. For this purpose, so-called "Call System" was used. Namely, the mother station can collect various data by a scanning technique. Total duration of this scanning is about one minute. Accordingly, at the mother station, various records can be obtained nearly continuously.

Recently a similar telemetering receiving set was installed at Sendai Meteorological Observatory. The tidal station is located at Ayukawa and the record obtained by an ordinary Fuess-type tide gauge installed in a well is transferred to the wireless signal by voltage transformation. At Sendai, the received data is recorded on the Fuess-type tide gauge recorder.

By this method the record of tsunami due to the Tokachi-oki Earthquake (May 16, 1968) was received at Sendai. This shows that this method will be valuable for tsunami research.

JMA is currently sending the tide gauge data obtained at Mera to the Tateyama Observatory by a digitized transmission system. This equipment is especially designed as a Tide-Level Telemetering system by using ultra-high frequency (UHF) (450 MHz.) band radio waves. In order to augment the accuracy of transmission, the parity check system is introduced. Therefore, these tide records at two stations coincide within a range of error of ± 1 cm. or so. This system also has sufficient accuracy to be useful for tsunami research.

As early as 1965, a submerged long-wave recorder of the Snodgrass-type was developed by Dr. Takahashi and others of the University of Tokyo.

INTRODUCTION

In Japan, tsunamis are generally experienced along the seashore facing the Pacific Ocean and caused by the earthquakes whose epicenters lie off the coast of Sanriku; but sometimes tsunamis occur at the seashore of the southern part of Japan. Furthermore, earthquakes in the Aleutian Islands or Chile sometimes cause tsunamis on Japanese coasts.

Hitherto the observations of such tsunami were done with ordinary tide-gauges installed at tidal stations on various coasts of Japan. Namely, sea-level change is recorded on an ordinary recorder, corresponding to the motion of a float in a well. Therefore the record of tidal variation is available only at tidal stations generally located far from a meteorological observatory. Such tidal station data was very inconvenient for the warning of tidal variation in cases of typhoons and earthquakes.

TELEMETRY

Several attempts have been made to find a way of remote display of this tidal-station data. In 1955, Nagasaki Marine Observatory succeeded in obtaining tidal data of the seashore through telemetering cable.

In 1960, this technique was improved to provide good linearity between the displacement and the output voltage. As is shown in the schematic diagram of Fig. 1, the measuring head is a cylindrical metallic tube having a diameter about 140 mm., a length about 240 mm., and, at the end of the tube, the water pressure acts on the bellows through the teflon membrane. Inside the teflon there is distilled water so that the bellows might not be subject to corrosion due to direct contact with sea water.

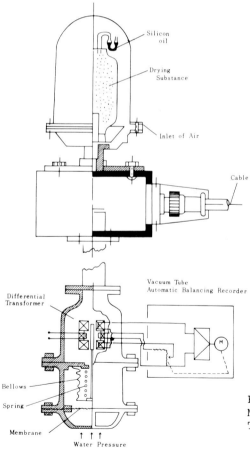

Fig. 1:
Measuring Part and Head of Tide Gauge.

The spring inside the bellows is in equilibrium with the water pressure, and the variation of water pressure causes the lateral displacement of the core between the coils of differential transformer. The transformer secondary output voltage change is designed to be perfectly linear to the displacement of the core, hence to the water pressure change. This core displacement, for the measuring range from 0 to 10 meters, is 2 mm. and the corresponding output voltage change is 63 mv. A.C.

In order to eliminate the effect of the atmospheric pressure on the bellows, the inner part of the bellows is connected with the atmosphere through the pipe, and at the upper part of the pipe a U-tube is attached. At the bottom of this U-tube some oil is inserted to eliminate the effect of the atmospheric pressure variation. Between this tube and the pipe a small chamber is installed as the drying head. This small chamber is designed to have the capacity to dry up the air of the inner part perfectly; due to this drying substance the electric materials are perfectly protected from bad effects on the electric circuit. A self-balancing recorder of A.C. potentiometer system is used. This recorder has the capacity of recording 6 elements; the duration of the recording paper is two months at the speed of 25 mm. per hour. The power source is 100 v. A.C. ± 10% (50 or 60 Hz.).

With wireless equipment, the record of a remote station can be obtained. After several trial experiments, such tide gauges and robot-type telemetering systems have been installed at main ports of Japan with satisfactory results.

A similar method was applied for the record of the abrupt attack of tsunami due to the earthquake.

The warning system was extended to include storm surge. Because of the abrupt increase of the sea level due to typhoon, the warning should be made very quickly by obtaining the instantaneous record of the tide level. Though the said robot-type gauge is a suitable apparatus, various data along the coastal line of Bay of Tokyo are collected and the so-called "Call System" adopted.

CALL SYSTEM

For warning of meteorological phenomena, the data collection need not be continuous. The telemetering system of central calling should be utilized. The central station calls a daughter station for collection of the suitable weather data.

After this daughter station has sent the whole data to the center, the next daughter station starts to transmit by a central calling signal. If this scanning system is suitable, the collected data at the center will act as if the continuous data are obtained and such data can be utilized for the warning. This principle of "Call System" may be utilized for the remote collection of various weather elements at various stations. Four stations and three elements of tide, wind force and direction, are selected, but this system has the capacity to handle at least 10 stations with 6 elements at each station.

The central station acts as a mother station and is installed in the JMA; four daughter stations are installed at Kawasaki, Yokohama, Yokosuka, and Chiba along the coastal part of Bay of Tokyo. The primary measuring instrument of the daughter station is a converter used for changing such measured values as wind force and direction and tide-level into electric signals convenient for telemetering. Its principle lies in changing all measured values first into mechanical displacements and then into A.C. voltages with a differential transformer. This daughter station is adjusted to send a digital wireless signal after receiving a wireless signal from the mother station. The wireless signals of the daughter station are the signals of tide-level, wind direction and wind force; the equipment is adjusted to send these signals one by one.

The time necessary to complete the sending of signals of the above 3 elements is approximately 14 seconds. The mother station can receive the necessary data of 4 stations within approximately one minute. Thus by continuous calling, various data can be obtained at every minute, one by one.

The central station has control equipment, radio set, printing and indicating equipment, while the daughter station has control equipment, radio set, and primary measuring apparatus.

The control equipment acts to control the sending the "Call Signal" to each daughter station at a definite time, and to print in a definite form and to indicate on the indicating board panel all the data measured at every daughter station. The equipment uses switching transistors and is composed of printed plates.

The radio equipment is very high frequency (VHF) FM one-channel wireless telephone for composing a signal transmission route between the mother station and each daughter station.

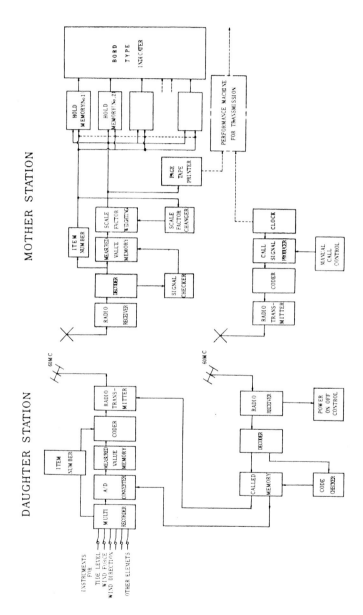

Fig. 2: Mother and Daughter Station Block Diagram.

The receiver being totally transistorized, has a very long life. Its frequency is 60 m. Hz. and its transmitting output 25 or 50 W. On the board, the data of various stations are shown with nixie tubes and the forecaster may be able to draw the suitable line on the map of Bay of Tokyo and the warning of the storm surge (Figure 2).

There is another reason for developing this call system technique for the warning of storm surge.

Ten years ago a typhoon named Vera struck Nagoya city and severe damage due to high tide occurred. After this disaster, Dr. Miyazaki of JMA succeeded in calculating the distribution of abnormal elevation of sea water in the bay of Nagoya, utilizing the IBM 704 computer. The distribution of sea level elevation due to storm surge can be estimated by assuming the intensity and path of the typhoon. The storm surge at various parts of Japan can be calculated assuming the path and intensity of typhoon. The calculated results are compiled for the coming season.

If a certain typhoon is forecast within several hours, the calculated values of abnormal tidal value at various places are obtained from the file by selecting a similar typhoon model already calculated by the computer. The value of abnormal tide can be forecast but the real typhoon is not the same with the model typhoon. Some discrepancies between the values of the forecast tide and the really observed tide arises.

To minimize such discrepancies the tidal variation at various stations along the coast of the bay must be watched and the forecaster should check and adjust the calculated value.

For this purpose continuous records of tide at coastal area are absolutely necessary. The above call system was developed to fulfill this requirement. Now the forecasters of JMA are utilizing the above theoretical methods and the call system technique to make the best possible forecasts.

Similar telemetering receiving set was installed recently at Sendai Meteorological Observatory. Tidal station is located at Ayukawa in the southern part of Ojika Peninsula and the record, obtained by an ordinary Fuess-type tide gauge installed in a well, is transferred to the wireless signal by voltage transformation and this signal is sent to the Sendai Observatory through the relay station at Koma Mountain, 324 m. high. In this case the observed data modulates a low-frequency FM wave of 960 Hz. and this low-frequency wave modulates the

UHF FM wave of 400 MHz.: the multimodulated 400 MHz. FM wave is used for transmission (Figure 3).

At Sendai, the ordinary robot-type radio wave (400 MHz.) is utilized, and the received signal is transmitted to the transistorized automatic converter and the data is recorded on the Fuess-type tide gauge recorder.

Figure 4 shows the records of Ayukawa and the received records at Sendai. These two records coincide fairly well.

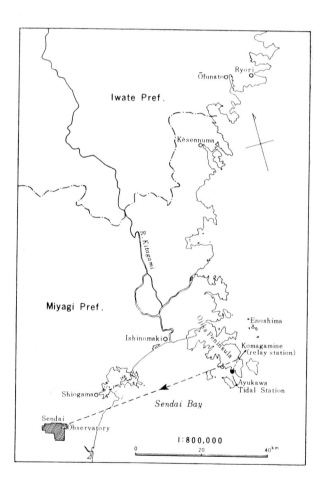

Fig. 3: Transmission Link from Ayukawa Tidal Station to Sendai Observatory via Koma Mountain Relay Station.

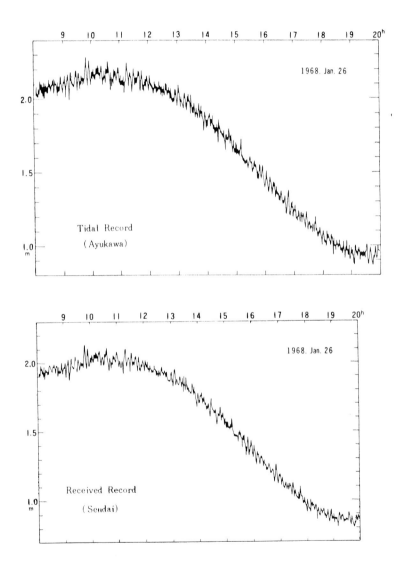

Fig. 4: On-Site Recording and Telemetered Recording for Qualitative Evaluation of Telemetry System.

The record of tsunami due to Tokachi-oki Earthquake (16 May 1968) was received at Sendai by this method. This shows that this method will be useful for tsunami research (Figure 5).

There is a tide-gauge station belonging to the University of Tokyo located at Enoshima Island, just to the north of Ojika Peninsula. JMA is now planning to install a tide gauge similar to that of Ayukawa to have tsunami records simultaneously at Sendai. JMA is also planning to collect records of tsunami at various places.

SEICHE STUDY

As early as 1965, a submerged long wave recorder of Snodgrass-type was developed by Dr. Takahashi, Dr. Aida, and Dr. Nagata of the University of Tokyo in order to study the tsunami. They installed four instruments in the depths varying from 10 to 40 m. at four points from bay head to bay mouth of Ofunato Bay. By using them, seiches were observed to learn the characteristics of the oscillation of Ofunato Bay where the Chilean Tsunami in 1960 caused severe damage. The spectral analyses of the records were made for each observation points. In Ofunato Bay, two modes with periods of 40 min. and 15 min. are predominant. The maximum amplitude of the seiche of 40-min. mode occurs at the bay head and that of 15-min. mode occurs at the intermediate point of the bay. These tendencies correspond to the height distribution of the Chilean Tsunami in 1960 and that of the Sanriku Tsunami in 1933, respectively. The observed amplitude distribution of the seiche of 40-min. mode shows good agreement with the calculated distribution.

DEEP-SEA TIDE GAUGE

Recently the Hydrographic Office, Maritime Safety Agency, has developed a deep-sea tide gauge and this tide gauge is now enplaced in the Sea of Japan at a depth of 200 m. The principle of this tide gauge is to record the change of tidal level due to the pressure change. The apparatus consists of a tire tube, bellows, and a recording drum. The inner pressure is adjusted beforehand at about 7 kg./cm.; when the apparatus reaches the bottom, the valve is closed and the inner pressure of this apparatus is in equilibrium with the ambient water pressure. (See Figure 6.)

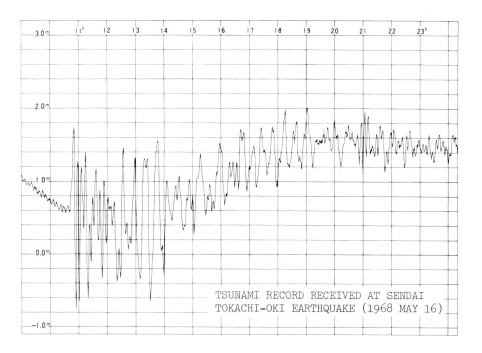

Fig. 5: The Telemetered Record of the Tokachi-oki Earthquake, as Received at Sendai.

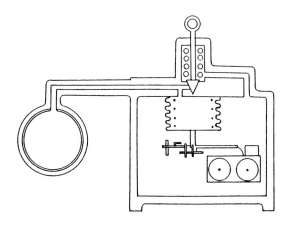

Fig. 6: Schematic Diagram of the Deep-Sea Tide Gauge.

Change of tidal level causes a minute pressure change in the tire; this small change is transmitted to the recording pen through the bellows. The drum is rotated by a battery at a speed of 10 mm./hr., so the continuous record of about one month can be obtained.

The apparatus developed by the Hydrographic Office is used to measure tidal variation over a range of 1 meter, but any amplitude of tidal variation can easily be recorded by modifying the instrument slightly. Therefore this instrument can be used for observing tsunami.

Port & Harbour Bureau, Ministry of Transportation, has also developed a pressure-type tsunami gauge. This gauge is dipped in the water, the variation of pressure is transmitted to the shore station by cable, and an electric circuit is used to filter periods less than 30 sec. Such tsunami recorders have been installed at Ofunato and Hachinohe.

JMA is currently sending the tide-gauge data obtained at Mera to the Tateyama Observatory by digitized transmission system. This equipment is specially designed as a Tide-Level Telemetering system by using UHF band (450 MHz.) radio wave. The variation of the tide level measured at the pick-up site on the beach is sent to the relaying station. The radio wave sent from the pick-up station is automatically transferred to the observatory or monitor/control site from the relaying station on another radio channel. The forecast/warning data included in the radio wave is analyzed and recorded on paper at the monitor/control site. The up-and-down variation of the tide level at the pick-up site is converted into the rotation angle (analog-to-digital) of the pulley connected with the float on the water surface, and is automatically recorded by pen-recorder. At the same time, the variation of rotation angle is also converted into an electrical signal (digital-to-analog) and is sent by transmitter as the "time-interval signal". This equipment has a very reliable characteristic due to the silicon transistorized circuit.

To augment the accuracy of transmission, the parity check system is introduced. Therefore, the tide records at the two stations coincide within ±1 cm.

JMA is now extending this system to Osaka Observatory and, in this case, the same is done for the records of tide and wind force and direction.

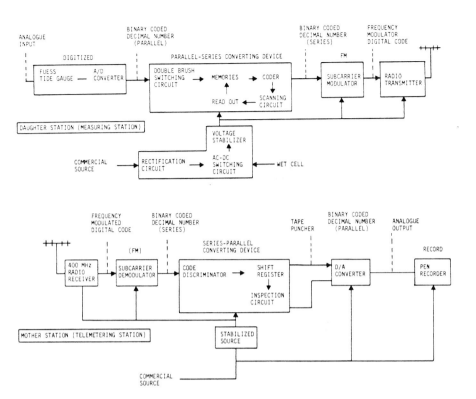

Fig. 7: Block Diagram of the Digital Transmission System for Telemetering the Mera Tide Gauge Data to the Tateyama Observatory.

OCEAN-WAVE METER

In 1968, a capacitance-type ocean wave meter was developed at Hiratsuka Branch of National Research Center for Disaster Prevention. This branch is located at the shore of Hiratsuka to the southwest of Tokyo and a marine observation tower about 20 m. high above the sea surface is installed 1 km. offshore where the water depth is about 20 m. This capacitance-type meter is attached to the tower.

The sensor of the meter is an iron pipe coated with epoxy resin or copper wire coated with resin; this forms a coaxial

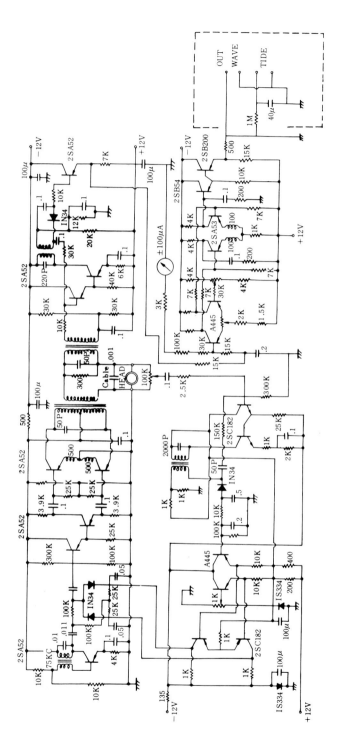

Fig. 8: Circuit Diagram of the Capacitance-Type Wave Meter.

condenser with sea water surrounding the pipe. The meter gives an instantaneous wave height through the electric circuit. This system was extended for the study of tide and tsunami by attaching a circuit as shown in Figure 8. The time constant was taken as 40 sec. The observed curve indicates that the wave meter can also be utilized as a tsunami recorder.

In Japan, especially at the inner part of a V-shaped bay, the tsunami wave sometimes reaches more than 20 m. high. In order to observe such a severe tsunami, the ordinary tide gauge is not suitable. Therefore, a new tsunami gauge which is capable of recording very high tides is needed but such a gauge is not yet developed.

For tsunami warning, the information of earthquake should be collected and analyzed automatically, and this principle is now under planning at JMA. If such a technique should be combined with a tsunami recorder, tsunami warning system would become more accurate and valuable.

BIBLIOGRAPHY

Inada, W. and Watabe, I. 1969. Capacitance type ocean wave meter (Japanese with English abstract). *Report of the National Research Center for Disaster Prevention No. 2*, pp. 57-68.

Sendai District Maring Observatory 1968. *Robot-type Tide Gauge at Ayukawa* (in Japanese).

Takahashi, R., Aida, I., and Nagata, Y. 1965. The submerged long wave recorders and the observation of the seiches in Ofunato Bay. *Journal of the Oceanographical Society of Japan, Vol. 22*, pp. 7-16.

Terada, K. and Fujiki, A. 1955. Electromagnetic telemetering tide gauge, Oceanographic research of Ariake Bay (II) (in Japanese). Pp. 61-65.

Terada, K., Futi, H., and Miyazaki, M. 1962. Storm surge forecasting method and telemetering, recording, printing and indicating system of tide and other elements -- so called "Call System" -- in Japan. *Oceanographical Magazine, Vol. 14*, pp. 43-62. Abbreviated paper is seen in *The 20th Anniversary Volume of the Oceanographical Society of Japan*. 1962, pp. 432-435.

15. Bourdon-Tube, Deep-Sea Tide Gauges

J. H. FILLOUX
Gulf General Atomic Incorporated
San Diego, California

ABSTRACT

Multiturn Bourdon tubes are remarkable pressure transducers that strongly decouple pressure effects, which consist mainly of rotations, from temperature effects, which consist mainly of translations. The small residual thermally induced rotations are related to coefficients of both thermoexpansion and thermoelasticity. A high degree of cancellation of the thermally induced rotations can be achieved with Ni-Span-C alloy. The thermoelastic coefficient of the latter can be adjusted by heat treatment to low positive or negative values. The use of a slightly negative value permits the counteraction, at least partially, of the small thermoexpansion effects. Simultaneously, the heat treatment greatly increases, by age hardening, the yield strength of this alloy. The plastic flow is thus reduced to very low level, even when the transducers are subjected to oceanic abyssal pressure. The creep rate is found to follow very closely Andrade's beta-creep law, in which plastic deformation accumulates proportional to the one-third power of the time elapsed after stressing. Drift can then be eliminated upon precise and legitimate ground.

Optical readout of the small rotations induced by pressure changes associated with tides permits frictionless, hence hysteresis-free, readout and high resolution without complexity. Multiple nulling provides linearity of response and continuous calibration.

The performance of a Bourdon-tube, deep-sea tide gauge is illustrated by an analysis of a one-week record obtained at a point one-third of the distance from San Diego to Hawaii, during October 1967.

INTRODUCTION

The pressure transducer presently described has originally been intended as a means to detect, on the sea floor, the pressure signature associated with the passage of a tsunami. The instrument has shown adequate sensitivity to satisfy the requirements but, by necessity, has been used so far solely to record tidal fluctuations on the open ocean.

Three main objectives are sought in this study. First, we review the main mechanisms that contribute to the temperature dependency of an ordinary Bourdon tube. From this understanding, we deduce the manner in which such a pressure transducer can be optimized. Since rotation and torque associated with the pressurization of a Bourdon tube are small quantities, the readout system must be very sensitive and frictionless. So, the second objective is to describe an optical readout scheme that satisfies these requirements. Finally, a deep-sea tide record is examined to provide an idea of the achieved performances.

In Fig. 1, we review the mechanisms of temperature dependency that are associated with the temperature coefficient of expansion C_{exp} of the material constituting the Bourdon tube. On the upper left of the figure, the continuous circle represents a ring made of a homogeneous material at temperature t. As the temperature rises to a value $t + \Delta t$, the diameter of the ring increases, but its circular shape is preserved. If the ring were split, as shown just to the right, the situation would remain essentially the same and the angular size of the missing segment would not change. From this observation, it may be deduced that a multiturn helix made of a homogeneous material would constitute an unsuccessful thermometer if one would attempt to measure the temperature by observing the rotation of its free end.

On the upper right of Fig. 1, we have represented an element of a Bourdon tube. We assume that the pressure applied is constant but that the temperature is rising. As the tube expands, its cross section A, as well as its moment of inertia I, increases. The forces F applied to the ends of the tube element increase as the linear dimensions to the second power, while the bending moment Mf, which tends to straighten the tube element, increases as the linear dimensions to the third power. Since the moment of inertia of the cross section increases as the linear dimensions to the fourth power, a stiffening of the tube results. This stiffening is linear with the relative changes in dimensions, hence linear with the changes in temperature. Therefore, the resulting temperature dependency is proportional to the relative changes in temperature and to the applied pressure p. This proportionality with applied pressure is of great importance.

The lower left diagram of Fig. 1 describes another mechanism of temperature dependency associated with the temperature coefficient of expansion. During the forming of the tube, a slight difference in the cold-work state of the inner and outer walls is generated. A small bimetal effect results, with an equivalent water head of a maximum of a few centimeters per degree centigrade.

In the lower-right diagram of Fig. 1, we show that a temperature-compensated multiturn Bourdon tube is also inherently insensitive to temperature gradients. Because of the small dimension of the cross section of the tube, temperature equilibrium across the tube is established very rapidly. It is conceivable, however, that a temperature difference may exist between the two ends. In this case, the Bourdon helix can be considered to be made of a succession of small elements 1,2,....n, at temperature $t + n \Delta t$. Since each element is assumed to be temperature independent, their ensemble should also be temperature insensitive.

We study now the effect of the thermoelastic coefficient C_{el}. (The thermoelastic coefficient describes the change of elastic modulus with temperature.) The left part of Fig. 2 represents a Bourdon tube at temperature t, free from internal pressure. In the center part, pressure p is applied to the tube with the temperature being unchanged. The pressurization causes a rotation of the end of the tube by an amount $\Delta \theta$ that is proportional to the pressure p and inversely proportional to the elastic modulus E at temperature t. On the right, the pressure is maintained constant but the temperature is increased to $t + \Delta t$. If the modulus of elasticity E at this new temperature decreases, as is common for most materials, the Bourdon tube rotates by an additional angle $\Delta \theta'$. The resulting temperature effect is then inversely proportional to the thermoelastic coefficient, and proportional to the relative temperature change and, again, to the applied pressure.

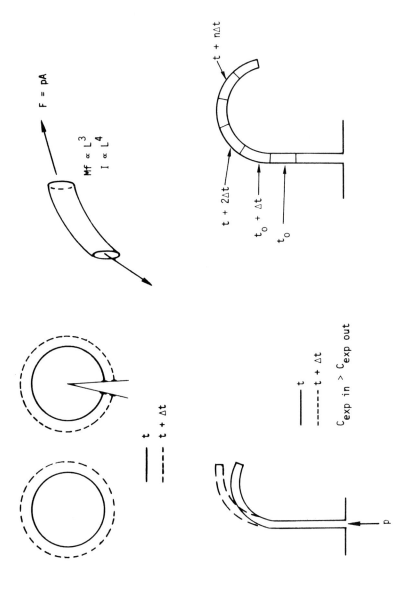

Fig. 1: Analysis of the Mechanism of Temperature Dependency Associated with Temperature Expansion.

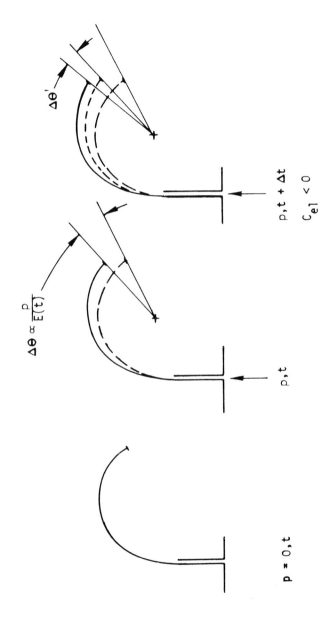

Fig. 2: Mechanism of Temperature Dependency Associated with the Changes of Modulus of Elasticity E with Temperature.

This suggests that appropriate combinations of nonzero thermal expansion and thermoelectric coefficients can eliminate most of the temperature dependency of a Bourdon tube. We shall see that this is indeed practical. This opportunity is provided by ferro-nickel alloys of adequate composition.

The two curves of Fig. 3 represent the dependency upon the iron-nickel proportion of the temperature coefficient of expansion (upper curve) and of the thermoelastic coefficient (lower curve) of ferro-nickel alloys. Optimization by control of the composition is difficult because of the steepness of the thermoelastic curve in the neighborhood of the zero crossing. A more satisfactory solution consists of lowering the hump of that curve toward zero by an appropriate addition of impurities. When this is done, the unavoidable variations with composition become much more tolerable, and carefull selected impurities can provide additional desirable physical properties. Ni-Span-C alloy, the composition of which is shown on the right part of Fig. 3, constitutes a highly satisfactory Bourdon-tube material. By means of an appropriate and relatively simple heat treatment, it is possible to perform final adjustment of the thermoelastic coefficient after the Bourdon tube has been formed. The same heat treatment permits age hardening of the cold-worked alloy to the very high yield strength of 13,000 kg./sq. cm. It follows that Ni-Span-C, properly cold worked and heat treated, is highly resistant to plastic flow. Nevertheless, because of the very high pressure at abyssal depth, creep is almost unavoidable. The analysis of the deep-sea tide record presented later shows that the plastic deformation of the Ni-Span-C Bourdon tube used in the tide gauge, very closely follows Andrade's beta creep law (Andrade, 1910). This law, originally empirical, has been justified recently for polycrystalline metals by Mott's dislocation dynamics model (Mott, 1953). The law states that plastic flow accumulates proportionally to the time elapsed after stressing to the one-third power. Since creep is slow, smooth, and featureless and can be filtered out using legitimate laws of physics, it is less detrimental than temperature effects that may be correlated with tides.

The instrument shown in the photograph of Fig. 4 is a deep-sea tide gauge that uses a 42-turn Bourdon-tube pressure transducer and an optical readout system. The tube is seen hanging vertically from the right flange of the pressure case, its lower end dipping in a small damping cup. The optical lever and the centering mechanism occupy the center. The fluctuations of the height of the water column are recorded on the small chart recorder seen to the left. The instrument is con-

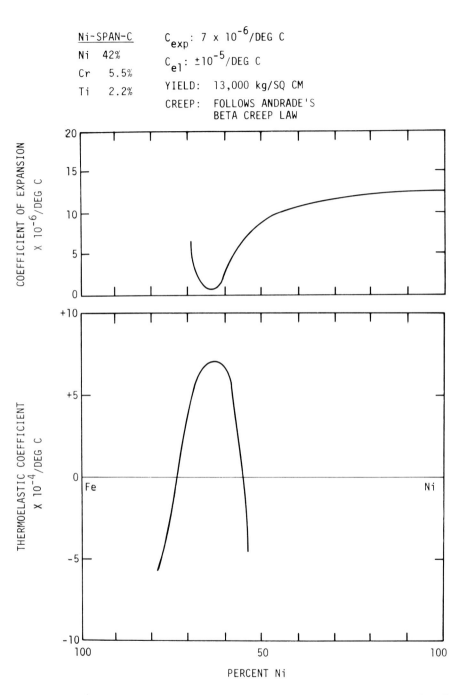

Fig. 3: Thermoexpansion and Thermoelastic Coefficient of Iron-nickel Alloys (left), and Properties of Ni-Span-C Alloy (right).

Fig. 4: Photograph of Self-Contained, Bourdon-Tube-Type, Deep-Sea Tide Gauges.

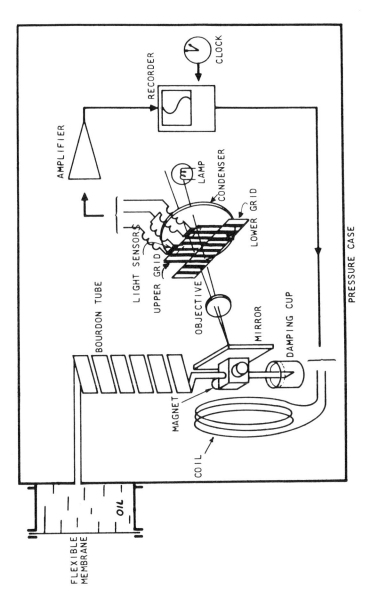

Fig. 5: Functional Diagram of the Tide Gauge.

tained within an aluminum pressure case about 18 cm. in diameter and 36 cm. long. The power supply is made up of four sealed cartridges, each containing five 1-1/2-V alkaline batteries, one of which is seen in the foreground.

Figure 5 illustrates the operating mechanism of the system. The lower end of the Bourdon tube is prolongated by a finger that carries a small mirror and a magnet. The reading mode is discontinuous. At reading time, which is controlled by a precise clock, a small incandescent lamp illuminates the grid located on the focal plane of the objective, which, itself, lies just in front of the mirror. The light is reflected back by the mirror and returns through the objective to form an image of the lower half of the grid onto the upper half. As the mirror rotates, the light transmitted through the upper grid is modulated. Small light-sensitive detectors convert fluctuations of illumination into varying electrical signals. If the grid pitch is made very small, minute rotations of the mirror induce rapidly varying signals. However, in this case, diffraction degenerates the linearity of response of the system.

The linearity is restored by the following scheme. After large amplification, the photocell signal is applied to a coil that creates a magnetic field, which, in turn, interacts with the magnet to rotate the Bourdon tube. The rotation, however, stops when the input signal decreases to a negligible nulling or error signal. Because of the periodicity of the grid, there is a multiplicity of such nulls, and because of the interruption of the illumination between readings, the system is always seeking the closest null position. One can take advantage of the situation to match the recorder impedance and the coil current in such a way that the recorder is automatically prevented from running off-scale by resetting to one side of the chart when the trace reaches the other side. The resulting discontinuity depends strictly on invariant quantities, namely, the elastic constant of the Bourdon tube, the focal length of the objective, and the pitch of the grids. The discontinuity, therefore, constitutes frequent and useful calibrations, which permit one to check the performance of the instrument. It also greatly increases the equivalent width of the chart of the recorder. These facts are exemplified in Fig. 6, which covers two days of tidal fluctuations in shallow water.

A seven-day record of the deep-sea tide at a position one-third of the way from San Diego to Hawaii was obtained during October 1967 (see Fig. 7). The distance from the nearest coast, that of Northern Baja California, was about 1250 kilometers.

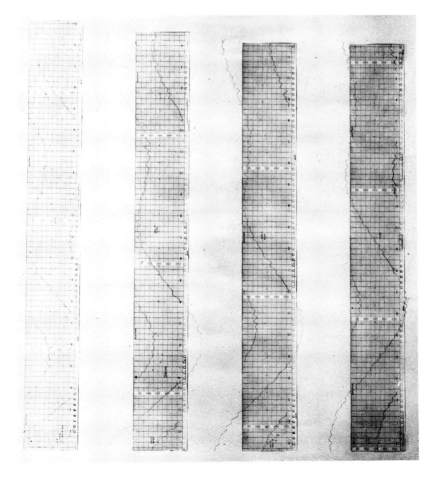

Fig. 6: Original Two-Day Tide Record Obtained in Shallow Water.

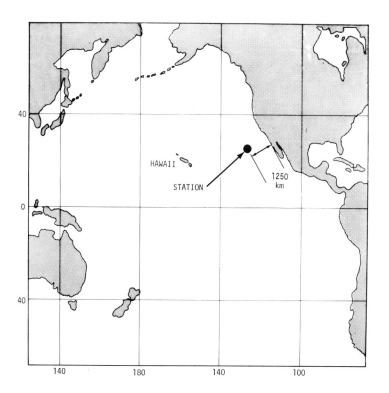

Fig. 7: Location of the Deep-Sea Tide Station.

On this occasion, the instrument was lowered to the sea floor and moored at a depth of 4.4 kilometers.

The upper fluctuating trace of Fig. 8 is a plot of the reading of the original record. The slow drift of this trace is an indication of the plastic flow in the Bourdon tube mentioned earlier. By means of least-squares analysis, the record has been split into several components: the creep component obeying Andrade's beta creep law; the tidal component, including only frequencies present in the gravitational driving forces; and the residual noise. The creep component is the smooth, monotonic curve across the upper trace. The tidal component is the lower fluctuating curve, and the background noise is the nearly straight line across the lower fluctuating trace.

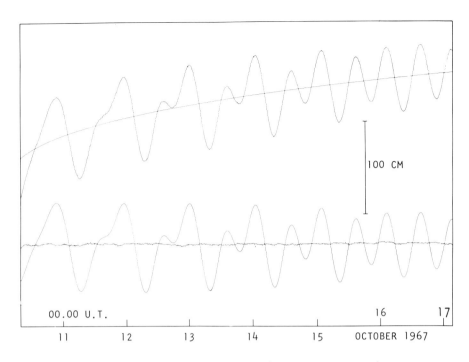

Fig. 8: One-Week Deep-Sea Tide Record (upper wavy curve), Creep Component (smooth upper curve), Pure Tidal Component (lower wavy curve), and Residual Noise (trace with horizontal trend across pure tidal component).

Table 1 gives the amplitude and phase with respect to the Greenwich meridian of the four main diurnal and semidiurnal constituents. To permit comparison with coastal values, the amplitudes and phases for the Scripps Institution of Oceanography tide gauge are also given. The most notable fact is probably that the offshore amplitude decay is very slow for the diurnal species and much larger for the semidiurnal one. The calculated amplitudes for the terdiurnal constituents S_3 and M_3, 0.2 cm. for each one, are indeed very small. From these observations, we may conclude that the response of the ocean to tidal forces, as well as the instrumental response, is very linear. The fact that the amplitude of S_3 barely rises above the noise level is an indication that nongravitational driving is probably very small. From Bogdanov's computation of the tidal amplitudes and phases over the Pacific Ocean, we

Table 1: Constituent Estimates and Other Pertinent Comparative Data

	Frequency (deg./hr.)	Station cm.	Station deg.	Scripps cm.	Scripps deg.	Bogdanov cm.	Bogdanov deg.
K1	15.04107	26.3	221	32.8	207	36.9	221
P1	14.95893	8.5	220	10.6	192	11.1	220
O1	13.94304	14.3	201	21.0	183	25.6	210
Q1	13.39866	2.5	187	3.8	204	4.8	204
M2	28.98410	22.4	115	48.9	143	34.7	172
S2	30.00000	10.4	97	19.9	136	13.0	160
K2	30.08214	2.8	96	5.6	131	4.5	159
N2	28.43973	4.5	124	12.1	119	7.4	178
S3	45.00000	0.2					
M3	43.47616	0.2					

OCTOBER 11, 1967

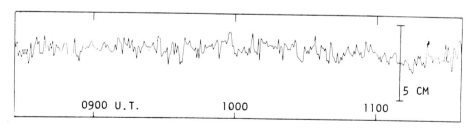

Fig. 9: Sample of Background Noise.

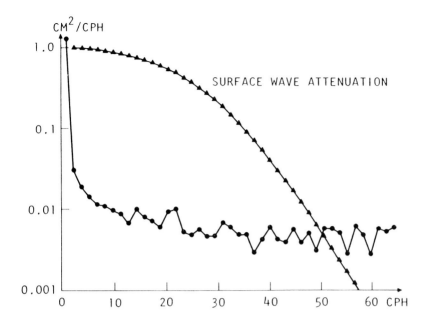
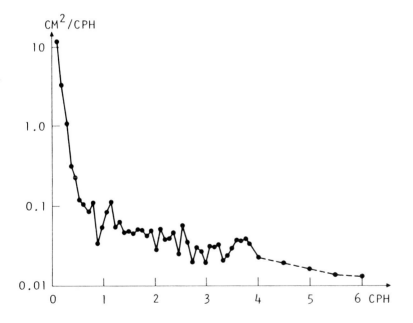

Fig. 10: Residual Noise Spectrum for Entire Record at 5-Min. Sampling Interval (upper), and Residual Noise Spectrum for 11 Hr. at 0.5-Min. Sampling Interval (lower).

have picked up the predicted values at the point of our observations. These values are shown in the right-hand column of Table 1. The predicted amplitudes are conspicuously larger than the observed ones, say by approximately 30 to 40%. This result suggests that in attempting to compute the open ocean tidal patterns, it is necessary to take into account the up-and-down fluctuations of the sea floor associated with the solid earth tides.

A sample of the background noise with a reading interval of 30 sec. is shown in Figure 9. The rms value of this noise is 0.8 cm.

The spectrum of the background noise for the entire record is shown on the upper part of Figure 10. The spectral-density fall sharply away from the tidal bands to a level of 0.05 sq. cm./cycle/hr. at 3 cycles/hr. It decreases much more slowly thereafter, as seen on the lower part of Fig. 10, to level off at approximately 0.006 sq. cm./cycle/hr. at around 40 cycles/hr. The attenuation of the pressure energy at the sea floor associated with surface waves is also shown on the lower part of Figure 10. The spectral density curve for the background noise shows no similarity to this attenuation curve, indicating that within the area of interest, say 10 to 30 cycles/hour, surface waves contribute very little, if any, noise energy. The pressure fluctuations in this interval could be generated by vertical seismic motion with an amplitude of the order of 20 microns. The fluctuation could also be associated with the instrumental noise or it could be generated by aliasing of higher frequency fluctuations.

BIBLIOGRAPHY

Andrade, E. N. 1910. On the viscous flow in metals and allied phenomena. *Proc. Roy. Soc. (London), Ser. a 84*, 1.

Mott, N. F. 1953. A theory of work-hardening of metals, II. Flow without slip-lines, recover, and creep. *Phil. Mag. 44*, 742.

16. An Instrumentation System for Measuring Tsunamis in the Deep Ocean

MARTIN VITOUSEK
Hawaii Institute of Geophysics
Honolulu, Hawaii
Contribution No. 298

GAYLORD MILLER
Environmental Science Services Administration
Joint Tsunami Research Effort
Honolulu, Hawaii

ABSTRACT

One of the major voids in the knowledge about tsunamis is knowledge of the nonlinear transformation of the waves as they interact with the shoreline. Theoretical work on this topic is difficult, and to date we have not had actual input data on the problem. There are no measurements of tsunamis in the open ocean. Even the smallest island has base dimensions commensurate with tsunami wavelengths, and probably deforms the waves which are recorded at the shoreline.

An instrumentation system has been developed which is capable of measuring tsunamis in the open ocean in depths to about 6 kilometers. Six systems have been built and tested; five of these were used in a recent experiment in the Aleutian Islands (Oct. 2, 1969).

As there also is interest in open-ocean tides, the recording systems have been designed to operate over long periods of time. For this reason non-corrosive materials were used. Eight glass spheres about 24 cm. in diameter house the various subsystems. These spheres are housed in two connected sections of polyvinyl chloride pipe and as a unit are buoyant. The buoyant unit is connected to a releasable battery and anchor unit. The instrument package is deployed by dropping it to the ocean floor; the buoyant package is retrieved at a later time by acoustic recall. There is a back-up timed release.

The spheres variously contain: acoustic transponder, computer-compatible tape recorder, timed release, radio beacon, flashing light, vibrotron battery, and associated electronics. The system is capable of recording any FM input signal. In this application a vibrotron absolute-pressure transducer is used.

The data are recorded digitally as the twelve least-significant bits of a period count of a 5-MHz crystal gated over 2^n cycles of the vibrotron. The exponent, n, may be any convenient integer; for tsunami work, 17 or 18 is the value used. The vibrotron has a center frequency of approximately 10.5 kHz and a response of about 0.2-Hz change per meter of water pressure change. The resolution of an individual reading is less than 0.5 millimeter.

The acoustic read-out is recordable at the surface on a standard depth recorder. A ping from the surface ship initiates an immediate response ping from the bottom unit and starts a slow count-down of the 12-bit register containing the last data reading. Upon completion of this count-down a second ping is sent from the bottom unit. For the surface ship the trace on the depth recorder appears as (1) bottom reflection, (2) first ping nearly simultaneously with the bottom reflection, and (3) second ping trace separated from the first by an amount proportionate to the water depth, full scale being about 2 meters.

The recent Amchitka experiment did not provide an opportunity to measure tsunamis in the open ocean, but it did provide an excellent test of the instrumentation system. A new series of measurements in the North Pacific is planned.

INTRODUCTION

Although there is wide agreement that long-period waves are transformed as they interact with the shoreline, there is very little agreement as to the exact nature of this transformation. Early studies showed that coherence between (two) tsunami recordings obtained at any two adjacent recording stations was poor. Recently, Adams (1969) demonstrated that for several linear models no significant transformation parameters appear which would let one predict run-up from current real-time seismic data. In summary, it is generally agreed that some nonlinear transformation of waves occurs as a tsunami interacts with a shoreline. Clearly, the transformation from the general pattern shoreward is simpler than the transformation from the inshore areas back seaward. An offshore observation is widely representative; an inshore observation is representative of purely local topography.

Evidence that the actual run-up is a local nonlinear transformation process stems from the fact that there is little correlation between nearby wave records (Continental Shelf, California) from tsunamis (Miller, 1964). Evidence that understanding of the run-up problem may be possible stems from the fact that the run-up ratios are similar, frequency by frequency, from place to place. A relatively low-frequency tsunami is universally low-frequency; similarly, a relatively high-frequency tsunami is widely high-frequency (Miller, 1964).

A need exists for tsunami observations made offshore. It may be possible to calculate the run-up of a tsunami given the offshore or deep-water input to the local topography.

GENERAL MEASUREMENT SCHEMES

Four possible methods of measuring a tsunami in the open ocean exist: (a) free-drop recoverable instrument package, (b) under-ship instrument, (c) under-buoy-mounted instrument, and (d) cable-mounted gauge. The last, while technically quite feasible, may be eliminated (except for existing cables) on the basis of cost. A durable cable might be expected to cost a substantial fraction of a dollar per foot, of the order of two-thousand dollars per mile. Thus, the laying of special cables is out of the realm of realistic costs for meaningful warning times or distances. Each hour of warning time would correspond to a cable cost of approximately one-half of a million dollars. It may well be possible to place a geophysical measurement package on existing trans-Pacific cables such as the cable between Fanning Island and Vancouver Island.

VIBROTRON SYSTEM

The free-drop system serves as an example of electronics which are essentially the same in all of the proposed systems.

A vibrotron transducer converts pressure into an FM signal. A taut tungsten wire is connected to a solid base at one end and to a metal diaphragm at the other. The wire is kept in oscillation at its resonant frequency by permanent magnets and an electrical connection through a feedback loop to an amplifier which takes its signal from the wire (Figure 1). A change in pressure in the diaphragm changes the tension and thus the free period of oscillation of the wire in this vibrotron. In practice, the center frequency of a 7000-meter full-range transducer is about 10.5 kHz at depth with a change in frequency of about 0.2 Hz per meter change in water pressure.

The basic logic clock for the sytem is the vibrotron itself. Changes in pressure due to tides or tsunamis are of the order of a meter. Thus, as a time base, the vibrotron is good to better than one part in 10^4.

To make a measurement, a precision 5-MHz crystal is counted over a duration of 2^n oscillations of the vibrotron. The output of the vibrotron is fed into a counter, and the interval between the start of a count and count number 2^n is used as a gate for the counting of the 5-MHz crystal. In this way a very precise period measurement is made of the vibrotron. A typical convenient value of n is about 18, corresponding to an averaging time for the recording of about 1/2 minute. Storing of this data, writing it on tape, and resetting the counters takes some additional time, and thus a reading about one per minute is obtained. The resolution or "least count" of each data point is a fraction of a millimeter, but the precision is less because the vibrotron is temperature-sensitive.

A 12-bit register is of adequate length for counting the 5-MHz output of the crystal; this corresponds to a 2-meter water-level change, depending on the center frequency and response of the particular vibrotron used and on the value of n which is appropriate.

The actual recorded value is that of the 12-bit period count of the crystal-count register. An incremental, computer-compatible tape recorder is used to store this data. Figure 2 schematically depicts the various components of the free-drop system. To summarize, a very precise period measurement is made of the FM vibrotron pressure transducer signal, and it is recorded on computer-compatible tape.

In addition to the stored data, a real-time read-out of the data from the free-drop system is possible. Upon command

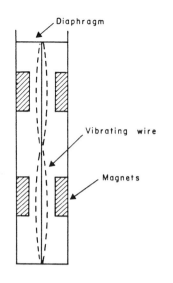

Fig. 1:
Vibrotron schematic.

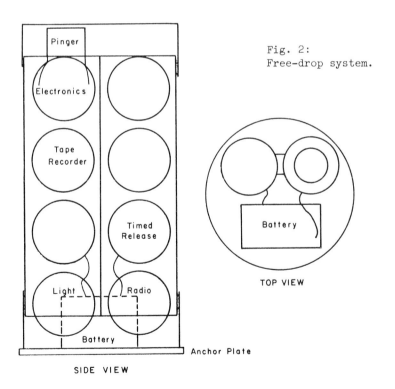

Fig. 2:
Free-drop system.

from a ship over the free-drop system -- consisting of a ping from a standard depth recorder, an immediate response ping is transmitted, and a slow count-down of the 12-bit register is initiated. At the termination of the count-down of the data register, a second ping is transmitted. The count-down requires up to three seconds. The recording made at the surface ship consists of: the outgoing ping, the bottom return, the immediate response, and the second response delayed by an interval of from 0 to 3 seconds, proportioned to the water pressure. For a ship positioned nearby the instrument on the bottom, the depth recorder shows: the surface trace, the bottom reflection, the immediate response trace, and a trace whose separation from the immediate response trace is proportional to the contents of the 12-bit register. Figure 3 depicts this output schematically.

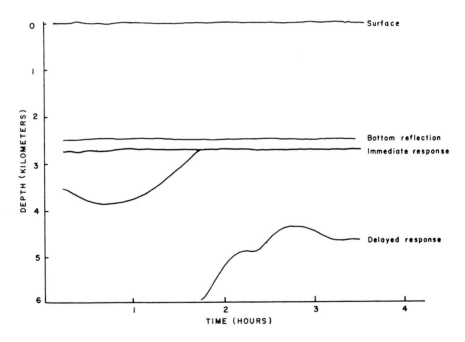

Fig. 3: Echo sounder trace schematic.

GENERAL SYSTEMS CONFIGURATIONS

Free-drop System

This system (Fig. 2) is constructed of individual components or subsystems housed in glass spheres. Six such systems have been designed and tested. The reasons for choosing glass over other materials are that glass is inexpensive, light, easy to handle, and not corrosive. The 9 1/4 inch (23.5 cm.) spheres are housed in sections of PVC (polyvinyl chloride) pipe. Six of the spheres contain subsystems, each performing a specific function. The flashing light, used in night-time recovery, is completely self-contained. The battery, which is sufficient for four days' use, is housed within the sphere. The normal position of the free-drop system when on the bottom is to have the flashing light and radio-beacon spheres at the end that is attached to the anchor baseplate. When released, the system inverts and floats with the light and radio above the sea surface. A mercury switch turns the flashing light on when the system is in the floating position.

The radio beacon, similarly activated, broadcasts at 246.8 MHz, producing a locatable signal when the unit is floating at the surface. A simple direction finder lets one "home in" on the unit from distances of several kilometers for surface-recovery ships, or several tens of kilometers from search aircraft.

A timed-release device is housed in a third sphere. This timer is used to trigger a cable cutter which releases the system in the event that it has not been successfully recalled by the acoustic recall subsystem.

The tape-recorder unit and the associated digital logic circuits are housed together in the fourth sphere. Tape formating is standard 200 BPI computer-compatible except that no inter-record gaps are made. Most computers can accept gapless tape up to record lengths of about 10^5 bits. After that length, multiple-load points must be placed on the tape and the tape read in overlapping sections.

An Accutron clock provides an independent minute count which is recorded on the data tape. This minute count is recorded modulo 64 in a single frame or 6-bit count across the standard 7-channel (6 data plus one parity) 200-bit-per-inch computer-compatible tape. Independently, hour marks are recorded by switching the 12-bit register from the least significant (most sensitive) bits of the crystal counter to the most significant (total pressure) bits of the counter. Thus the tape recording of a data value usually has the minute count at the time of a data sample as 6-bits followed by the 12-bit data value. Once per hour the 12-bit data value contains the absolute pressure to the nearest few meters.

An acoustic transponder is attached to a fifth sphere which contains its associated electronics. This unit receives the 12-kHz acoustic signal from the surface, initiates the first return pulse and count-down of the 12-bit register, and produces the second acoustic pulse. The transponder is also useful in locating the system after it is released from the bottom anchor and battery case.

Acoustic recall is accomplished by a series of pings spaced at a specified interval. We have used half the standard 600-fathom depth-counter repetition rate or 6 seconds. Upon receipt of a series of such pings, and only such pings, a cable cutter severs the power cable and hold-down wiring which connect the free-floating glass-sphere section of the system to the base-plate anchor and lead-acid battery case.

The sixth sphere contains mercury batteries to power the vibrotron transducer and the remaining two glass spheres are empty and provide buoyancy. If any one sphere should fail, there is sufficient buoyancy in the remaining seven to bring the unit back to the surface.

Common automotive-type batteries are used as the major power supply. These need not be pressure-protected, requiring only that insulating oil separate the cells and that the connections to the terminals be well insulated (Snodgrass, 1969).

The sequence of events during a measurement is, typically: deployment over the side of a ship of the system firmly attached to its anchor and battery case base plate, recording by this system of water pressure as it sits on the bottom, and recall of the floating glass-sphere portion of the system by acoustic command from the surface. If the acoustic command system fails, the unit surfaces according to the schedule pre-set in the time release.

Under-Ship System

If one takes a sub-set of the free-drop recoverable system described above which consists only of those components required for the acoustic-data transmission, then a unit which would provide data to a surface ship is described. In this usage, permanent bottom units (non-recoverable) are to be emplaced at the sites of the ocean-station weather ships. The units consist of only the data acquisition electronics and acoustic transponder systems. Special long-life lead-acid batteries provide the required power. The first unit, Fig. 4, will be placed under ocean-station "VICTOR" by early 1970. Allowing for variability in acoustic transmission-path travel times, transducer noise, and the motion of the receiving ship, a resolution of about 1 cm. in water level is expected.

Total capacity of the battery pack will be approximately one month of continuous output at the standard 1-per-minute data-repetition rate. Plans for the use of this capacity are to record this tide once per hour to obtain a record of several months' duration and to record all tsunamis occurring during the three years that the batteries will retain their charge.

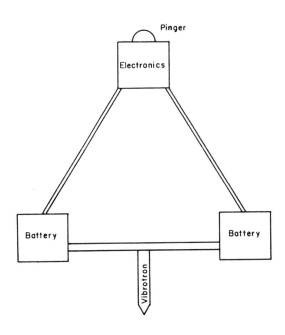

Fig. 4: Permanent Under-Ship or Under-Buoy System.

Buoy System

A bottom unit similar to that described above could be used to communicate pressure data to a surface buoy. Components for a buoy system have been acquired and emplacement is planned during 1970. The first buoy system (Fig. 5) is of spar type with a two-component mooring line (Isaacs, 1967). The upper metal portion of this line is designed for protection against fish bite. The lower part is made up of three sections of synthetic line and acts as a shock absorber and, being partially buoyant, keeps the line from dragging on the bottom.

First tests will be made about 140 km. north of Maui Island. Line-of-sight radio communications from Haleakala Mountain will be used. Until such time as the Geo-Orbiting-Environmental Satellites (GOES) system becomes available (probably in 1972), it will not be practical to emplace the small, low-powered buoy system in the open ocean.

Data transmission of tsunami data requires relatively low power. The bit rate is only of the order of one bit per second. Power at the buoy is generated by a wave-activated generator.

The sequence of events for a data sampling from the buoy system will involve the following steps: radio query from shore to the buoy receiver, acoustic query to the independent bottom unit, immediate acoustic reference response from the bottom unit, generation of the count-down of the data register, return delayed acoustic response to the surface, and repetitions (probably three times) of this sequence. The buoy transmitter will respond with either the acoustic pulses in analog form or with a time count in digital form of the interval between pulses.

If a durable buoy system can be developed, data can be provided from areas where there are no islands between tsunamigenic regions and endangered shores. One of the first operational emplacements is planned 1800 km. north of Hawaii.

Cable Systems

A cable-mounted unit similar to the bottom unit designed for the buoy system has been built and "life-tested" in the laboratory. Three major differences exist between it and the standard independent bottom unit. Temperature is measured as well as pressure. Power is transmitted down the cable. The time-delay data pulses are transmitted by cable. Plans are to place this instrument at some future time on the abandoned Vancouver Island-to-Fanning Island cable about 700 km. north of Fanning Island. In order to reduce the temperature noise on the standard bottom units, the vibrotron will be on a probe inserted into the sediments (Figure 4).

TESTS AND MEASUREMENTS

Tests of the free-drop tsunami recording system have been made in waters off the south shore of Oahu. Records of up to

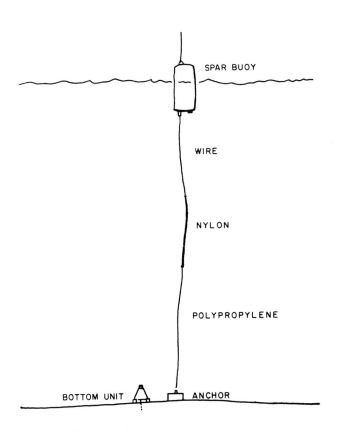

Fig. 5: Buoy system.

a week's duration have been made in these tests. Cold tests were conducted in the laboratory to insure that the various electronic and mechanical systems would operate in the near 0°C environment at the sea bottom.

The first actual use of the systems was made in October 1969 offshore from Amchitka Island in the Aleutian Islands. On October 2, 1969, an underground nuclear test was conducted on Amchitka Island. It was expected that a small water-wave would be generated as the region near the test site was deformed by the generation of the cavity. Close inshore, there existed the possibility of measuring any net ground motion by recording sea level. There was also the very small probability that the shock wave from the event could trigger an earthquake which might, in turn, generate water waves.

Five of the glass-sphere systems were deployed (Fig. 6) in depths between 470 m. and 4100 meters. An older model

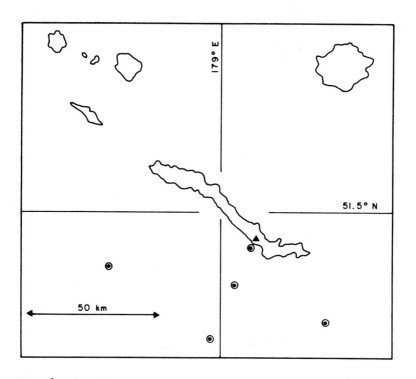

Fig. 6: Location of Instruments Relative to Amchitka Island.

free-fall system with electronics similar to the glass-sphere system but housed in an aluminum sphere was tethered in about 30 m. water depth by divers. Its purpose was to measure any net ground deformation.

One difficulty which occurred during emplacement was the premature firing of four of the explosive cable-cutter releases as the helicopter carrying the systems hovered near a radar unit on the ship. It is believed that some rectification of the electromagnetic radar radiation produced enough current to trigger the cutter. The required 6-sec. repetition rate was produced by the rotating radar antenna.

Four of the glass-sphere systems were successfully recalled acoustically, and data tapes were obtained. Figure 7 shows a portion of the tide record obtained from the instrument at 1800 m. depth. The fifth unit was recovered using the time release, but a malfunction spoiled data recovery.

Figure 8 shows the small-amplitude long-period waves generated by the ground deformation. The period of the two most prominent waves is about 10 minutes; the wave heights are a few centimeters. These waves have wavelengths which are long compared to the depth of water and thus propagate at the speed of any other tsunami.

FUTURE TESTS

The apparent background of water-pressure fluctuation at the bottom of the open ocean is about 10^{-3} cm.^{2}/MHz. This may be in the form of very small amplitude gravity waves generated by atmospheric fluctuations, pressure changes partially generated by internal waves, turbulence, atmospheric pressure changes,

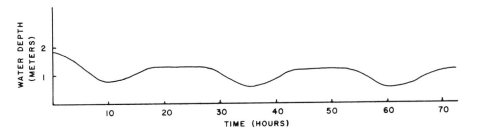

Fig. 7: Tide record at depth of 1800 meters.

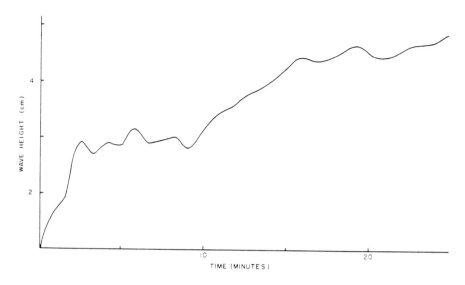

Fig. 8: Waves generated by Amchitka tests.

and nonlinear wave-wave interactions. In order to study this portion of the spectrum we plan to emplace a triangular array of free-fall systems about 100 km. north of Oahu Island. Instrument separation will be about 10 km., which will insure that there will be good coherence between the records even for the highest frequency of the shallow-water gravity waves.

BIBLIOGRAPHY

Adams, Wm. Mansfield. 1969. *Prediction of Tsunami Inundation from Existing Real-Time Seismic Data.* State of Hawaii.

Isaacs, John D. 1967. Some Remarks on Present and Future Buoy Development, *Transactions, Second International Buoy Technology Symposium,* Marine Technology Society.

Miller, G. R. 1964. *Tsunamis and Tides.* Ph.D. Thesis, 120 pp., University of California, San Diego.

Snodgrass, Frank. 1968. Deep Sea Instrument Capsule. *Science,* 162:78-87.

17. Estimating Earthquake Rupture Length from T Waves

R. H. JOHNSON
Hawaii Institute of Geophysics
Honolulu, Hawaii
Contribution No. 299

ABSTRACT

Earthquake T waves may find application to the tsunami warning problem by being interpreted in terms of the length of rupture. This dimension is a more direct indication of the potential of an earthquake to generate a tsunami than is earthquake magnitude. Earthquake swarms within a confined region may be recognized as such by the repetition of their signature. The length of rupture of a major earthquake can be estimated from the duration of the high-level portion of the T phase at two or more stations. It is speculated that equivalent information may be obtained from one recording station by azimuthally tracking the incoming T waves.

INTRODUCTION

Earthquake T waves, in common with tsunamis, are generated by movements of the ocean floor. In the case of T waves the movement is oscillatory, while for the tsunami it more closely resembles a moving-step function. Although the oscillatory waves are not likely to indicate the magnitude and direction of permanent displacement of the ocean floor, they may find application in delimiting the lateral extent of the affected region. From the standpoint of tsunami generation this dimension is probably more diagnostic than an estimate of earthquake magnitude.

BACKGROUND

In a study of the earthquake sequence initiated on 4 February 1965, Johnson and Norris (1968) showed that T-phase source solutions clustered about points, called T-phase radiators, along the Aleutian Ridge. Each aftershock which originated in the vicinity of a particular radiator generated a T phase with a signature characteristic of that limited source region. Thus an earthquake swarm from a confined region, though it may include events of magnitude 6 or 7, is readily recognized as such by the repetition of the logarithmic power-level signature (Johnson, et al., 1963).

For a major earthquake, however, T-phase radiators are excited sequentially as the rupture proceeds along the fault zone. The duration of the T-phase is longer and its signature is perhaps a composite of those for the myriad of aftershocks which follow. Using the duration of the high-level portion of the T phase, plus the length of the aftershock region, Johnson and Norris (1968) estimated the speed of rupture for the main shock of the Rat Islands series. The value obtained, 3.5 km./sec., was in good agreement with values derived for other earthquakes by other means. This result indicates that the duration of the T phase can be used, as suggested by Eaton, et al. (1961), to estimate the length of the fractured region.

DISCUSSION

An example in which the present day level of T-phase interpretation might have forestalled a false tsunami alarm is the Fox Islands earthquake swarm of 21 December 1962. A 24-hour section of the Oahu hydrophone record containing this swarm is shown in Figure 1. The original record shows a total of 19 earthquake T phases, each with the same characteristic double-peaked signature. In this case the first peak is due to abyssal generation at the epicenter (Johnson et al., 1968) followed 75 seconds later by the peak radiated from the slope. The regularity of the signature here implies that each earthquake of the series excited the same region of the ocean floor, although with varying strength. The disturbed region must, then, have been too confined to have generated a dangerous tsunami. In fact, none was detected.

In cases such as the foregoing, only an upper limit on rupture length can be inferred from the T-phase duration. However, when the rupture length is greater than the radiator spacing, say 100 km., this fact should be evident from differences

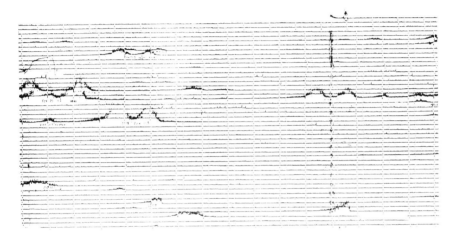

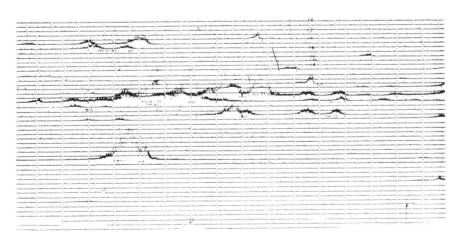

Fig. 1: Twenty-four Hour Section of the Oahu Hydrophone Record Containing the Fox Islands Earthquake Swarm of 21 December 1962.

in signature between the main shock and its aftershocks as well as differences among the aftershocks. The duration measured at two or more stations can then be used to estimate rupture length as the distance between the first and last radiator excited.

The delay and confusion encountered in attempting to gather data from widely separated stations under emergency conditions would probably render such a technique impracticable. An alternate method would be to use an array of hydrophones at one recording station. The azimuth of the received signal could then be computed by cross-correlation.

Figure 2 shows how this technique might have been applied in the case of the 4 February 1965 Rat Islands earthquake. Assuming a tracking array at R, Oahu, the signal would presumably have swept counter-clockwise from ray A to ray B. As epicentral coordinates are normally available early from the seismograph network, only a differential measurement in azimuth is needed, ray A being centered on the epicenter O.

The T-phase duration is measured as the time between arrivals from the first and last radiator excited by the main shock. Although the time for the first radiator is generally quite clear, selection of the last radiator may be confused by the presence of aftershocks. Figure 3 shows the signals received at four hydrophone stations for the main shock of the 1965 Rat Islands series. The choice of peaks for the termination time at Oahu appears open to question and, in this case, was based on comparisons with the signals at the other stations. However, a correlogram might resolve this difficulty as one would expect a monotonic progression in azimuth of the peaks.

If the duration from Fig. 3 is taken as 7 m. 18 s. and the speed of rupture as 3.5 km./sec., the termination of rupture would lie on the elliptical locus in Figure 2. The curve is drawn with a width corresponding to ±0.26 km./sec. rupture velocity. Termination of rupture is indicated by the intersection of the elliptical curve with ray B at T. The other intersection would have required that the signal sweep clockwise from B to A.

An array which might be suitable for such tracking could consist of three hydrophones suspended in the sound channel at the corners of a 1000-meter triangle. According to Born and Wolf (1959), the maximum aperture across which an extended incoherent source can be studied by cross-correlation techniques is given by

$$\Delta = \rho \, \bar{\lambda}/(L \cos \Theta)$$

where Δ is the maximum aperture in radians, $\bar{\lambda}$ is a mean wave length, and $L \cos \Theta$ is the hydrophone separation normal to the direction of the source. The coefficient depends inversely on the correlation coefficient and may reasonably be taken as equal to 0.2 (Hardy, 1969). At a frequency of 30 Hertz, Δ is about 0.6° for the hypothetical array. In Fig. 2, rays A and B have been drawn with this breadth. At the range of Oahu from the Rat Islands the angle subtended by any radiator spacing is less than 0.6°; therefore, correlation techniques may be expected to operate satisfactorily.

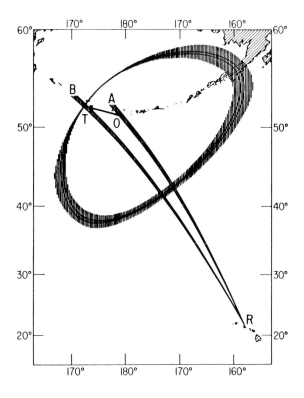

Fig. 2: Diagram Illustrating Proposed Method of Using an Array of Hydrophones at Oahu to Estimate Rupture Length.

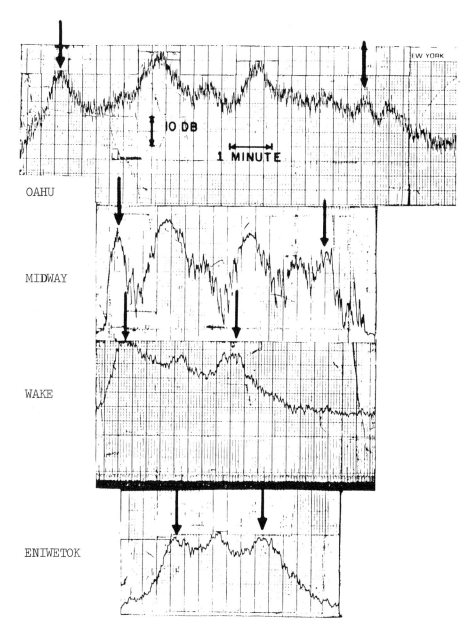

Fig. 3: Hydrophone Records from Four Stations Containing the Main Shock of the 1965 Rat Island Earthquake Series.

A related experiment which bears upon this aspect was conducted by Hardy (1969). He used correlation between pairs of hydrophones suspended from a ship to obtain azimuths of echoes from seamounts. His sound sources were explosives detonated under the same ship. He observed that while correlations were obtained for dozens of features, they were rarely obtained from distances less than 30 miles (56 km.), the aperture being too large. This was interpreted as corroboration of the foregoing relationship for estimating the maximum aperture.

Although no permanent hydrophone installation is presently available for azimuthal studies of T waves, limited results could be obtained from a temporary array. This could be moored, suspended, or drifting and tended by a surface vessel. In any 24-hour period tens of T phases would be expected although it would be very unlikely for a major earthquake to occur. However, worthwhile tests could be made of the ability of such a system to resolve adjacent radiators excited by a single earthquake and for establishing the degree of correlation of the underwater signals for varying hydrophone spacing and frequency.

BIBLIOGRAPHY

Born, M. and Wolf, E. 1959. *Principles of Optics*. Pergamon Press, N. Y.

Eaton, J. P., Richter, D. H., and Ault, W. U. 1961. The tsunami of May 23, 1960, on the island of Hawaii. *Bull. Seism. Soc. Am.*, *51*, pp. 135-157.

Hardy, W. A. 1969. Seamount detection by cross correlation. *Hudson Laboratories Tech. Rept. 169*.

Johnson, R. H. and Norris, R. A. 1968. T-phase radiators in the Western Aleutians. *Bull. Seism. Soc. Am.*, *58*, pp. 1-10.

Johnson, R. H., Northrop, J., and Eppley, R. 1963. Sources of Pacific T phases. *J. Geophys. Res. 68*, pp. 4251-60.

18. Developments and Plans for the Pacific Tsunami Warning System

L. M. MURPHY and R. A. EPPLEY
Environmental Science Services Administration
Coast and Geodetic Survey
Rockville, Maryland

ABSTRACT

The Pacific Tsunami Warning System is continuing to expand in both the number of participants and the scope of warning responsibility. Plans are now underway for addition of six new seismic and eight new tide stations to the Warning System. New visible recording seismic systems and electronic recording tide gauges are being supplied to many participating stations. Earlier warnings to residents of Hawaii for locally generated tsunamis is the objective of the Experimental Tsunami Warning System in Hawaii, now being installed by the University of Hawaii under an Environmental Science Services Administration (ESSA) contract. Several recent tsunami operations of the Alaska Regional Warning System are described as well as a new research study to determine the effectiveness of the Alaska Warning System.

INTRODUCTION

Twenty-one years ago the Seismic Sea Wave Warning System was organized with four seismological observatories and nine tide stations. These initial seismic stations were College, Sitka, Tucson, and Honolulu; the tide stations were located at Attu, Adak, Dutch Harbor, Sitka, Palmyra, Midway, Johnston Island, Hilo, and Honolulu. All but one of these participating stations, the Palmyra tide station, are still part of the Warning System. In 1948 the primary objective of the Warning System was to supply tsunami-warning information to civil authorities in Hawaii and to various military centers for dissemination to military bases throughout the Pacific.

From this rather modest start, the Tsunami Warning System (TWS) has increased considerably both in number of participants and in scope of warning responsibility. The seismic stations now number twenty-one and the number of tide stations has increased to forty-one. (See Figure 1.) This expanded participation has permitted extending the dissemination of tsunami warnings to eleven countries bordering the Pacific Ocean. The System is now an effective one and has undoubtedly saved many lives, but it can be improved. These improvements may be described as:

(1) increase the number of participating seismic and tide stations;
(2) improve the instrumentation at certain seismic and tide stations;
(3) develop faster communication methods for the transmission of seismic and tsunami data and for dissemination of tsunami warning information; and,
(4) develop analytical techniques to permit faster and more accurate epicenter determinations, extent, and direction of faulting, magnitude determinations, tsunami expected times of arrival (ETA's), and evaluation of tsunami heights.

EXPANDED INSTRUMENTATION

A program to carry out the first two improvements has begun. Seismic observatories at Wellington, Port Moresby, Suva, Tacubaya, Antofagasta, and Easter Island will join the TWS. Data from these stations are expected to expedite epicenter determinations in certain areas, as well as provide a greater degree of accuracy. New tide stations expected to join the TWS include: Galapagos; Marsden Point, New Zealand; Gambier Island, Tuamotu Arch; Okinawa; Amchitka; Salina Cruz and Manzanillo, Mexico; and Talara, Peru. The increased coverage provided by these stations should help to reduce tsunami detection times, as well as provide additional data for tsunami evaluation

Upgrading of instrumentation at a number of TWS stations recently became possible through the purchase of new visible-recording seismic systems and new electronic recording tide

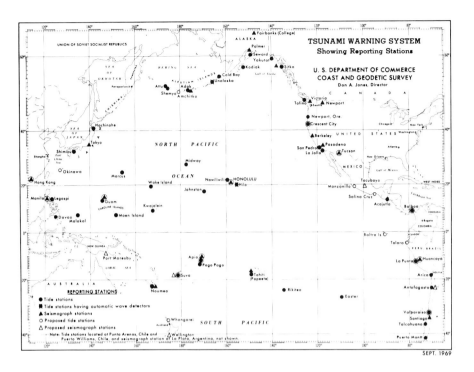

Fig. 1: Map Showing the Locations and Type of Reporting Stations in the Pacific Tsunami Warning System.

gauges. The seismic equipment consists of a short-period vertical seismometer, usually a Johnson-Matheson, a Geotech photo tube amplifier, a Geotech Helicorder, a Geotech crystal-controlled timing system, and a standby power supply. (See Figs. 2 and 3.) This equipment replaces visible systems at a number of stations which are not particularly suitable for recording "P" phases. The timing system permits scaling of phases to an accuracy of a tenth of a second. Although this precision is not required at present for the epicenter determination procedures, the anticipation of computer-derived epicenters emphasizes the need for greater accuracy in station readings. The battery power supply allows a station to continue recording for several hours in the event of a loss of line power.

Fig. 2:
Sensor Portion of the Seismic Equipment Planned for Installation at Some Reporting Stations. Seismometer on Left; Phototube Amplifier in Center, Power Supply at Right.

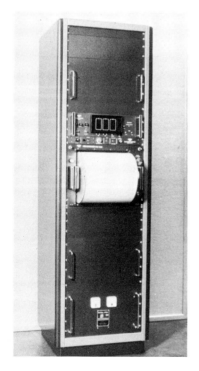

Fig. 3:
Timing and Display Portion of the Seismic Equipment Planned for Installation at Some of the Reporting Stations. From Top to Bottom: Timing System; Helicorder Amplifier; Helicorder; and Power Supply.

For several years some of the participating tide stations have been supplied with remote-recording systems. These systems are to be furnished to all TWS tide stations who have a need for remoting their tidal data. The use of the remote recorder eliminates the necessity of visiting gauge sites to visually inspect the marigram. This reduces the hazard to station personnel and the time required to respond to requests for wave information. Such remote-recording systems also provide readily available wave data for use in determining the time to safely sound the "all clear" following tsunami warnings. The instrumentation required consists of a Bristol Metameter transmitter mechanically linked to the tide gauge (see Fig. 4) and a Bristol Dynamaster recorder containing a Metameter receiver. The signals are usually transmitted over hard wire in the form of timed electrical pulses whose duration is proportional to the sea level. Because the receiver measures only the time duration of these pulses, rather than the magnitude of any electrical quantity, system accuracy is unaffected by wide variations in circuit resistance and other transmission-line characteristics.

HAWAII REGIONAL WARNING SYSTEM

A contract was awarded to the University of Hawaii to establish a new experimental tsunami warning system for locally generated tsunamis in Hawaii. This system, being installed under the direction of Drs. William Adams and Martin Vitousek, consists of seismic and hydraulic gauge stations on various islands. Data from these stations are telemetered in real time by radio to the Honolulu Observatory (HO). Four seismic signals are transmitted from the Big Island of Hawaii. Signals from two existing seismic stations, Naalehu and Pahoa, will be telemetered from the Hawaii Volcano Observatory (HVO) through the co-operation of the U. S. Geological Survey. Those stations are now being telemetered via phone line to HVO. Radio telemetry carries the signals from HVO to a relay station at Mauna Kea, where they are transmitted to HO. New Big Island seismic stations have been established at Kona and Mauna Kea. The Kona data are transmitted with the hydraulic gauge signals to Haleakala, Maui, where it is relayed to HO. The new station at Mauna Kea is telemetered to HO with the Naalehu and Pahoa data. The other seismic station in the System is located at Haleakala, Maui, with direct radio transmission to HO. (See Figure 5.)

Because of the high seismicity of the Island of Hawaii and the probability that any local tsunami generation would occur there, two hydraulic stations are part of the Warning System, one at Kona, on the Kona Coast; and the other at Honuapo, on the southeast coast. If this experimental system proves to be successful, additional hydraulic gauges will probably be established in the Hawaiian Islands.

With all data visually recorded at HO (see Fig. 6), potentially tsunamigenic shocks should be located within 10 to 15 minutes. Procedures for issuing warnings may be based on a threshold magnitude so that areas near the epicenter can be warned whenever the earthquake's magnitude exceeds the threshold value. The records from the hydraulic gauges may then confirm or refute tsunami generation.

Fig. 4: Sensor Portion of the Tide Gauge Equipment Planned for Installation at Some Reporting Stations. Bristol Sensor at Left, Conventional Tide Gauge at Right.

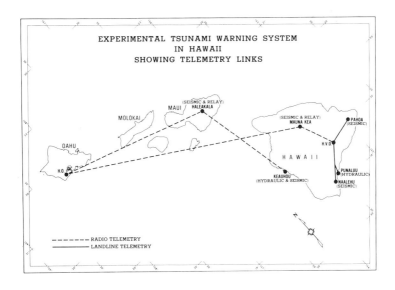

Fig. 5: Map Showing the Proposed Telemetry Links in the Experimental Tsunami Warning System in Hawaii.

Fig. 6: View of Partially Assembled Display Racks Planned for Installation at Honolulu Observatory for Read-Out of Telemetered Real-Time, Seismic and Water Level Data on Islands of Hawaii and Maui.

ALASKA REGIONAL WARNING SYSTEM

The Alaskan Regional Warning System (ARWS) has been in operation a little over two years. Although no tsunami activity in Alaska has occurred during this time, several earthquakes occurred this year which served to test the procedures that have been developed for the ARWS. On May 14 of this year a magnitude 7 earthquake occurred in the Andreanof Islands about 250 kilometers west of Adak. Within eight minutes, Adak Observatory had determined the distance, direction, and magnitude of the shock. Because of their local warning responsibility, Adak Observatory issued appropriate warnings to officials on Adak and Amchitka Islands about 15 minutes after the time of the earthquake. Meanwhile, ARWS had determined the epicenter and, 22 minutes after the origin time, issued a tsunami warning to the Alaska Disaster Office, Alaska Command, and the Federal Aviation Administration for the area from Shemya to Adak. After monitoring the tide gauges at Shemya and Adak and observing no tsunami activity, ARWS cancelled the warning.

A smaller earthquake of magnitude 6.6 in the same area on September 12 brought the ARWS into action again. A warning for the coastal area from Attu to Adak was issued by ARWS about 18 minutes after the time of the shock. When no unusual activity was observed on the tide gauges, the warning was cancelled. Although the technical capability to provide tsunami warnings to all but the immediate epicentral area in Alaska appears to exist, no way to measure the effectiveness of the Warning System has been developed.

System Effectiveness

How well the warnings are disseminated to the public and how the public responds to the warnings should be determined. Answers to these questions may result from a study now underway by the Department of Sociology of the University of Colorado. Dr. J. Eugene Haas, Professor of Sociology, is the principal investigator for this study, which is being supported by ESSA. Dr. Haas plans to select four or five towns in south central and southern Alaska which appear to have high tsunami hazard risk. Then, using a probability sample of households, he will interview the general public concerning their understanding of tsunami hazards and knowledge about appropriate action to be taken in response to warnings.

He then plans to test two contrasting approaches to educating the public. One approach will emphasize the use of the mass media such as radio, newspapers, TV, while in the other, emphasis will be placed on face-to-face presentations and discussions. These educational methods will be used in two of the sample communities. Following the public education efforts, he will again contact the respondents in each sample to determine the extent to which there has been any change in their knowledge of tsunami hazard and appropriate response to warnings. It is hoped that the results of this study will be used as a basis for a general public educational program, either by ESSA or by disaster-warning agencies in Alaska.

SATELLITE COMMUNICATIONS

The proposed expansion in the number of seismic and tide stations previously described is the beginning of what will eventually be an automated international warning system. About 30 seismic stations and about 120 tide stations will be required for complete coverage of the Pacific. Data telemetry of this vast network has become feasible with the use of the geostationary satellite as a communication relay station.

The geostationary satellite maintains a fixed position 22,300 miles above the equator on any selected meridian. It receives and relays information, such as seismograms or marigrams, from a surface area of $70°$ radius. During several years of test operation of a geostationary satellite, there has never been a communication failure. The useful life has been calculated at five years. The satellite is not subject to failure due to earthquake action, as are submarine cables, and it is not susceptible to the varying ionospheric conditions of high-frequency radio. Also it does not have the very large power requirements of other techniques. The geostationary satellite should provide an excellent vehicle for tsunami warning communications.

As presently planned, each of the 25 land-based seismographs and five ocean-bottom seismometers will record into a memory bank which will continuously accept new data and discard old data. Whenever a large signal is received by a seismometer, its transmitter will be activated and the recording will be sent to Honolulu in proper time sequence starting with the earliest data in the memory. The delay will assure that the signal, prior to the alarm, will be transmitted and will assist in the signal identification at HO. In addition, an interrogation pulse will be sent to each of the other seismometers in the net so each may begin to transmit to Honolulu.

The volume of data and the need for rapid analysis precludes the usual methods of analysis so an on-line computer at HO must be programmed to identify the seismic signals by mathematical analysis, extract the arrival times, compute the hypocenters, calculate origin time and determine the estimated times of arrival of the tsunami, if generated, at all points. Once the estimated times of arrival have been obtained, a watch can be issued and the computer can program a series of queries to a selected segment of the 120 tide stations based on the progress of the tsunami. Each tide station in turn may then begin transmission at the appropriate time. A voice channel will be used to broadcast the watch and warning information.

The first operational geostationary satellite is scheduled tentatively for launch in 1971 and will be positioned at 100° W. longitude. This will provide coverage for most of North and South America and several of the island stations in the Eastern Pacific. By 1975, additional satellite coverage should include the Western Pacific, and an automated tsunami warning system should be operational.

19. Problems of the Tsunami Warning Service in the U.S.S.R.

Z. K. ABOUZIYAROV
Hydrometeorological Service
Moscow, U.S.S.R.

ABSTRACT

Tsunamis are generally caused by seaquakes in the Pacific Ocean. Earthquakes with epicenter on the continental slope of the Kuril-Kamchatka trough are the primary threat. The travel time of the tsunamis to the shore is only about 30 minutes. The U.S.S.R. tsunami warning system was started after the disastrous 1952 Kamchatka tsunami. Three autonomous seismo-tsunami stations were established. Each station alerts a section of the coast. From 1958 to 1964 no tsunami was missed: of the 13 warnings issued, only 5 cases were not accompanied by a significant tsunami. Special sensors have been designed and installed, and are now being improved.

Long-term, statistical tsunami forecasts have been made for the Kuril-Kamchatka area. The generation, propagation, and run-up are included in the forecasting analysis. Travel-time charts for the Kuril-Kamchatka area have been completed. A long-term periodicity (140 to 160 years) of great seaquakes with increased seismicity for the twenty years just prior to each great seaquake has been noted. The exact time of a great seaquake cannot yet be forecast. A historical catalog for the period 1734-1958 has been completed.

INTRODUCTION

Tsunami waves are one of the menacing elemental forces of nature and usually arise as a result of seaquakes.

Meteorological phenomena (storms and cyclones in middle latitudes), eruptions of submarine volcanoes, or landslides are seldom the cause of tsunami origin.

The chronicle of the most disastrous tsunamis indicates that their sources are in regions with high seismic activity. Such an area is the Pacific Ocean Basin. The most seismically active regions in the Pacific Ocean are the Japanese, Kuril-Kamchatka, Aleutian, Guatemalan, Atakhama, and other deep-sea troughs. The slopes of these troughs (mainly on the side of continents) are epicentral zones of seaquakes and regions of tsunami generation.

It has been established that 90 to 95 percent of all tsunamis in the Pacific Ocean arise as a result of seaquakes, i.e., they are of seismic origin. However, not every seaquake is accompanied by a tsunami.

The main cause of tsunamis is considered to be tectonic seaquakes associated with a sudden displacement of large volumes of the earth-crust. The release of high stresses causes the appearance of large faults, breaks of the crust, and the formation of mud flows. As a result of lowerings and raisings of the ocean bed, rapid changes of volume and pressure occur. The whole column of water in the source region begins to move and a tsunami is generated.

Tsunami waves travel in all directions from the source of the seaquake, but the wave energy is distributed nonuniformly. Most of the energy is concentrated, as a rule, along the normal to the fracture zone of the crust. At the coast, where waves propagate from the deep sea into shallow water, the wave energy is transformed. A portion of kinetic energy is converted into potential energy, resulting in an increase of the height of a wave and steepness of its front slope. The breaking of waves occurs, then massive withdrawal of water back into the ocean.

The main danger on the Pacific Ocean coast of the Far East U.S.S.R. is tsunamis from seismic sources over the continental slope of the deep-sea Kuril-Kamchatka trough. Computations show that the travel time of these tsunamis to the nearest shore is customarily 20 to 30 minutes. During this period of time, the inhabitants of populated areas exposed to floods from tsunamis need time to take shelter in safe places.

Based on the repetition rate of recorded tsunamis, the Kuril-Kamchatka zone is divided into three regions. The most violent tsunamis were observed on the eastern coast of Kamchatka between Cape Lopatka and Cape Kamchatka, as well as on the Comandor Islands. More moderate tsunamis were recorded opposite the South Kuril Islands, beginning from Urup. In the third region, covering the area opposite the Middle Kuril Islands (between the Straits of Krusenstern and the Straits of Boussol), tsunamis were not recorded.

Two types of tsunamis are segregated in the Far East as follows:

a) Disastrous tsunamis generated near the deep-sea trough formed at the depth of 5000 to 6000 m. in places of main longitudinal breaks. These tsunamis cover a wide sector of the coast (up to 1000 kilometers).

b) Tsunamis generated near the shore are characterized by limited spreading and, in many cases, by a moderate or small force evidently associated with lateral faults, and the motion of coastal blocks. These tsunamis usually cover the coast for 300 to 400 kilometers.

As a rule, the greatest portion of tsunami energy is propagated mainly along a normal to the fracture zone. Therefore, Aleutian tsunamis do not reach the shores of the U.S.S.R. but frequently damage the coasts of the Hawaiian Islands. The tsunami recorded on April 1, 1946 is a classic example of the direction of tsunami travel.

The causes and mechanism of tsunami origin and travel are still poorly known, and the science cannot successfully make long-term predictions of or prevent them. The short-term prediction of a tsunami is the warning of an immediate, real danger from a tsunami which will soon swoop down on to the coast.

Specific operations services for warning the population of the tsunami menace are established in those countries which have been most damaged by tsunamis (the U.S.S.R., Japan, the U.S.A.). The task of these services is to determine the occurrence of tsunamis and to notify official bodies so the population will have time to evacuate to safe places.

On the Far East coast of the U.S.S.R., the tsunami warning service was established after the disastrous Kamchatka tsunami in 1952. The service was set up on the basis of three specialized seismo-tsunami stations at Petropavlovsk-on-Kamchatka, Kurilsk, and Uthno-Sakhalinsk. This service has autonomy as separate seismic stations, i.e., if there is danger of a tsunami, each seismic station may announce the alarm independently from other stations.

The supervision of the tsunami warning service is performed by the Hydrometeorological Service of the U.S.S.R. The institutes of the Academy of Sciences, U.S.S.R., and a number of other institutions render effective assistance. Tsunami headquarters are organized at central points to evacuate the population in the event of tsunami danger. For this purpose all means are maintained in constant readiness.

The establishment of the tsunami warning service stimulated research and experimental investigations. At the same time, many problems of methods and organization had to be solved.

In this paper, only the problems concerning tsunami warning and tsunami zoning of the Kuril-Kamchatka region are considered.

When the tsunami warning service was organized in the U.S.S.R., it was important to take into account the isolation of islands included in the Kuril-Kamchatka ridge, the close position of the tsunamigenic zones, and the inadequate scope of instrumental seismic and marigraphic observations taken in the past.

The experience of the service in warning and notifying the population of tsunamis in past years has shown that despite shortcomings, the available service is able, in principle, to correctly inform the population which tsunamis are dangerous and what measures of protection should be taken. Information on the activity of the tsunami warning service for the period 1958-1964 is given in Table 1.

During these years, no tsunami was missed, though of thirteen issued warnings, five cases proved to be false. This is because the mechanism of tsunami origin is not fully studied. The science is still not able to give simple answers to the questions of when and where the seaquake will take place, whether it will cause a tsunami, and what dimensions these waves will be. The tsunami warning service is based on information about the epicenter location and force of the seaquake. Then tsunamigeneity of the seaquake is determined with special nomograms; if it is tsunamigeneous, the alarm of tsunami is announced.

All means for observing the development of the tsunami phenomenon are put into operation; level stations make frequent observations of the sea level; information of the tsunami development, broadcast by warning services of Japan and the U.S.A., are received. Success depends to a considerable extent on the effectiveness and prompt cooperation between all systems and sections of the warning service.

In order for the warnings of the tsunami to be made in good time, three special instruments were constructed to determine quickly the position of the epicenter. The UBOPE-1 instrument in normal operative conditions is designed to record violent seaquakes: for weak seaquakes, the UBOPE-2 instrument is used. Since these instruments are too sensitive, the UBOPE-0 instrument was constructed with the same scale of records. All these instruments satisfactorily record tsunami and dangerous seaquakes with magnitudes of 7.0 to 8.5 at epicentral distances of 150 to 2000 kilometers.

To obtain objective data on level variations resulting from tsunami and to determine regions and duration of floods near the coasts, a method of separate recording of tsunami waves was proposed by the Research Institute of Hydrometeorological Instruments (Shenderovich, 1961). At present the warning service uses a float-type GM-23-I tsunami recorder, a hydrostatic GM-23-II tsunami recorder, and a GM-30 tsunami detector signal system.

The Sakhalinsk Composite Research Institute of the Academy of Sciences, U.S.S.R., is also engaged in designing special sea-level sensors installed in the open sea near origins of tsunami. Setting several such sensors in the most important source areas will considerably improve the quality of tsunami warnings.

One of the main tasks confronting scientists is the preparation of a map showing zones of the seismically active Kuril-Kamchatka region, according to their tendency to cause tsunamis. This work is essential not only for rational positioning and building of shore structures and settlements but also for tsunami forecasting.

The main problems to be solved for this objective are as follows:

1. To estimate from seismic data the probability of vertical motions of various intensities and various lengths

Table 1: Activity of the Tsunami Warning Service in the Far East for the Period 1958-1965.

Date (local time)	Who and on the basis of which data announced the alarm of tsunami	Time of processing seismic data (in min)	Characteristic of tsunami
Nov 7, 1958	Sakhalinsk Administration of the Hydrometeorological Service on the basis of broadcast by Japan Meteorological Agency	—	Moderate tsunami in the south of the Kuril Islands
May 4, 1959	Seismic station Petropavlovsk	5	Weak tsunami in the area of Cape Shipunsky and the Bay of Avachinsk
Mar 20, 1960	Tsunami station Uthno-Sakhalinsk	10	No tsunami
May 24, 1960	Sakhalinsk Administration of the Hydrometeorological Service on the basis of information from a man on duty of the port post Severo-Kurilsk	—	Destructive tsunami on the Pacific coast of Kamchatka and the Kuril Islands
Jul 29, 1960	Tsunami station Uthno-Sakhalinsk	10	No tsunami
Feb 13, 1961	Tsunami station Uthno-Sakhalinsk	7	Very weak tsunami in the south of the Kuril Islands
Oct 13, 1963	Seismic tsunami station Uthno-Sakhalinsk	10	Destructive tsunami on the Pacific coast of Urup and Iturup islands

Date	Station		Result
	Seismic post of tsunami station Kurilsk	7(20)	
	Seismic station Sevezo-Kurilsk	10	
Oct 20, 1963	Sakhalinsk Administration of the Hydrometeorological Service on the basis of information from hydrometeorological station Van-Der-Lind	–	No tsunami
Mar 28, 1964	Seismic tsunami station Uthno-Sakhalinsk	12	Very weak tsunami on the Pacific coast of Kamchatka and the Kuril Islands
Dec 24, 1964	Seismic tsunami station Uthno-Sakhalinsk	14	Very weak tsunami on Matua
Dec 25, 1964	Sakhalinsk Administration of the Hydrometeorological Service on the basis of information from seismic tsunami station Uthno-Sakhalinsk	10	No tsunami
Oct 16, 1964	Tsunami station Petropavlovsk	10	No tsunami
Feb 4, 1965	Sakhalinsk Administration of the Hydrometeorological Service on the basis of information from seismic tsunami station Uthno-Sakhalinsk	26	Very weak tsunami on the Pacific coast of Kamchatka and the Kuril Islands

at all points of the Pacific Ocean between the continental shelf and the axis of the deep-sea Kuril-Kamchatka trough.

2. To relate bottom deformation to deformation of the ocean surface.

3. To construct refraction diagrams and to relate a probable characteristic of water lift in the region of sources to that of tsunami height on some contour line near the coast (for example, 200 meters).

4. To take into account features of the coastal relief and to obtain prediction of the water height and width of the flooded zone.

In 1958, an elementary map of seismic zoning on the Pacific Ocean coast of the Kuril Islands and Kamchatka was composed by a group of experts of the Academy of Sciences, U.S.S.R., under supervision of E. F. Savaransky (Figure 1). The map was prepared on the basis of data about the epicenter position of past seaquakes accompanied with tsunami, the wave travel theory, the geomorphological features, and the relief of the coastal zone. The map shows areas of most probable origin of seaquakes accompanied with tsunami, computed relative heights of waves on the coast for several epicenters, and the extent of wave inundation.

The most complicated part of the task with regard to the zoning is the establishment of relationships between parameters of seaquakes and possibility of tsunami generation.

Such relationships are determined mainly on a statistic basis, though lately original theoretical solutions of this task are also obtained (G. S. Podyapolsky, 1968).

Fig. 1: Scheme-Map of the Degree of Danger from Tsunami on the Kuril-Kamchatka Coast. 1. Epicenters of Seaquakes with Violent Tsunami (with indication of the year) and Zones of Probable Generation of Seaquakes Accompanied with Tsunami (shaded sections) 2. Low Shores Dangerous During Tsunami 3. Shores with Moderate Tsunami Wave Climb 4. High Shores Comparatively Harmless During Tsunami 5. Considerable Increase of Heights Due to Features of the Bottom Contour 6. Maximum Heights of Waves Observed During Tsunami on November 4-5, 1952 (6 m).

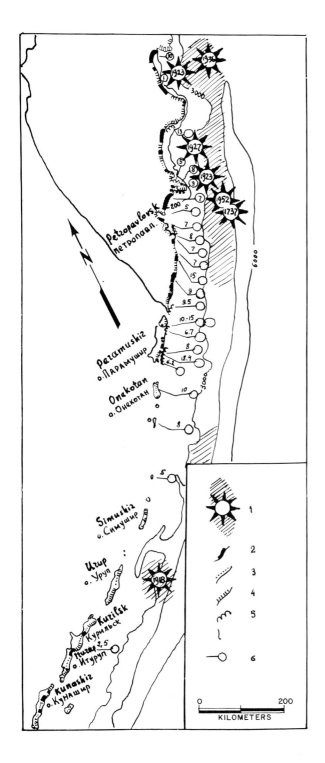

An attempt at tsunami zoning of the Kuril-Kamchatka region according to features of seismicity was made by V. N. Averyanova on the basis of statistic analysis of data on seismic activity and direction of motions in a source of the seaquake.

L. N. Ikonnikova performed computations of maximum tsunami wave heights on the coast of the Kuril Islands and Kamchatka.

As a basis of the computations, Brovikov's formula describing the change of waves as they move to shallow water was taken. The wave refraction was computed using the method proposed by academician V. V. Shuleikin. The wave travel rate was computed with Lagrange-Eure's formula ($C = \sqrt{gh}$) and in shallow water with Skott-Rassel's formula ($C = \sqrt{g(H \pm h)}$). The height of an initial wave in the origin source was found on the basis of studies made by seismologists. The intensity of the tsunami over the seaquake source was determined from the intensity of the seaquake. A relationship was obtained which permits determination of the height of tsunami waves in bays of any shape.

According to the computations made by L. N. Ikonnikova, the possible maximum height of tsunami on the Pacific Ocean coast of the U.S.S.R. can vary, depending on the relief, within 10 to 18 meters (extreme values are 1 and 27 meters).

For the purpose of tsunami warning, maps showing the tsunami travel time from a seaquake epicenter are of great importance. Numerous approximate methods based on the theory of long waves are applied for this. Bathymetric maps for the basins under consideration are usually used, and by averaging depths between nearest contour lines, the computation of travel time for a given section is performed with Lagrange-Eure's formula. All intervals of time are computed in a similar manner for separate sections of some path, added together, and the total time of wave travel from the origin source to the point of observation is obtained. Such maps for regions from the Kuril-Kamchatka trough to coasts of Kamchatka and the Kuril Islands were composed by Z. K. Grigorash and R. N. Samuseva.

In the Far East Research Institute, refraction maps of tsunami paths are composed for different positions of seaquake epicenters by R. A. Yaroshena. These maps show that if the seaquake occurs in the trough, divergence of wave paths is observed. If the wave comes over a knoll, its travel rate is decreased and water masses are inhibited; in consequence, the wave front curves, wave paths converge, and an increase of the wave height occurs over the knoll. The Pacific Ocean coast of Kamchatka and the Kuril Islands is divided into regions according to the predicted intensity of tsunami; this intensity varies due to features of the bottom contour between tsunami source and a coastal line.

There are 6 regions:

(1) Kamchatka and the Kronoki Bay,
(2) the Cape of Shipunsky,
(3) the South-Kamchatka region,
(4) the North-Kuril region,
(5) Rashuwa, Ketoi, Simushir islands, and
(6) the South-Kuril region.

Visible divergence of wave paths is observed in the fifth region and convergence in the first region. In other regions, the shape of the paths is rather simple.

All regions mentioned above are of great importance for the tsunami zoning of the Kuril-Kamchatka region. Since tsunamis are closely associated with seaquakes, investigations on the development of long-range seaquake forecast are of great interest. In this respect, an interesting investigation made by S. A. Fedotov should be noted. Fedotov has developed a statistical method for forecasting variations of seismicity with time. He noticed the fact that violent seaquakes are repeated in the same place. This suggested that gradually increasing tensions in the crest are released at some critical moment, i.e., the seaquake occurs. After this, an abatement of seismicity and stabilization of the regime is observed for several tens of years. Then an increase of seismicity occurs about 20 years prior to the next violent seaquake. On the basis of statistical processing of seaquakes observed for a long period of time, Fedotov determined a definite periodicity of seaquakes according to their size. From his computations, this period is equal to 140 to 160 years. Fedotov considers that it is possible to forecast probable variations of seismicity tens of years ahead if it is known what moment of the cycle a given region goes through at present. Unfortunately, it is impossible to determine this is most cases. But that certainly does not mean forecasting of the exact time of seaquakes for one period or another.

A report of data about tsunami in the U.S.S.R., composed by S. L. Soloviev and M. D. Ferchev, is also important for the tsunami zoning of the Kuril-Kamchatka region and studies of other problems related to the problem of tsunami. In this report, descriptions of tsunami observed on the coast of seas bordering the Soviet Union are given for the period 1734-1958.

Investigations of the theory of tsunami waves and modeling of tsunami have not been discussed here since these problems are analyzed in detail in the review of S. S. Voit.

This review does not cover all the investigations related to the tsunami zoning of the Kuril-Kamchatka region. As mentioned above, a positive solution of this problem is closely associated with a qualitative improvement of tsunami forecasting. The seismic method now applied in the tsunami warning service still has a low reliability and for the present there is no way to improve this method notably. Therefore, at present it is proposed to improve the system of warning by designing instruments for direct determination of tsunami waves in the open ocean. However, the quality of tsunami forecasting can be improved using only seismic, marigraphic, and hydrophysical methods in combination. But it is necessary to add a good network of seismic and water-level stations, and a reliable communications service.

BIBLIOGRAPHY

Podyapolsky, G. S. 1968. Relationship between a tsunami wave and the buried source generating it. *Problema Tsunami*. Izd-vo Nauka, pp. 51-62.

Shenderovich, I. M. 1961. Tsunami recorders built by the Scientific Research Institute of Hydrometeorological Instrument Building. *Byulleten' soveta po seysmologii, problemy tsunami*. Moscow, No. 9, pp. 67-73.

Soloviev, S. L. 1965. The earthquakes and tsunamis of October 13 and 20, 1963, on the Kuril Islands. Academia Nauk S.S.S.R., 101 pages.

TSUNAMI PROPAGATION AND RUN-UP

20. Tsunami Propagation over Large Distances

R. D. BRADDOCK
Queensland University
Queensland, Australia

ABSTRACT

The Grid Refinement Technique is an iterative procedure which employs discrete methods to obtain sequences of approximate solutions to continuous optimal problems. This technique is applied to the propagation of tsunamis over large distances, and the results of calculations on the propagation of the Alaskan tsunami of 1946 demonstrate its capabilities.

The ray paths of the Alaskan tsunami of 1964 to various localities along the west coast of America were calculated and the propagation of the wave system into this region is discussed. In the southwest Pacific, the wave system was dispersed by the complex bathymetry northeast of Australia. However, some of the wave energy entered the Tasman Sea by way of the Fiji Basin, and its subsequent reflection by the Undulla Deep explains the late arrival times in New Zealand and at Macquarie Island.

INTRODUCTION

The travel time $T(S)$ of a tsunami along a curve S is given by the functional (Green, 1961)

$$T(S) = \int_S \frac{ds}{c} , \tag{1}$$

where $c = \sqrt{gh}$ is the long-wave phase speed in water of variable depth h. Applying Fermat's Principle to (1) leads to an optimization problem which can be handled using variational methods. The resulting differential equations which govern the curvature of the ray path contain derivatives of the phase speed and hence derivatives of the water depth. On the continental shelves, the bathymetry is sufficiently well documented that these derivatives may be estimated accurately and the ray path determined by numerical integration (Arthur, et al., 1952). In the deeper basins which constitute a large percentage of the oceans, the bathymetry is not well known, the derivatives cannot be accurately estimated, and the ray equations cannot be integrated accurately.

The travel times for many tsunamis have been studied by evaluating (1) along the appropriate great circle arcs; such studies can be useful (Green, 1946), but the results require careful interpretation since anomalies can arise. (See later discussion on 1946 Aleutian Tsunami). The wave-front method or Huygens' Construction may also be used to study the propagation, although this technique necessitates enormous calculations if accurate answers are required. Further, the accuracy of the final answers depends heavily on the accuracy of the preceding computations and cannot be improved without re-running the full calculation. A modified form of this technique calculates travel times along sets of great circle paths and connects points of equal travel time to obtain the approximate wave fronts (Gilmour, 1961). Charts constructed by this method are obviously incorrect if the great circle arcs do not closely approximate the ray paths.

The grid refinement (GR-) technique (Braddock, 1968) is a new method for systematically calculating convergent sequences of approximations to an optimal path and its corresponding travel time. The first terms in these sequences are usually provided by the corresponding great-circle approximations and the subsequent terms are obtained by an iterative sampling of the velocity field using discrete grids. The convergence of the approximating sequences has been established, and it has been shown that the functional (1) satisfies the conditions for convergence. The technique has been applied to the propagation of tsunamis over the deep oceans, where bathymetric data are scanty, and unlike the wave-front method, it requires only moderate calculations which are easily performed by computer.

The application of the GR-technique to the propagation of the Alaskan tsunami of 1946 to Sitka demonstrates its capabilities, and a remarkable correlation with the observed travel time is obtained. Some of the ray paths for the Alaskan tsunami of 1964 are obtained and the observed and estimated travel times discussed. Finally, the propagation of this tsunami into the southwest Pacific Ocean is considered, and many of the problems presented by the tide gauge records are resolved.

THE GR-TECHNIQUE

Assume that a travel time T_1 and corresponding path S_1 have been obtained as first approximations to the optimal travel time T_{min} and corresponding path S_{min} between two endpoints A and B. Further approximations to these optimal values may be obtained as follows:

For a suitable grid spacing DS_1, divide S_1 by the points E_j, $j = 1, 2, \ldots, m$, into m-1 intervals of length less than but approximately equal to DS_1. (See Fig. 1). Select a normal grid spacing DN_1 and define the grid parameter

$$U = DS_1 / DN_1 . \qquad (2)$$

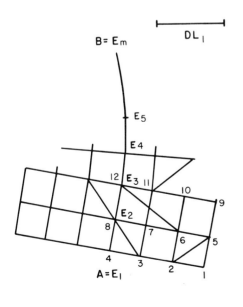

Fig. 1: Illustration of Typical Construction of Finite Discrete Network of Paths in the Field of Travel.

At each point E_j, $j = 1, 2, \ldots, m$, construct a normal to S_1 and mark off nodes or points at intervals of DN_1 along each normal. Now select a control length DL_1 such that

$$DL_1 \geq \max(DS_1, DN_1),$$

and define a second grid parameter by

$$Q = DL_1 / DS_1. \tag{3}$$

Using arcs or lines of length less than DL_1, complete the grid by constructing links between the nodes. The travel time along each arc is then estimated using known or interpolated bathymetric data.

The above construction has provided a finite discrete network of paths in the field of travel such that the travel times along the individual arcs are known. Using a suitable algorithm (Dantzig, 1960), the shortest time path S_2 through this discrete grid is calculated together with the corresponding travel time T_2.

Next, select a new tangential grid spacing $DS_2 \leq DS_1$, define the grid refinement parameter by

$$R = DS_2 / DS_1, \tag{4}$$

and, by interpolation, add extra nodes to S_2 so that the maximum spacing between any two adjacent nodes which form S_2, is less than DS_2. Determine DN_2 and DL_2 from U and Q, respectively, construct a new grid about S_2, and hence obtain further approximations to the optimal values.

This process generates a monotonically decreasing sequence of travel times $\{T_i\}$ and a sequence of paths $\{S_i\}$ which converges to a limit curve (Braddock, 1968). Actually, each S_i is the center-line of the succeeding construction and hence a possible path through the grid. Thus

$$T_i \geq T_{i+1},$$

for all $i \geq 1$. In constructing the grids, it is not necessary to use arcs of great circles between the nodes since, as the refinement proceeds, the arcs become shorter and straight line approximations are sufficiently accurate, certainly more accurate than the bathymetric data.

Usually the above computations require the use of a computer containing a high-speed access memory of moderate capacity. The number of nodes marked off along the normals to each approximating path is determined mainly by the capacity of the memory bank although the values of the parameters U and Q must also be considered.

PROPERTIES OF THE GRID PARAMETERS

The parameters U, Q, and R govern the magnitude of the calculations and the machine storage requirements needed to generate the approximating sequences. They also control the accuracy obtained at each step. Obviously, $0 < R \leq 1$, from (4). For small R, i.e., $R \lesssim 1/2$, the tangential spacing DS_i decreases rapidly and large numbers of nodes are required to specify the approximating paths. This leads to storage difficulties which usually halt the refinement before sufficient accuracy is obtained. (See Braddock, 1968). Alternatively, large values of R, i.e., $R \gtrsim 3/4$, produce a slow refinement and hence very many calculations.

In general, the number of calculations may be reduced by restricting the number of arcs in each grid, i.e., by controlling the value of Q. Obviously, $Q \geq 1$ from (3), since $Q < 1$ implies that

$$DL_i < DS_i ,$$

and hence, that S_i may not be included in the set of possible paths through the succeeding grid. Then $\{T_i\}$ is not necessarily monotonic and may even be increasing.

Now consider a grid constructed using $U = 1.0$, i.e., the normal and tangential grid spacings are equal. If $Q = 1.0$, this grid contains only first order normal and tangential arcs, cf., the arcs 1-2 and 5-9 in Figure 1. If Q is increased to $\sqrt{2}$, the first order diagonals are included, cf., the arc 3-8 in Fig. 1, while larger values include the higher order arcs. It has been found that $\{T_i\}$ and $\{S_i\}$ are relatively insensitive to the use of the higher order arcs (see Braddock, 1968).

Finally, it has also been found that greater accuracy may be obtained by setting

$$U > 1, \quad \text{i.e.,} \quad DN_i < DS_i.$$

As the refinement proceeds, the normal distance between the exact solution S_m and the approximation S_i decreases, and smaller mesh spacings are required to obtain higher accuracy. Hence, either R is small or U must be large, thus permitting small normal variations in the approximating paths.

For large U, the number of arcs in each grid is also large since, in this case,

$$DL_i \gg DN_i.$$

If the grid is restricted to contain only first order normal, tangential, and diagonal arcs, then

$$DL_i \gtrsim (DS_i^2 + DN_i^2)^{1/2}.$$

Hence, on dividing by DS_i,

$$Q \gtrsim \{1 + \frac{1}{U^2}\}^{1/2}. \tag{5}$$

However, there is one disadvantage to the use of the GR-technique. There are certain bathymetric distributions such that several curves yield minimum values of (1). The curve S_{min} for which (1) attains its minimum value T_{min} with respect to all possible paths between the two end-points, is called *the optimal path*. Other curves S_e for which (1) attains a minimum value with respect to curves which are close to S_e, are called *locally optimal paths*. In the GR-technique, it is possible that $\{S_i\}$ and $\{T_i\}$ may converge to a locally optimal solution. This situation can usually be avoided by using broad initial grids which thoroughly sample the full velocity field.

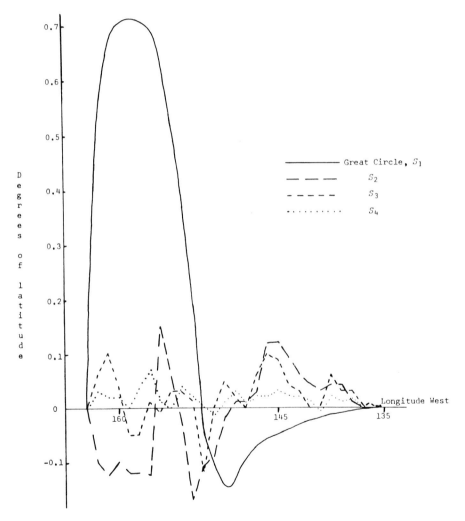

Fig. 2: Relative Map of Approximating Paths. The Ordinate is the Difference Between S_5 and the Particular Approximating Path, Measured in Degrees of Latitude.

PATH OF THE ALEUTIAN TSUNAMI OF APRIL 1, 1946, TO SITKA

On April 1, 1946, an earthquake generated a large tsunami on the continental shelf near the Aleutian Islands. This tsunami is particularly interesting since there is a large discrepancy between its observed travel time of 2 hours 42 minutes and its computed travel time of approximately 6 hours along the corresponding great-circle arc to Sitka, Alaska. Zetler (1947) observed that this arc passes through shallow water on the continental shelf and consequently yields a large travel time. He further observed that a path to the southeast and then along the Aleutian trench to Sitka exploits the bathymetry and yields a travel time of approximately 3 hours. See Figure 3.

By starting from the great-circle arc as the first approximation and using R = 0.6 and U = 3.0, where Q is determined from (5), the GR-technique generated the approximating travel times: 5 hours 59 minutes, 3 hours, 2 hours 49 minutes, 2 hours 45 minutes, and 2 hours 43 minutes. The approximating paths S_1, S_2, ..., S_5 are shown in Fig. 2, where the abscissa is the approximating path S_5 and the ordinate is the difference, in degrees of latitude, between S_5 and S_1, S_2, S_3, and S_4. It can be seen that the paths converge rapidly.

The last three approximate travel times obtained from the GR-technique, are all less than Zetler's observed value of 2 hours 55 minutes, possibly because of an error in estimating the arrival time of the tsunami. The Sitka tide record exhibits two small waves (travel time: 2 hours 42 minutes) followed by a train of larger waves (travel time: 2 hours 55 minutes). Theory and experiment show that the initial front of a tsunami dominates as the waves propagate over large distances (Wilson, 1962), whereas at small distances from the epicenter, the first few waves are frequently quite small and easily overlooked. It is suggested here that these two small waves were the initial front of the tsunami. Figure 3 shows that the computed path closely follows the path suggested by Zetler (1947).

PROPAGATION OF THE TSUNAMI OF MARCH 28, 1964, IN THE EASTERN PACIFIC

In their analysis of the Alaskan tsunami of March 28, 1964, Spaeth and Berkman (1967) assumed that the epicenter was situated at 60° N., 147° W. This assumption, to be adopted in this study, is justified since the regions of large scale earth movement, which generated the tsunami, are located near or to

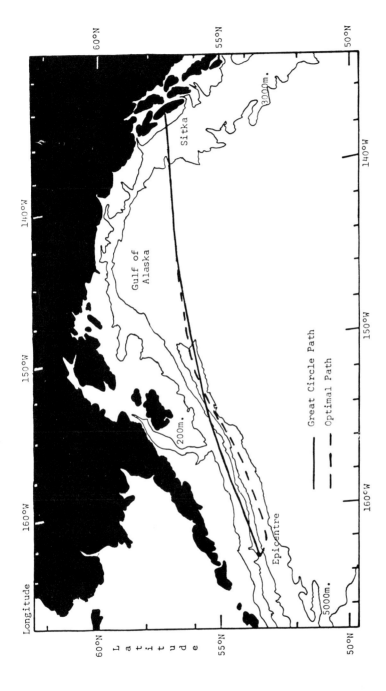

Fig. 3: Bathymetry of the Gulf of Alaska. The chart illustrates the relative positions of the Aleutian Trench and the epicenter of the Alaskan earthquake of April 1, 1946. The great circle arc and the optimal path obtained using the grid refinement technique are also shown.

the north of this point. The eight localities (Sitka, Astoria, Ensenada, Manzanillo, Punta Arenas, Talara, Arica, and Talcahuano) were selected since they are approximately equally spaced along the west coast of America. The corresponding observed travel times of the tsunami to these localities are given in Table 1.

The calculated travel times to these localities were obtained from the GR-technique by using great circles as first approximations. Except for Sitka and Astoria, these arcs passed over the American continents, and broad initial grids, including a section of the Pacific Ocean, were employed. As a consequence, early approximations were inaccurate and the sequences converged relatively slowly. Table 1 shows that the computed travel times to Astoria and Punta Arenas are less than the observed values and that the largest difference of 19 minutes occurs for Arica. These differences are probably due to inaccuracies in the Bathymetric data employed in the calculations.

The calculated paths to Sitka and Astoria pass seaward of the corresponding great circles, thus avoiding the shallow waters at the head of the Gulf of Alaska. (See Figs. 4 and 5). These paths do not reach the Aleutian trench and hence, the propagation of this particular tsunami in this region, differs from the example discussed in the previous section. North of latitude 20° N., the paths to Ensenada and Manzanillo pass through deeper water west of the United States and then curve across the continental shelf to the respective harbors, while the path to Punta Arenas continues on around the coast of Mexico. The path to Talara also skirts the coast of Mexico but then moves east of the Galapagos Islands to Talara, on the western tip of South America. The paths to Arica and Talcahuano exploit the deeper waters west of the Galapagos Islands and then curve across the continental shelf to the South American coast.

PROPAGATION OF THE TSUNAMI OF MARCH 28, 1964, IN THE SOUTHWEST PACIFIC

The complex bathymetry and large distance between the southwest Pacific and Alaska is reflected in the regional tide gauge records for this tsunami. The McMurdo Sound and Brisbane tide records indicate that both the pack ice surrounding Antarctica, and the Great Barrier Reef and island chains northeast of Australia, prevented the tsunami from reaching these localities. The observed travel times (see Table 1) for Nelson

and Greymouth are some four to five hours greater than for
Coff's Harbour, Sydney, and Lyttleton, all of which are approximately equidistant from Alaska. Further, the observed value
for Sydney is less than that for Coff's Harbour, which is
nearer to the epicenter, whereas the Macquarie Island record
shows evidence of recurring tsunami activity over a period of
at least seven hours. This activity is in three sections with
the travel times 26 hours 39 minutes, 30 hours 39 minutes. and
31 hours 54 minutes. However, the timing on this record is
obviously incorrect.

Preliminary Calculations

Except for two particular cases, great circles were again
employed as first approximations for the GR-technique. The
calculated values for Coff's Harbour, Sydney, and Hobart agree
with the observed travel times (see Table 1), and each path
passes eastward of the corresponding great circle to the
Fiji Basin. From there, the tsunami was refracted by the
Norfolk Island Ridge and Lord Howe Rise and proceeded across

Table 1: Comparison of Computed and Observed Travel Times
of the 1964 Alaskan Tsunami.

Tide Station	Observed Travel Time	Calculated Travel Time
Sitka	1 hour 30 mins.	1 hour 31 mins.
Astoria	4 hours 20 mins.	4 hours 19 mins.
Ensenada	6 hours 6 mins.	6 hours 11 mins.
Manzanillo	8 hours 39 mins.	8 hours 46 mins.
Punta Arenas	12 hours 47 mins.	12 hours 39 mins.
Talara	14 hours 20 mins.	14 hours 21 mins.
Arica	16 hours 54 mins.	17 hours 13 mins.
Talcahuano	18 hours 39 mins.	18 hours 51 mins.
Brisbane	Indefinite	18 hours 25 mins.
Coff's Harbour	17 hours 39 mins.	17 hours 37 mins.
Sydney		
(a) Camp Cove	16 hours 54 mins.	17 hours 7 mins.
(b) Fort Denison	17 hours 9 mins.	Not Calculated
Hobart	20 hours 19 mins.	20 hours 30 mins.
Canton Island	8 hours 39 mins.	Not Calculated
Pago Pago	10 hours 15 mins.	Not Calculated
Nelson	22 hours 20 mins.	18 hours 55 mins.
Greymouth	23 hours 20 mins.	18 hours 5 mins.
Lyttleton	18 hours 34 mins.	17 hours 2 mins.
McMurdo Sound	Indefinite	Not Calculated

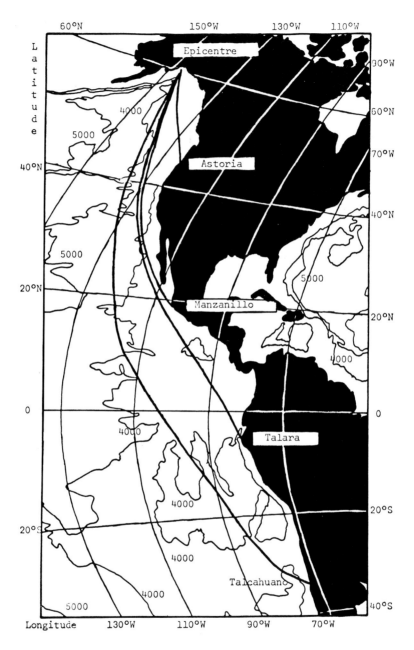

Fig. 4: Calculated Ray Paths for the Tsunami of 28/3/64 to Astoria, Manzanillo, Talara, and Talcahuano.

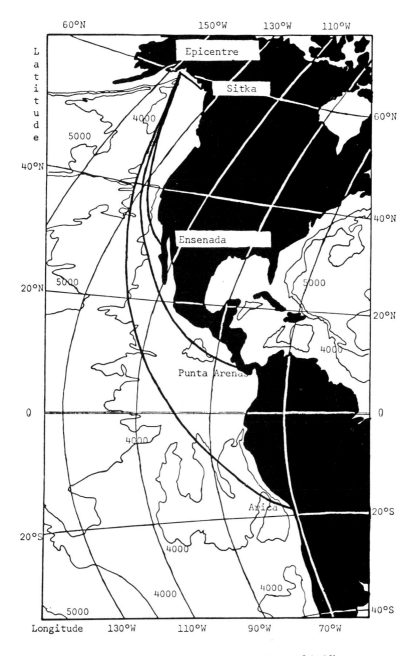

Fig. 5: Calculated Ray Paths for the Tsunami of 28/3/64 to Sitka, Ensenada, Punta Arenas, and Arica.

the Tasman Sea to the east coast of Australia. (See Fig. 6). Two refinements were made of the path to Brisbane: the first starting with a great circle from Alaska to Brisbane; the second starting with a composite path from Alaska to the Fiji Basin and hence to Brisbane. However, these failed to converge properly, the approximating paths continually changing position in the region of islands and reefs northeast of Australia. This confirmed the earlier deduction from the Brisbane tide record.

The great circle arc from 60° N., 147° W. to Macquarie Island passes through the Tasman Sea west of New Zealand and, when employed as the first approximation, a calculated travel time of 20 hours 33 minutes was obtained. When a composite path consisting of great circle arcs from 60° N., 147° W. to 40° S., 170° W. and then to Macquarie Island was used, as the initial path for the GR-technique, a travel time of 23 hours 32 minutes was obtained. The corresponding calculated path passes east of New Zealand around the base of the New Zealand Plateau. As very little wave energy would pass this way, a more probable route is a larger time path across the plateau.

The calculated and observed values for Lyttleton differ by approximately 1.5 hours, and the corresponding path passes near the Tonga-Kermadec trench system. (See Fig. 6). Wave energy travelling along these trenches would be scattered, and another section of the tsunami may be recorded as the initial wave. The Lyttleton record shows strange oscillations before the main waves arrived, but these cannot be definitely attributed to the tsunami. The estimated travel times for Nelson and Greymouth compare favorably with the value for the east coast of Australia, but they do not agree with the observed values. The tide records do not show any unusual activity at the estimated arrival times. (Table 1).

The Reflected Wave

The estimated travel times to Macquarie Island, already corrected by altering the timing of the record by 6 hours (Table 2), are 6 to 7 hours less than the observed values. The second corrected value does not agree with the corresponding calculated value, but the tsunami may have followed an alternate path across the New Zealand Plateau. The observed values for Nelson and Greymouth and the third corrected value for Macquarie Island indicate that a section of the tsunami passed down the eastern side of the Tasman Sea, but these observations do not correspond to any direct route from Alaska.

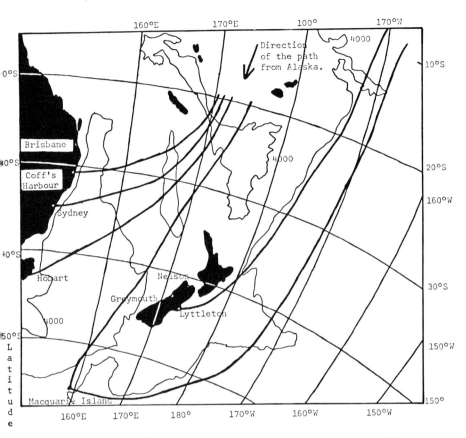

Fig. 6: Approximate Travel Paths for the Tsunami of 28/3/64 in the Southwest Pacific.

Bathymetric features, such as the western edge of the Undulla Deep on Australia's east coast, can reflect appreciable amounts of wave energy. Lamb's formula (Hilaly, 1967) indicates that approximately 50 percent of the wave energy could be reflected by the Undulla Deep, but the lack of suitable tide records has prevented the calculation of an experimental reflection coefficient. This hypothesis was tested by calculating travel times (Table 3) and approximating paths (Fig. 7) from the edge of the Undulla Deep near Sydney to Coff's Harbour, to Nelson, Greymouth, and Macquarie Island.

299

The results indicate that the point of reflection for Greymouth is nearer to Coff's Harbour than Sydney, and that, in the case of Macquarie Island, reflection also occurred north of Sydney. However, consideration of the angle of incidence (about 30°) indicates that the path to Macquarie Island was reflected from a point about 120 miles south of Sydney. Table 3 does not fully explain the value for Nelson but indicates that this particular path may have been reflected from the wider continental margin north of Coff's Harbour, or possibly from the Lord Howe Rise. Since precise data are not available, the points of reflection could not be estimated more accurately.

Table 2: Comparison of Observed, Calculated and Corrected Travel Times for the Arrival of the 1964 Tsunami at Macquarie Island.

Path followed to Macquarie Island	Calculated	Travel Times Observed	Corrected
Via the west side of New Zealand	20 hrs 33 mins	26 hrs 39 mins	20 hrs 39 mins
Via the east side of New Zealand	23 hrs 32 mins	30 hrs 39 mins	24 hrs 39 mins
Indefinite	...	31 hrs 54 mins	25 hrs 54 mins

Table 3: Calculated Travel Times for the Reflected Wave.

	Nelson	Greymouth	Macquarie Island
Observed travel times	22 hrs 20 mins	23 hrs 20 mins	25 hrs 54 mins
Computed paths: Reflected from Sydney	23 hrs 46 mins	21 hrs 59 mins	24 hrs 4 mins
Reflected from Coff's Harbour	25 hrs 16 mins	24 hrs 47 mins	26 hrs 5 mins

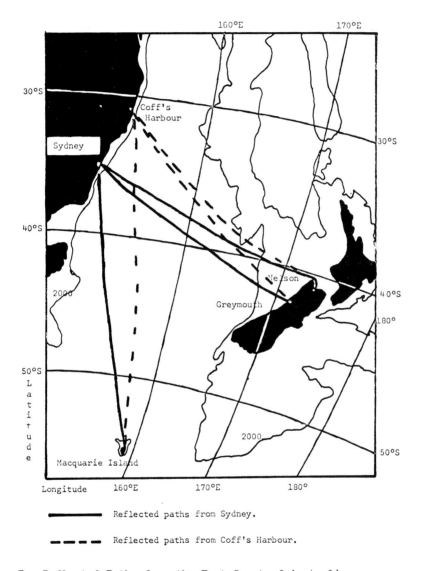

Fig. 7: Reflected Paths from the East Coast of Australia.

CONCLUSION

The GR-technique provides a new method of studying the propagation of tsunamis across the ocean depths. The example of the propagation of the 1946 tsunami to Sitka illustrates the high degree of accuracy in calculated travel paths and times obtained from the method. Even badly chosen initial approximations, such as those offered by the propagation of the 1964 tsunami in the eastern Pacific, do not affect the final accuracy. It has also been used successfully to discuss the propagation in regions with a complex land and bathymetry distribution, so much so that a complete picture of the propagation of the 1964 tsunami into the southwest Pacific can be presented.

The 1964 tsunami propagated toward the southwest Pacific along ray paths closely approximating great circle arcs. One section of the wave system passed into the Solomon and New Hebrides Island, where it was dispersed or dissipated by the complex bathymetry. The section of the wave train which travelled into the Fiji Basin was refracted by the Norfolk Island Ridge and Lord Howe Rise and then passed on to the east coast of Australia. Some of the wave energy was reflected by the Undulla Deep, recrossed the Tasman Sea, and was recorded in New Zealand and on Macquarie Island. A final section passed down the east coast of New Zealand and travelled on into the Antarctic seas.

BIBLIOGRAPHY

Arthur R. S., W. H. Munk, and J. D. Isaacs 1952. The direct construction of wave rays. *Trans. Am. Geophys. Union.* Vol. 33, p. 855.

Braddock, R. D. 1968. Optimal problems in physical oceanography. Ph.D. thesis. South Australia: Flinders University.

Dantzig, G. B. 1960. A shortest path algorithm. *Management Sc.* Vol. 6, p. 187.

Gilmour, A. E. 1961. Tsunami warning charts. *N.Z. J. Geol. Geophys.* Vol. 4, p. 132.

Green, C. 1946. Seismic sea wave of 1/4/46 as recorded at tide gauges. *Trans. Am. Geophys. Union.* Vol. 27, p. 490.

Green, R. 1961. The sweep of long waves across the Pacific. *Aust. J. Phys.* Vol. 14, p. 120.

Hilaly, N. 1967. Diffraction of water waves over bottom discontinuities. *Hydraul. Eng. Lab., Univ. Calif., Tech. Report,* Hel 1-7.

Spaeth, M. G., and S. C. Berkman 1967. The tsunami of 28/3/64 as recorded at tide stations. E.S.S.A., *U. S. Coast and Geodetic Survey, Tech. Report No. 33.*

Wilson, B. D. 1962. *The Nature of Tsunamis.* Nat. Eng. Sc. Co., Tech. Report, SN 57-2.

Zetler, B. D. 1947. Travel times of seismic sea waves to Honolulu. *Pacific Sc.* Vol. 1, p. 185.

21. Some Hydrodynamic Models of Nonstationary Wave Motions of Tsunami Waves

S. S. VOIT and B. I. SEBEKIN
Institute of Oceanology
Moscow, U.S.S.R.

ABSTRACT

For the development of the hydrodynamical models of tsunami wave generation and their consequent propagation, the waves originating from the concentrated point source with an undetermined law of function are investigated. The influence of the Coriolois force on the basic wave parameters is investigated. The estimate of the long-wave theory approximations for the tsunami phenomena is given.

The reflection of the concentrated-source waves from the straight shore is explored, and the calculating formulas for the unsteady wave amplitudes and wave motion velocities are given.

The reflecting wave asymmetry due to Coriolis force is shown.

INTRODUCTION

For the successful application of physico-mathematical methods to geophysical problems, in particular the investigation of tsunami waves, one constructs models revealing the main features of the phenomenon that is of interest for the investigator. The simplest model of generation and distribution of tsunami waves is the displacement of the bottom in a basin filled with ideal liquid. The same kind of model was thoroughly investigated in Lamb (1947) where it was shown that earthquake waves incited by bottom vibrations are slightly different from waves incited by a point hydrodynamic source located in a liquid. Therefore, in the present work waves generated by motion of the source with an arbitrary law chosen on the basis of seismology data will be studied. The influence of Earth rotation on the tsunami waves and the effects on wave reflection from rectilinear shore will be studied. This investigation is carried on in terms of the linear theory of long waves, widely used in dynamic oceanology. In spite of simplifying suppositions, the theory explains well the main physical phenomena occurring in ocean. The comparative simplicity of the linear-theory equations of long waves permits obtaining analytical decisions of model tasks taking into consideration different effects. Their introduction into exact science results in insoluble mathematical difficulties. Being approximate, the linear theory of long waves considerably narrows the range of the physical parameters that can be investigated. But the equation of long-wave theory, both linear and nonlinear, can be obtained from the exact equations with the help of only one supposition of hydrostatic pressure change. These equations can also be obtained as the first approach in perturbing the exact equations by a small parameter.

This small parameter is the ratio of fluid depth to some typical horizontal dimension, for example, the wave length.

Here the linear theory of long waves is considered as a first approach to exact linear theory in dividing by a small parameter, h/λ where h = fluid depth and λ = wave length. Meanings of physical parameters when the linear theory of long waves is adequately exact are evaluated. An harmonic source operating in the flat nonrotary basin is assumed. The result of this problem is well known in the term of long-wave theory. The elevation of the free surface is determined by the formula

$$\tilde{\zeta}(r,t) = \frac{Q\sigma}{4c^2} e^{i\sigma t} H_0^{(2)}(\sigma\frac{r}{c}) \ , \quad c = \sqrt{gh},$$

where σ = source frequency,
Q = second consumption,
g = acceleration of gravity strength,
h = fluid depth,
t = time,
r = distance from the source to investigation point, and

$H_o^{(2)}(z)$ = Hankel function of the second kind, zero order.

In the terms of three-dimensional theory, the solution of this problem is obtained (Sebekin, 1967). Analyzing the solution in degrees of small parameter h/λ, we have the asymptotic line:

$$\zeta = \tilde{\zeta} \; [1 - 2\pi^2(h/\lambda)^2 + C(h/\lambda)^4 - \ldots] .$$

We shall require the second term of the series to be 10^n times less than the first term. Hence we find conditions connecting admissible length of the wave and the period with the depth of the fluid:

$$\lambda \geq 10^{n/2}\pi\sqrt{2h} , \qquad T \geq 10^{n/2}\sqrt{2h/g} .$$

These formulas permit a minimum period and length of the waves to be defined, provided the depth of the basin is given and the maximum error of the solution obtained with the help of the theory of long waves is indicated. For example, with 4 km. ocean depth and an acceptable error of 1%, the minimum length of the wave is 180 km., and the minimum period 15 minutes. With an acceptable error of 10%, λ = 55 km., T = 4.5 minutes: decreasing the basin depth increases the accuracy of the theory of long waves. When the depth is 1 km. and the acceptable error is 1%, λ = 45 km., T = 7.5 minutes.

These results have been obtained without taking into consideration the rotation of the basin. Rotation of the basin will not decrease, at least, the accuracy of the theory of long waves since for a fixed frequency of the source the presence of rotation increases the wave length.

$$\lambda = 2\pi c^2/\sqrt{\sigma^2 - 4\omega^2} .$$

The flat model restricts the size of the area where the wave motion is investigated. The global shape of the Earth need not be taken into consideration for dimensions less than 2000-3000 kilometers.

Introducing the hydrodynamic source as wave generation factor is equivalent to the work of concentrated strength of pressure (Lamb, 1947) and the complete strength of pressure is connected with the displacement of the source, according to the formula:

$$P(t) = \frac{Q\rho c^2}{h} \int_0^t f(\xi) d\xi,$$

where ρ = density of fluid, and function $f(t)$ indicates the time dependence of the source function. The source imitates the work of concentrated pressure quite exactly, if the field of pressure application is much less in typical horizontal dimension than the length of the radiating waves.

For the above mentioned qualitative reasons the results of the solution of our model problem can be applied to investigating wave motion within 2000-3000 km. distance, with minimum length of waves ~ 50-100 km., periods ~ 5-15 minutes and disturbances situated in regions which have typical horizontal dimension about a kilometer.

Let us suppose that the plane boundless basin is flooded with the heavy, incompressible, ideal fluid. The depth of the basin is h. The basin rotates around the vertical axis. The angular velocity of the basin is ω. The axis z of the cylindrical coordinate system coincides with that of the rotation of the basin. Let us write the linear equations of the long-waves theory for the movements which have axial symmetry (Stretensky, 1936).

$$\frac{\partial V_r}{\partial t} - 2\omega V_s + g \frac{\partial \zeta}{\partial r} = 0$$

$$\frac{\partial V_s}{\partial t} + 2\omega V_r = 0$$

$$\frac{\partial \zeta}{\partial t} + \frac{h}{\tau} \frac{\partial}{\partial r} (V_r) = 0 ,$$

where V_r is the radial component of the velocity,
V_s is the tangential component of the velocity,
ζ is the elevation of the free surface from the level of the relative equilibrium,

r is the distance between the point of observation and the origin of the coordinates, and
t is time.

Let us assume that at the initial moment of time $t = 0$ at the point $r = 0$ the source within the fluid starts acting. The displacement of the source is $Q \cdot f(t)$ where $f(t)$ is an arbitrary, non-dimensional function of time. When the solution of the system (1) must satisfy initial conditions

$$V(r,0) = V_s(r,0) = \zeta(r,0) = 0, \qquad (2)$$

and the restriction:

$$\lim_{r \to 0} (2\pi r h V_r) = Q f(t). \qquad (3)$$

For solving the boundary value problem (1) to (3), let us perform the transform of Laplace at variable t. Solving the equations for Laplace transform, we get

$$\zeta(r,p) = \frac{Q}{2\pi c^2} G(p) p K_0(\frac{r}{c}\sqrt{p^2+4\omega^2})$$

$$V_r(r,p) = \frac{Qg}{2\pi c^2} G(p) \frac{p^2}{\sqrt{p^2+4\omega^2}} K_1(\frac{r}{c}\sqrt{p^2+4\omega^2}) \qquad (4)$$

$$V_s(r,p) = -\frac{Qg\omega}{\pi c^2} G(p) \frac{p}{\sqrt{p^2+4\omega^2}} K_1(\frac{r}{c}\sqrt{p^2+4\omega^2}).$$

where p is the parameter of Laplace transform, and $K_0(z)$ and $K_1(z)$ are Hankel's functions.

$$G(p) = f(p)[1 + 4\omega^2/p^2]$$

is the Laplace transform of the function

$$G(t) = f(t) + 4\omega^2 \int_0^t (t - \xi) f(\xi) d\xi.$$

Using the theorem about the convolution, we get

$$\zeta(r,t) = \frac{Q}{2\pi c^2} \frac{\partial}{\partial t} \int_0^{t-r/c} \frac{\cos[2\omega\sqrt{(t-\xi)^2-(r/c)^2}]}{\sqrt{(t-\xi)^2-(r/c)^2}} G(\xi) d\xi \qquad (5)$$

$$V(r,t) = \frac{Qg}{2\pi c^2} \frac{\partial}{\partial t} \int_0^{t-r/c} \frac{\cos[2\omega\sqrt{(t-\xi)^2-(r/c)^2}]}{\sqrt{(t-\xi)^2-(r/c)^2}} (t-\xi) G(\xi) d \qquad (6)$$

$$V_s(r,t) = -\frac{Qg^0}{2\pi c^2 r} \frac{\partial}{\partial t} \int_0^{t-r/c} \sin[2\omega\sqrt{(t-\xi)^2-(r/c)^2}] G(\xi) d\xi . \qquad (7)$$

If $t < r/c$ then $\zeta(r,t) = V(r,t) = V_s(r,t) = 0$.
Formulas (5) to (7) give the solution of the boundary value problem (1) to (3).

If $\omega = 0$, the solution (5) to (7) coincides with the well-known solution of the problem about radial waves in the immovable basin. In this case the tangential component of the velocity is zero.

Let us suppose that $f(0) = 0$, then we can write the elevation of the free surface in the following form

$$\zeta(r,t) = \zeta_1(r,t) + \zeta_2(r,t),$$

where

$$\zeta_1 = \frac{Q}{2\pi c^2} \int_0^{t-r/c} \frac{\cos[2\omega\sqrt{(t-\xi)^2-(r/c)^2}]}{\sqrt{(t-\xi)^2-(r/c)^2}} \frac{df(\xi)}{d\xi} d\xi \qquad (8)$$

and

$$\zeta_2 = \frac{2Q\omega}{\pi c^2} \int_0^{t-r/c} \frac{\cos[2\omega\sqrt{(t-\xi)^2-(r/c)^2}]}{\sqrt{(t-\xi)^2-(r/c)^2}} \phi(\xi) d\xi \qquad (9)$$

$$\phi(t) = \int_0^t f(\xi) d\xi . \qquad (10)$$

Consequently, $Q\phi(t)$ is the total quantity of the fluid which the source emits for the moment t. Formula (8) describes the elevation determined by the derivative of the displacement with respect to time. If $\omega = 0$, then ζ_1 coincides with the elevation of the free surface in the immovable basin. Term ζ_2 is the effect of rotation. If the basin is immovable, then ζ_2 is zero. The supplementary elevation ζ_2 is determined by the integral of the displacement of the source and appears due to the variation of the total quantity of the fluid emitted to the moment of time $t-r/c$.

Using the solution (5) to (7), we can investigate reflection of the long unsteady waves from a rigid wall situated in the rotating basin. In the rotating Cartesian coordinates, long-wave equations have the following form

$$\frac{\partial u}{\partial t} - 2\omega v + g\frac{\partial \zeta}{\partial x} = 0$$

$$\frac{\partial v}{\partial t} + 2\omega u + g\frac{\partial \zeta}{\partial y} = 0 \qquad (11)$$

$$\frac{\partial u}{\partial x} + \frac{\partial v}{\partial y} + \frac{\ell}{h}\frac{\partial \zeta}{\partial t} = 0 ,$$

where u and v are the components of the velocity. Let us suppose that the source, studied above, starts acting at the moment $t = 0$ in the point which have coordinates (x_0, y_0). Then the initial conditions must be satisfied:

$$\zeta(x,y,0) = u(x,y,0) = v(x,y,0) = 0. \qquad (12)$$

Besides that, let the fluid occupy the space $y > 0$. On the rigid wall $y = 0$ the condition of not leaking must be satisfied:

$$v(x,0,t) = 0 \qquad (13)$$

To solve the boundary value problem (11) to (13) we set the long unsteady wave source at the point $(x_0, -y_0)$. Because the problem is linear, the total elevation of the free surface has the form

$$\zeta_1 + \zeta_2 + \zeta \qquad (14)$$

The components of the velocity have the similar form: $u_1 + u_2 + u$, $v_1 + v_2 + v$. The letters having the subscript "one" correspond to the elevation and the velocity from the source at the point (x_0, y_0). The letters having the subscript "two" correspond to the elevation and the velocity from the source at the point $(x_0, -y_0)$. The functions ζ, u, and v must satisfy the tidal equations (11), the initial conditions (12), and the boundary condition

$$v(x,0,t) = -v_1(x,0,t) - v_2(x,0,t) \qquad (15)$$

We can find the magnitudes of the functions $v_1(x,0,t)$ and $v_2(x,0,t)$ using formulas (6) and (7). The boundary condition (15) becomes

$$v(x,0,t) = -\frac{2(x-x_0)}{\sqrt{(x-x_0)^2+y_0^2}} \, (V_s)_{y=0} \qquad (16)$$

One can solve the boundary value problem (11), (12), (16) using the Laplace transform on variable t and the Fourier transform on variable x. After calculation, we get

$$\zeta(x,y,t) = -\frac{Q\omega}{2\pi^2} \int_{\varepsilon-i\infty}^{\varepsilon+i\infty} \frac{f(p)}{p} e^{pt} dp \int_{-\infty}^{\infty} \frac{s(p\sqrt{s^2+(p^2+4\omega^2)/c^2} - 2i\omega s)}{(p^2+c^2s^2)\sqrt{s^2+(p^2+4\omega^2)/c^2}}$$

$$\exp\left[-(y+y_0)\sqrt{s^2+(p^2+4\omega^2)/c^2} + is(x-x_0)\right] ds \qquad (17)$$

$$u(x,y,t) = -\frac{Qg\omega}{2c^2\pi^2} \int_{\varepsilon-i\infty}^{\varepsilon+i\infty} \frac{f(p)}{p^2} e^{pt} dp \int_{-\infty}^{\infty} \frac{s(2\omega p - isc^2\sqrt{s^2+(p^2+4\omega^2)/c^2})}{(p^2+c^2s^2)\sqrt{s^2+(p^2+4\omega^2)/c^2}}$$

$$\exp\left[-(y+y_0)\sqrt{s^2+(p^2+4\omega^2)/c^2}+is(x-x_0)\right] ds \qquad (18)$$

$$v(x,y,t) = -\frac{Qg\omega}{2c^2\pi^2} \int_{\varepsilon-i\infty}^{\varepsilon+i\infty} \frac{f(p)}{p} e^{pt} dp$$

$$\cdot \int_{-\infty}^{\infty} \frac{s \exp[-(y+y_0)\sqrt{s^2+(p^2+4\omega^2)/c^2} + i(x-x_0)]}{\sqrt{s^2+(p^2+4\omega^2)/c^2}} ds \quad (19)$$

The total wave field is described by the formulas (17) to (19) plus the fields of two sources at the points $(x_0, y_0$ and $(x_0, -y_0)$. When the basin is immovable, the motion described by formulas (17) to (19) is zero.

The integrals in formulas (17) to (19) are unsuitable for investigation and calculation. The integrals (17) to (19) can be transformed to a more suitable form using the following algebraic identities.

$$\frac{s(p\sqrt{s^2+(p^2+4\omega^2)/c^2} - 2i\omega s)}{(p^2+c^2s^2)\sqrt{s^2+(p^2+4\omega^2)/c^2}} \equiv \frac{ps}{p^2+c^2s^2} - \frac{2i\omega}{c^2} \frac{1}{\sqrt{s^2+(p^2+4\omega^2)/c^2}}$$

$$+ \frac{2i\omega}{c^2} \frac{p^2}{(p^2+c^2s^2)\sqrt{s^2+(p^2+4\omega^2)/c^2}}$$

$$\frac{s(2\omega p - isc^2\sqrt{s^2+(p^2+4\omega^2)/c^2})}{(p^2+c^2s^2)\sqrt{s^2+(p^2+4\omega^2)/c^2}} \equiv -i + \frac{ip}{p^2+c^2s^2}$$

$$+ \frac{2\omega ps}{(p^2+c^2s^2)\sqrt{s^2+(p^2+4\omega^2)/c^2}}.$$

The solution may be written in the form

$$\zeta(x,y,t) = \frac{Q}{\pi}[2\omega\frac{\partial N}{\partial t} + c^2\frac{\partial^2 N}{\partial x \partial y} - \frac{4\omega^2}{c^2} I] \quad (20)$$

$$u(x,y,t) = \frac{Qg}{\pi}[\frac{2\omega}{c^2} \frac{\partial I}{\partial y} - \frac{\partial^2 N}{\partial t \partial y} - 2\omega\frac{\partial N}{\partial x}] \quad (21)$$

$$v(x,y,t) = -\frac{2Qg\omega}{\pi c^2}\frac{\partial I}{\partial x}, \qquad (22)$$

where

$$I(x,y,t) = \frac{1}{4\pi i}\int_{\varepsilon-i\infty}^{\varepsilon+i\infty}\frac{f(p)}{p^2}e^{pt}\,dp$$

$$\cdot\int_{-\infty}^{\infty}\frac{\exp[-(y+y_0)\sqrt{s^2+(p^2+4\omega^2)/c^2}+is(x-x_0)]}{\sqrt{s^2+(p^2+4\omega^2)/c^2}}\,ds \qquad (23)$$

$$N(x,y,t) = \frac{\omega}{2\pi i\omega}\int_{\varepsilon-i\infty}^{\varepsilon+i\infty}\frac{f(p)}{p}e^{pt}\,dp$$

$$\cdot\int_{\infty}^{\infty}\frac{\exp[-(y+y_0)\sqrt{s^2+(p^2+4\omega^2)/c^2}+is(x-x_0)]}{(p^2+c^2s^2)\sqrt{s^2+(p^2+4\omega^2)/c^2}}\,ds. \qquad (24)$$

Then functions I and N can be determined by simple real integrals of convolution type. The internal integral in the formula (23) can be calculated (Gradstein and Ryzhic, 1963). Then

$$I(x,y,t) = \frac{1}{2\pi i}\int_{\varepsilon-i\infty}^{\varepsilon+i\infty}\frac{f(p)}{p^2}K_0(\frac{r}{c}\sqrt{p^2+4\omega^2})e^{pt}\,dp, \qquad (25)$$

where $r^2 = (x-x_0)^2 + (y+y_0)^2$. Calculating the Melline's integral (25) (Ditkin and Prudnikov, 1961) one gets that

$$I(x,y,t) = 0 \qquad \text{if } ct < r$$

and

$$I(x,y,t) = \int_0^{t-r/c}\frac{\cos[2\omega\sqrt{(t-\xi)^2-(r/c)^2}]}{\sqrt{(t-\xi)^2-(r/c)^2}}\phi(\xi)d\xi \quad \text{if } ct \geq r \qquad (26)$$

The function $\phi(t)$ is determined by the displacement of the source $f(t)$ by

$$\phi(t) = \int_0^t f(\xi)d\xi . \qquad (27)$$

For simplification of the integral (24) a theorem from the theory of the Fourier transformation (6) may be used.

$$\int_{-\infty}^{\infty} F(u) G(u) e^{-ixu} du = \int_{-\infty}^{\infty} g(\xi) f(x-\xi) d\xi ,$$

where $F(u)$ and $G(u)$ are Fourier transforms of the functions $g(x)$ and $f(x)$. In our case

$$F(u) = \frac{1}{p^2 + c^2 s^2} , \qquad G(u) = \frac{\exp[-(y+y_o)\sqrt{s^2+(p^2+4\omega^2)}/c^2 - isx_o]}{\sqrt{s^2+(p^2+4\omega^2)}/c^2} .$$

After calculation, we get the function N in the form

$$N = \frac{\omega}{2\pi i c^3} \int_0^{\infty} d\xi \int_{\varepsilon-i\infty}^{\varepsilon+i\infty} \frac{f(p)}{p^2} K_o (\frac{1}{c}\sqrt{p^2+4\omega^2} \sqrt{(x-x_o-\xi)^2+(y+y_o)^2}) e^{p(t-\xi/c)}$$

$$\cdot dp + \frac{\omega}{2\pi i c^3} \int_0^{\infty} d\xi \int_{\varepsilon-i\infty}^{\varepsilon+i\infty} \frac{f(p)}{p^2} K_o (\frac{1}{c}\sqrt{p^2+4\omega^2} \sqrt{(x-x_o-\xi)^2+(y+y_o)^2})$$

$$\cdot e^{p(t-\xi/c)} dp .$$

In the last expression one can perform both integrations. The internal integration can be performed using formulas (25) and (26). The external integration can be performed using a change of the variable of the integration. After this calculation, we get

$$N(x,y,t) = 0 \text{ if } ct < r$$

and

$$N(x,y,t) = \int_0^{t-r/c} \frac{(t-\xi)\sin[2\omega\sqrt{(t-\xi)^2-(r/c)^2}]}{c^2(t-\xi)^2-(x-x_0)^2}\phi(\xi)d\xi,$$

if $ct > r$. (28)

where $\tau^2 = (x-x_0)^2+(y+y_0)^2$. The function $\phi(t)$ is determined by formula (27).

Thus, the solution of the boundary value problem (11), (12), (16) is determined by expressions (20), (21), (22) where the functions I and N are determined by formulas (26) and (28). To obtain the total wave field we must add the motion described by the functions ζ, U and V, and the motions generated by the sources in the points (x_0,y_0) and $(x_0,-y_0)$. For the total elevation of the free surface we get

$$\zeta_\Sigma = \frac{Q}{2\pi c^2}[\frac{\partial^2 I(r_1)}{\partial t^2} + 4\omega^2 I(r_1)]$$

$$+ \frac{Q}{2\pi c^2}[\frac{\partial^2 I(r)}{\partial t^2} - 4\omega^2 I(r)] + \frac{Q}{\pi}[2\omega\frac{\partial N}{\partial t}+c^2\frac{\partial^2 N}{\partial x\partial y}], \quad (29)$$

where $r^2 = (x-x_0)^2 + (y+y_0)^2$, $r_1^2 = (x-x_0)^2 + (y-y_0)^2$. The first term in formula (29) describes the incident wave, the other terms describe the reflected wave. Two first terms determine the axial symmetric motions around the points (x_0,y_0) and $(x_0,-y_0)$ respectively. The third term does not have an axial symmetry. The destruction of the axial symmetry in the reflected wave is the effect of the Coriolis force.

The solution of this point-source problem can be used for calculation of the free-surface elevation when the disturbance is distributed over an arbitrary region.

BIBLIOGRAPHY

Ditkin, V. A. and Prudnikov, A. P. 1961. *The Integral Transformations and Operatory Calculations*. Moscow.

Gradstein, I. S. and Ryzhic, I. M. 1963. *The Tables of Integrals, Sums, Series, and Productions*. Moscow.

Lamb, H. 1947. *Hydrodynamics*. Moscow.

Sebekin, B. I. 1967. Diffraction of surface waves on a wedge. Izv. Academy of Sciences U.S.S.R. *Physics of Atmosphere and Ocean*, Vol. 3, N.8.

Sretensky, L. N. 1936. *The Theory of Wave Motions of a Fluid*. Moscow.

Voit, S. S. and Sebekin, B. I. 1969. Some hydrodynamical models of unsteady wave motions of tsunami type. *Sbornik Morskie Gidrofizicheskie Issledovania*, N.1, Sevastopol.

22. Some Problems of Hydrodynamics of Tsunami Waves

L. V. CHERKESOV
Nautical Hydrophysical Institute
Sevastopol, Crimea, U.S.S.R.

ABSTRACT

This paper presents the results of investigations of the effect of nonuniformity and viscosity of a fluid, and also the change of the bottom profile of a basin on waves caused by initial disturbances.

For tsunami-type waves:

1. Long surface and internal waves arising in a shallow edge of the basin as a result of initial disturbances in deep-water region are investigated. The fluid is ideal and two-layered. Cases of the sharp but continuous change in depth of the basin and the vertical step of the bottom are discussed separately. It is shown that in the first case the amplitude of internal waves in a shallow zone under real conditions of a sea is far greater than that of surface waves, whereas for the vertical step of the bottom there is the opposite effect.

2. Long surface waves arising in a shallow edge under the action of initial disturbances concentrated in a deep-water region are studied. The fluid is viscous and uniform; the depth of the basin is changing sharply but continuously. It is shown that viscosity of a fluid causes lowering of sea level before the arrival of the basic wave of the rise. A formula is obtained defining the time between the beginning of an ebb and the arrival of the maximum of the crest.

Expressions for decrement of decay and the decrease of wave velocity due to viscosity are found.

3. Under the assumptions of a general theory of waves, the effects of a density discontinuity and of a continuous change in density with depth on surface and internal waves arising from initial disturbances of a bottom section of the basin are investigated.

4. The influence of viscosity and density discontinuity upon surface and internal waves generated by initial disturbances is investigated. It is shown that viscosity of a fluid induces an ebbing wave on a density interface which precedes the arrival of the main internal wave. It is also found that the decrement of the decay of unsteady internal waves is many times greater than that of unsteady surface waves.

For all the problems discussed in the paper, precise (under assumptions of a linear theory) expressions for types of generated waves are obtained. A further analysis of these expressions by asymptotic and numerical methods is given.

INTRODUCTION

This report presents results of a study on the influence of heterogeneity and viscosity of fluids, as well as the changes in the basin bottom profile, on tsunami waves caused by initial disturbances.

CASE OF CONTINUOUS CHANGE
IN OCEAN BOTTOM

Long waves produced by initial disturbances in a nonuniform fluid contained in a rotating basin of variable depth were investigated. (See Figure 1.)

Let an unlimited basin filled with fluid have the depth $h_1 + h_3$ in the region $x < 0$, and the depth $h_1 + h_4$ in the region $x > 0$. On the line $x = 0$ an abrupt change in the basin depth occurs ($h_3 > h_4$). The upper layer of the fluid of h_1 depth has density ρ_1 and the fluid below this layer has density ρ (where $\rho > \rho_1$).

At the initial time $t = 0$ the fluid is static: The free surface and density interface in the deep-sea region ($x < 0$)

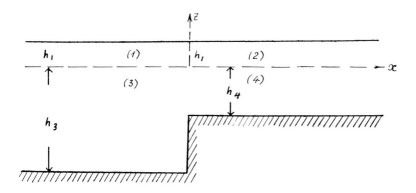

Fig. 1: Cross-Section of Basin with Vertical Step in Bottom. The Dashed Line Separates Two Fluids of Different Densities.

are disturbed and have forms ζ_1^0 (x,y) and ζ_1^0 (x,y), respectively; the free surface and density interface in the shallow-water region are horizontal. The problem investigated is that of initial disturbances on the free surface and interface, assuming that the waves are long and taking the effect of Coriolis force into consideration.

The system of equations for determination of the wave velocities and elevations in each of the four regions considered has the form

$$\frac{\partial u_K}{\partial t} - 2\omega v_K = -\frac{\partial P_K}{\partial x}, \quad \frac{\partial v_K}{\partial t} + 2\omega u = -\frac{\partial P_K}{\partial y}$$

$$\frac{\partial}{\partial t}(\zeta_K - \zeta_{K+2}) = -h_K \left(\frac{\partial u_K}{\partial} + \frac{\partial v_K}{\partial y}\right) \quad (K = 1,2,3,4)$$

(1)

the index of the velocity component coincides with the area number of the region in which the system of equations is discussed. Then $h_1 = h_2$, $\zeta_5 = \zeta_6$, and functions P_K will be

$$P_{1,3} = g(\zeta_{1,2} + h_1 - z), \quad \gamma = \rho_1 \rho^{-1}$$

$$P_{3,4} = g[(\zeta_{1,2} - \zeta_{3,4} + h_1)\gamma + (\zeta_{3,4} - z)] \tag{2}$$

Functions u_K, v_K, ζ_K at $t = 0$ correspond to the following initial conditions:

$$u_K = v_K = 0, \qquad \zeta_2 = \zeta_4 = 0$$

$$\zeta_1 = \zeta_1^0(x,y), \qquad \zeta_3 = \zeta_3^0(x,y) \tag{3}$$

At the step in the sea bottom at (x=0), the free surface and the density interface, as well as the mass of liquid, are continuous functions of x. Hence, at $x = 0$,

$$\zeta_1 = \zeta_2, \quad \zeta_3 = \zeta_4, \quad u_1 = u_2, \quad h_3 u_3 = h_4 u_4 \tag{4}$$

Applying the Laplace transformation to time (t) and the Fourier transformation to y, a precise solution is obtained for the system (equation 1) with initial and boundary conditions for arbitrary initial elevations concentrated in the shallow water region.

To simplify the final expressions ζ_K, expressions of the free surface ζ_2 and of the density interface ζ_4 in a shallow-water region are used for the case where

$$\zeta_1^0(x,y) = \zeta_3^0(x,y) = \zeta^0 = \text{const} \quad \omega = 0 \tag{5}$$

in the range $-b-a < x < -b+a$, $-\infty \leq y \leq \infty$, $0 < a < b$.

In this case, we have

$$\zeta_2 = \eta_2 + \eta_4, \quad \zeta_4 = \eta_2^1 + \eta_4^1 \tag{6}$$

where

$$\eta_2 = \zeta^0 \frac{\Delta_2}{\Delta}, \qquad \eta_2^1 = \zeta^0 \frac{h_4}{h_1+h_4} \frac{\Delta_2}{\Delta}$$

in the region $(t - t_o) v_2 > x > (t - t_o - t_1)v_2$ and are equal to zero outside this region; and

$$\eta_4 = \zeta_0 \frac{\Delta_4}{\Delta}, \qquad \eta_4^1 = -\zeta^0 \frac{\Delta_4}{\Delta} \frac{h_1+h_4}{\epsilon h_4}$$

in the region $(t - t_o) v_4 > x > (t - t_o - t_1)v_4$ and are equal to zero outside this region.

Here

$$t_o = \frac{b-a}{v}, \quad t_1 = \frac{2a}{v}, \quad v_4 = \sqrt{\frac{g\epsilon h_1 h_4}{h_1+h_4}}, \quad \epsilon = \frac{\rho-\rho_1}{\rho},$$

$$v = \sqrt{g(h_1+h_3)}, \qquad v_2 = \sqrt{g(h_1+h_4)}$$

and Δ, Δ_2, Δ_4, the known expressions are dependent upon the initial parameters of the problem.

Thus a surface wave, while passing over a sharp change in the sea bottom, generates a surface wave η_2 in the shallow-water region, travelling at velocity v_2, and an internal wave η_4^1, moving at velocity $v_4 \ll v_2$ (for $\epsilon \ll 1$) and representing a secondary wave as a result. In view of the well-known phenomenon of secondary tsunami waves, determination of the ratio of the amplitudes of these waves is of special interest. Since the dependence of their amplitudes on the initial parameters of the problem (h_1, h_3, h_4, ϵ) is rather complicated, the numerical calculations were carried out for the amplitudes of these waves and for their ratios for $h_1=64$ m, $h_5=3.5 \times 10^3$ m, and $\epsilon=3.6 \times 10^{-3}$ and for various values of h_4 ($h_3=h_5+h_4$) varying from 500 to 2 meters. Results of these calculations are given in Table 1, where A is the amplitude of the wave η_2 and B is the amplitude of the wave η_4^1 at $\zeta_o=1$ m.

Analysis of the numerical calculations shows that the amplitude of the internal wave in a shallow region can be many times greater (of the order of ten) than that of the surface wave. It follows from this that an abrupt change in the sea bottom together with a hiatus in density may be one of the significant causes of secondary tsunami waves.

CASE OF VERTICAL STEP IN OCEAN BOTTOM

Now consider the problem with the profile of the basin bottom at x = 0, having a vertical step, and with the normal velocity component equal to zero. In the previous case, the bottom profile had an abrupt but continuous change and the restriction of equality of flows. As in the previous long-wave theory, this condition cannot be satisfied in the previous system of equations, so the previous scheme must be modified to include the second surface of density discontinuity in the deep-water region, coinciding in the undisturbed state with continuation of the horizontal bottom of a shallow-water part of the basin (Figure 2).

The solution to this problem is obtained in the same manner as the previous problem. The expressions obtained for the free surface type ζ^2 and for interface type ζ^4 are analogous in form except that Δ, Δ_2, and Δ_4 now have different dependence on the initial parameters. As this dependence is very complex, the comparison of the surface-wave amplitude η_2 and the internal wave amplitude η_4^I were carried out by numerical calculations for $h_1=64$ m, $h_5=3.5 \times 10^3$ m, $\varepsilon=3.6 \times 10^{-3}$, and for various values of h_3, ranging from 500 m. to 2 meters. The results of these calculations are given in Table 2, where

Table 1: Amplitude in Meters of the Surface Wave (A) and of the Internal Wave (B) and Their Ratio (B/A) for Various Depths of the Higher Density Fluid on the Shelf. Continuous Change in Ocean Bottom.

h_4(m)	B	A	B/A
500	1.8611	0.7296	2.5507
300	2.6181	0.7666	3.4154
100	4.8089	0.8259	5.8228
50	6.4762	0.8468	7.6477
30	7.7333	0.8552	9.0422
20	8.6937	0.8588	10.1223
10	10.1763	0.8611	11.8171
2	12.6226	0.8598	14.6805

A is the amplitude of the surface wave, B is the amplitude of the internal wave, and $g\varepsilon = 1 - \rho_5 \rho_3^{-1}$.

The results of numerical calculations at given values of the parameters indicates that the ratio B/A reaches its greatest value when the density interface is at the distance of 40 meters from the bottom of the shallow-water region and the density difference is close to 0.3.

Comparison of results of these calculations with those for analogous values of initial parameters (Table 1) shows that the effect of secondary internal waves on the ratio B/A is about two orders less in the case of a vertical step in the sea bottom (other conditions being equal) than in the case of a sharp but continuously changing bottom profile.

CASE OF VARIABLE DEPTH OCEAN
WITH VISCOSITY INCLUDED

The influence of viscosity on long waves arising from initial disturbances in a basin of variable depth is next investigated.

Let a viscous, incompressible fluid fill a horizontally boundless basin. The depth of the basin in the region x<0

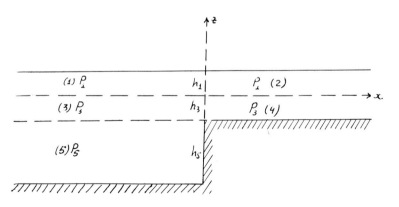

Fig. 2: Cross-Section of Basin with Vertical Step in Bottom. The Two Horizontal Dashed Lines Separate Three Layers of Two Fluids.

Table 2: Amplitude in Meters of the Surface Wave (A) and of the Internal Wave (B) and Their Ratio (B/A) for Various Depths of the Higher Density Fluid on the Shelf. Abrupt Vertical Step in Ocean Bottom.

h_3 (m)		A	B	B/A
500	10^{-2}	0.27171	0.01830	0.0674
400	10^{-2}	0.25515	0.02069	0.0811
300	10^{-2}	0.23497	0.02394	0.1017
200	10^{-2}	0.20926	0.02877	0.1373
100	10^{-2}	0.17392	0.03556	0.2045
50	10^{-2}	0.14901	0.03965	0.2661
40	10^{-2}	0.14335	0.04005	0.2794
30	10^{-2}	0.13691	0.03989	0.2912
20	10^{-2}	1.12988	0.03852	0.2965
14	10^{-2}	0.12538	0.03625	0.2892
8	10^{-2}	0.12041	0.03170	0.2633
6	10^{-2}	0.11862	0.02911	0.2454
2	10^{-2}	0.11466	0.01962	0.1711
40	10^{-1}	0.14298	0.03777	0.2642
30	10^{-1}	0.13662	0.03779	0.2767
20	10^{-1}	0.12964	0.03543	0.2732

equals h_1 and in the region $x>0$, equals h_2 ($h_1>h_2$). Along the line $x = 0$, the depth of the basin changes sharply but continuously. At initial time, $t = 0$, the fluid is static, the free surface is disturbed in the deep-water region, and is undisturbed in the shallow-water region. The type of waves generated is investigated, assuming that these waves are long.

In this case the system of equations of motion is in the form:

$$\frac{\partial u}{\partial t} = -g\frac{\partial \zeta}{\partial x} + \nu\frac{\partial^2 u}{\partial z^2}, \quad \frac{\partial v}{\partial t} = -g\frac{\partial \zeta}{\partial y} + \nu\frac{\partial^2 v}{\partial z^2}$$

$$\frac{\partial \zeta}{\partial t} = -\int_{-h}^{0} \left(\frac{\partial u}{\partial x} + \frac{\partial v}{\partial y}\right) dz \qquad (7)$$

with initial conditions (at $t = 0$),

$$u_1 = u_2 = 0, \quad \zeta_1 = af(x,y), \quad \zeta_2 = 0 \qquad (8)$$

and boundary conditions

$$u_\kappa = 0 \text{ at } z = -h_\kappa, \quad \frac{\partial u_\kappa}{\partial z} = \frac{\partial v_\kappa}{\partial z} = 0 \text{ at } z = 0 \qquad (9)$$

$$\zeta_1 = \zeta_2, \quad \int_{-h_1}^{0} u_1 \, dz = \int_{-h_2}^{0} u_2 \, dz \quad \text{at } x = 0 \qquad (10)$$

In equation 7 at u, v, ζ, h the index κ ($\kappa=1,2$) is not taken into consideration. Index "1" refers to the deep-water region and index "2" to the shallow-water region. Boundary conditions (equation 9) denote the adherence of the liquid to the bottom of the basin and the absence of tangential stress on the free surface; conditions (equation 10) express continuity of the free surface and of the flow of liquid at the step in the sea bottom.

The solution to the system of equations is found by applying the Laplace transformation to time t and the Fourier transformation to the variable y. In the plane case, for the function $f(x,y)$ depending only on x and having the form

$$f(x) = \begin{cases} \cos \frac{\pi}{2b} x^1 & |x^1| < b \\ 0 & |x^1| > b \end{cases}$$

$$x^1 = x + d, \qquad d > b > 0 \qquad (11)$$

the following expression is obtained for the expression of the free surface in the shallow-water zone ($x > 0$) of the basin:

$$0 < x < c_2 (t_2 - t_3) \qquad (12)$$

$$\zeta_2(x,t) = 0(\varepsilon)$$

$$c_2(t_2 - t_3) < x < c_2 t_2$$

$$\zeta_2 = \frac{a\sqrt{h_1}}{\sqrt{h_1} + \sqrt{h_2}} \cos\gamma \exp(-\xi) \qquad (13)$$

$$\gamma = \frac{\pi}{2b} \frac{c_1}{c_2} [x(1+\varepsilon_2 \frac{h_1-h_2}{h_2}) - c_2 t(1-\varepsilon_2) + d\frac{c_2}{c_1}]$$

$$\xi = \frac{\sqrt{\pi}}{4} \varepsilon_1 (t + \frac{x}{c_2} \frac{h_1-h_2}{h_2}),$$

$$\varepsilon_2 = \frac{b}{2\sqrt{\pi c_1}} \varepsilon_1, \qquad \varepsilon_1 = \nu^{1/2} c_1^{1/2} b^{-1/2} h_1^{-1},$$

$$c_\kappa = \sqrt{gh_\kappa}, \qquad t_2 = t - t_1,$$

$$t_1 = (d-b)c_1^{-1}, \qquad t = 2bc_1^{-1},$$

$$x > c_2 t_2$$

$$\zeta_2(x,t) = 0 \qquad (14)$$

where $0(\varepsilon)$ approaches zero as ε approaches 0.

It follows from previous equations that velocity u of propagation of the basic (solitary) wave is equal to $u = c_2(1-\varepsilon h_1 h_2^{-1})$, which is less than the velocity of propagation (c_2) of a similar wave in an ideal fluid.

It is also clear from this that at some points of the shallow-water region $x = \ell > 0$, beginning from

$$t = \frac{d-b}{c_1} + \frac{\ell}{c_2} \qquad (15)$$

and during the time interval

$$\Delta t = \varepsilon_2 \left(\frac{d-b}{c_1} + \frac{\ell}{c_2} \frac{h_1}{h_2} \right)$$

an ebb tide is observed ($\zeta_2 < 0$). This ebb tide preceding the arrival of the basic wave arises only because of forces due to viscosity. The time between the beginning of the ebb tide and the arrival of the maximum of the wave is

$$T = \frac{b}{c_2} \sqrt{\frac{h_2}{h_1}} + \Delta t \qquad (16)$$

The maximum value of the wave amplitude at the point $x = \ell$ decreases due to viscosity as $\exp(\xi_1)$, where

$$\xi_1 = \frac{\sqrt{\pi}}{4} \frac{\varepsilon_1}{c_1} \left[d + \ell \left(\frac{h_1}{h_2} \right)^{3/2} \right] .$$

This decrease, as can be seen, depends essentially upon the ratio of depths h_1 and h_2. Let

$$h_1 = 4 \cdot 10^3 \text{ m}, \qquad \ell = 3 \cdot 10^3 \text{ m, and}$$
$$b = 4 \cdot 10^4 \text{ m}, \qquad d = 4 \cdot 10^5 \text{ m}.$$
$$h_2 = 4 \cdot 10^2 \text{ m},$$

Hence the time during which the ebb tide at the point $x = \ell$ is observed for velocity values equal to 1 cm./sec., 200 cm./sec., and 10^3 cm./sec. is equal to 7.3 sec., 101.8 sec., 227.4 sec., respectively. Consequently, the time between the beginning of the ebb tide and the arrival of the maximum equals 207.3 sec., 301.8 sec., 427.4 sec., respectively.

SOURCE AT BOTTOM OF OCEAN

Now consider the problem of waves arising both in the free surface and density interface for two-layered liquid affected by disturbances prescribed at the bottom of the basin.

On the surface of a liquid of ρ_2 density and h_2 depth, let float a layer of some other fluid of ρ_1 density and h_1 depth. At initial time $t = 0$, both liquids are static and the free surface and density interface are undisturbed. The form of waves arising on the free surface and density interface under the influence of vertical oscillations of the basin bottom having velocity $W = W_o f(x,y) \psi(t)$ $[\psi(0) = 0]$ is investigated. Potential velocities of the upper and lower layers of liquid are designated by ϕ_1 and ϕ_2. For determination of these functions there are then two equations, $\Delta\phi_1 = 0$ ($0 < z < h_1$) and $\Delta\phi_2 = 0$ ($-h < z < 0$), with boundary and initial conditions

$$\frac{\partial^2 \phi_1}{\partial t^2} + g \frac{\partial \phi_1}{\partial z} = 0 \quad \text{at } z = h_1,$$

$$\frac{\partial \phi_1}{\partial z} = \frac{\partial \phi_2}{\partial z}, \quad \frac{\partial^2 \phi_2}{\partial t^2} - \gamma \frac{\partial^2 \phi_1}{\partial t^2} + g\varepsilon \frac{\partial \phi_1}{\partial z} = 0 \quad \text{at } z = 0,$$

$$\frac{\partial \phi_2}{\partial z} = W_o f(x,y) \psi(t) \quad \text{at } z = -h_2,$$

$$\frac{\partial \phi_2(x,y,0,0)}{\partial t} = \gamma \frac{\partial \phi_1(x,y,0,0)}{\partial t}, \quad \phi_2(x,y,z,0) = 0,$$

$$\frac{\partial \phi_1(x,y,h_1,0)}{\partial t} = 0, \quad \text{and} \quad \phi_1(x,y,z,0) = 0,$$

where $\gamma = \rho_1/\rho_2$ and $\varepsilon = 1 - \gamma$.

The forms of the free surface ζ_1 and the density interface ζ_2 are given by

$$\zeta_1 = -\frac{1}{g}\left(\frac{\partial \phi_1}{\partial t}\right)_{z=h_1} \quad \text{and} \quad \zeta_2 = -\frac{1}{g\varepsilon}\left(\frac{\partial \phi_2}{\partial t} - \gamma \frac{\partial \phi_1}{\partial t}\right)_{z=0}$$

The problem is solved by the method of integral transforms. The result is that the exact solution for any function $f(x,y)$ and $\psi(t)$ is obtained in the form

$$\psi(t) = \begin{cases} \cos \omega t & 0 < t_o \leq t \leq t_o + \tau \\ 0 & 0 \leq t < t_o, \ t > t_o + \tau \end{cases}$$

Results for the long waves and function are

$$f(x,y) = \exp(KR^2/H^2) \quad [R = \sqrt{x^2+y^2}] \quad (17)$$

$$t \geq t_o + \tau, \quad \zeta_1 = \zeta_{11} + \zeta_{12}, \quad \zeta_2 = \zeta_{21} + \zeta_{22}$$

$$\zeta_{1s} = \begin{cases} \frac{1}{\sqrt{R}} \eta_s & c_s(t_1-\tau) < R < c_s t_1 \\ 0 \ (R^{-3/2}) & R > c_s t_1 \end{cases} \quad (18)$$

$$\zeta_{2s} = A_s \zeta_{1s}, \quad \eta_s = N_s \cos \gamma_s \quad (s = 1,2)$$

where

$$N_1 = \frac{W_o \sqrt{\pi}}{2\sqrt{K}} \frac{p}{c} \sqrt{\frac{\omega}{c_1}} \left(1 + \frac{3}{2} \frac{h_1 h_2}{H^2} \varepsilon\right), \quad A_1 = \frac{h_2}{h_1+h_2}, \quad (19)$$

$$N_2 = \frac{W_o \sqrt{\pi}}{2\sqrt{K}} \frac{p}{c} \sqrt{\frac{\omega}{c_2}} \sqrt{\frac{h_1 h_2}{H^2} \varepsilon}, \quad A_2 = -\frac{h_1+h_2}{h_2 \varepsilon},$$

$$\gamma_s = \frac{\omega}{c_s} R - \omega t - \frac{\pi}{4}, \quad c_1 = c\left(1 - \frac{1}{2}\frac{h_1 h_2}{H^2}\varepsilon\right),$$

$$c_2 = c\sqrt{\frac{h_1 h_2}{H^2}\varepsilon}, \quad c = \sqrt{gH}, \quad H = h_1 + h_2, \quad t_1 = t - t_o.$$

From these equations it is clear that two systems of waves, $\zeta_{\kappa 1}$ and $\zeta_{\kappa 2}$ ($\kappa = 1,2$), moving with different velocities, are generated on the free surface and interface.

Basic disturbances are concentrated in annular areas. For wave systems of $\zeta_{\kappa 1}$, $c_1(t_1-\tau) < R < c_1 t_1$; and for wave systems $\zeta_{\kappa 2}$, $c_2(t_1-\tau) < R < c_2 t_1$, with $c_2 \ll c_1$.

Amplitudes of the main waves decrease as $R^{-1/2}$ with increasing R. The main disturbance is preceded by waves whose amplitudes decrease as $R^{-3/2}$ with increasing of R.

Waves ζ_{21} and ζ_{22} arising on the density interface differ from the corresponding waves ζ_{11} and ζ_{12} on the free surface only by their amplitudes. The wave ζ_{21} has the same phase as the wave ζ_{11} but lesser amplitude ($0 < A_1 < 1$); the wave ζ_{22} has opposite phase and a considerably larger amplitude than the wave ζ_{12} ($A_2 < 0$, $|A_2| \gg 1$).

The ratio of the amplitude of the internal wave ζ_{22} to the amplitude of surface wave ζ_{11} equals

$$\delta = h_1^{1/4} h_2^{-1/4} \varepsilon^{-3/4}$$

For realistic sea conditions, this ratio can be rather large. For instance, for $h_1 = 100$ m, $h_2 = 4.10^3$ m, $\delta = 25$. Thus the internal waves can have amplitudes which are many times greater than those of surface waves.

SOURCE AT SURFACE OF OCEAN WITH VISCOSITY INCLUDED

The problem of unsteady waves arising in a viscous two-layered liquid under the action of initial disturbances of the free surface and density interface is now considered with the general linear theory of long waves. The effects of viscosity and density discontinuity on surface and internal waves are investigated.

Assume that a liquid having density ρ_1, coefficient of kinematic viscosity ν_1 and depth h_1, floats on another liquid of density ρ_2, coefficient of kinematic viscosity ν_2 and depth h_2. At initial time $t = 0$, normal stresses are constant on the free surface, tangential stresses are absent, and the free surface ζ_1 and density interface ζ_2 are deflected from the equilibrium as

$$\zeta_s(x,0) = a_s f_s(x) \quad (s = 1,2).$$

Assuming that the generated waves are long and motions are slow, a system of equations of motion is obtained in the form,

$$\frac{\partial u_1}{\partial t} = -g\frac{\partial \zeta}{\partial x} + \nu_1 \frac{\partial^2 u_1}{\partial z^2}, \quad \frac{\partial u_2}{\partial t} = -g\gamma\frac{\partial \zeta_1}{\partial x} - g\theta\frac{\partial \zeta_2}{\partial x} + \nu_2 \frac{\partial^2 u_2}{\partial t^2}$$

$$\int_0^{h_1} \frac{\partial u_1}{\partial x} dz = \frac{\partial \zeta_2}{\partial t} - \frac{\partial \zeta_1}{\partial t}, \quad \int_{-h_2}^0 \frac{\partial u_2}{\partial x} dz = -\frac{\partial \zeta_2}{\partial t}$$

with the initial and boundary conditions

$$u_1 = u_2 = 0, \quad \zeta_s = a_s f_s(x) \quad \text{at } t = 0,$$

$$\frac{\partial u_1}{\partial z} = 0 \text{ at } z = h_1, \quad u_1 = u_2, \quad \rho_1 \nu_1 \frac{\partial u_1}{\partial z} = \rho_2 \nu_2 \frac{\partial u_2}{\partial z} \text{ at } z = 0,$$

$$u = 0 \text{ at } z = -h_2, \quad \gamma = \rho_1 \rho_2^{-1}, \quad \theta = 1 - \gamma.$$

Using the method of integral transforms to solve the problems with the functions

$$f(x) = \begin{cases} \cos \frac{\pi x}{2\ell} & |x| \leq \ell \\ 0 & |x| > \ell \end{cases}$$

the following expressions for the elevations of the free surface and the density interface are obtained:

$$\zeta = \xi_1 + \xi_3 \qquad \zeta_2 = \xi_2 + \xi_4$$

$$\xi_i = \begin{cases} A_i \exp(-\gamma_i t) \cos \frac{\pi}{2\ell}(x - v_i t) & c_i t - \ell < x < c_i t + \ell \\ 0(\epsilon) & \ell < x < c_i t + \ell \\ & x > c_i t + \ell \end{cases} \qquad (20)$$

$$\gamma_1 = \gamma_2 = c_1[\nu^{1/2} r_1 + \nu r_2], \quad \gamma_3 = \gamma_4 = c_3[\nu^{1/2} r_1 r_3 + \nu r_2 r_3^2],$$

$$\nu_1 = \nu_2 = c_1 (1 - \nu^{1/2} r_4), \quad \nu_3 = \nu_4 = c_3 (1 - \nu^{1/2} r_4 r_3),$$

$$r_1 = \frac{1}{4} \frac{\pi^{1/2}}{\ell^{1/2} g^{1/4} H^{5/4}} , \quad r_2 = \frac{1}{4g^{1/2} H^{5/2}} , \quad r_3 = \frac{1}{2} \frac{H}{h_1} \theta_1^{-1/4} ,$$

$$r_4 = \frac{\ell^{1/2}}{2 g^{1/4} \pi^{1/2} H^{5/4}}$$

$$c_1 = c_2 = c [1 - \frac{1}{2}\theta_1 - \frac{5}{8}\theta_1^2], \quad c_3 = c_4 = c \sqrt{\theta_1}(1 + \frac{1}{2}\theta_1),$$

$$c = \sqrt{gH}, \quad \theta_1 = \sqrt{\frac{h_1 h_2}{H^2}\theta} , \quad H = h_1 + h_2 , \quad \varepsilon = \nu^{1/2} g^{1/4} \ell^{-1/2} H^-$$

CONCLUSIONS

For a point with fixed coordinate x, the maxima of surface waves $\zeta_{1,2}$ are $\exp[\frac{\gamma_1 x}{v_1}]$ times less than those in an ideal liquid; maxima of the internal waves $\zeta_{3,4}$ are $\exp[\frac{\gamma_3 x}{v_3}]$ times less than the corresponding maxima in an ideal liquid. As $\frac{\gamma_3}{v_3} \gg \frac{\gamma_1}{v_1}$, it is evident that internal waves are attenuated by viscosity much more than surface waves. The ratio of attenuation decrements is

$$\Delta = \frac{\gamma_3}{v_3} \bigg/ \frac{\gamma_1}{v_1} = \frac{1}{2} (\frac{H}{h_1})^{5/4} \theta^{-1/4} \gg 1$$

So, for real conditions of the sea (H = 4,000 m, h = 100 m, $\theta = 4 \cdot 10^{-3}$) Δ = 200 meters. It is also seen from the formulas that both surface and internal waves will have lower speeds in a viscous liquid than in an ideal liquid ($v_1 < c_1$, $v_3 < c_3$). After passage of the main internal wave ξ_4, the density interface continues to vibrate, due to the effect of the viscosity. This vibration is a positive wave having the same shape as the surface wave ξ_1. The main positive waves $\xi_{1,2}$ and $\xi_{3,4}$ are preceded by negative waves. This phenomenon arises exclusively due to forces of viscosity. The width of the precursors to waves $\xi_{1,2}$, $\xi_{3,4}$ are equal to $\ell_1 = \nu^{1/2} c_1 r_4 t$ and $\ell_2 = \nu^{1/2} c_3 r_4 r_3 t$, respectively. The width of regions ℓ_1 and ℓ_2 increases with the increase in distance from the epicenter of the initial disturbances.

For a point with fixed coordinate x, the duration from the beginning of the ebb tide until the arrival of the maxima of the main positive wave is

$$t_1 = \frac{\ell(1 - \nu^{1/2}r_4) + \nu^{1/2}r_4 x}{c_1(1 - \nu^{1/2}r_4)^2}$$

for wave $\xi_{1,2}$ and

$$t_2 = \frac{\ell(1 - \nu^{1/2}r_4 r_3) + \nu^{1/2}r_4 r_3 x}{c_3(1 - \nu^{1/2}r_4 r_3)^2}$$

for wave $\xi_{3,4}$.

BIBLIOGRAPHY

Cherkesov, L. V. 1966. To the problem of tsunami in a nonuniform sea, 1. *Okeanologia*, t. 6, V. 5.

Cherkesov, L. V. 1965. To the problem of tsunami in a nonuniform sea, 2. *Izvestia AN SSSR*, Physika atmosphery i okeana, t. 1, N. 8.

Cherkesov, L. V. 1965. Long waves in a viscous fluid. *Izvestia AN SSSR*, Physika atmosphery i okeana, t. 1, N. 1.

Cherkesov, L. V. 1966. On the effect of viscosity on waves of tsunami type. *Izvestia AN SSSR*, Physika atmosphery i okeana, t. 2, N. 12.

Fedosenko, V. S. and Cherkesov, L. V. 1968. On internal waves resulting from underwater earthquakes. *Izvestia AN SSSR*, Physika atmosphery i okeana, t. 4, N. 11.

23. Transformation of Tsunamis on the Continental Shelf

A. V. NEKRASOV
Hydrometeorological Institute
Leningrad, U.S.S.R.

ABSTRACT

Some effects due to reflections accompanying entry of the tsunami wave into the shelf region are considered. Intensive reflection occurs at two places, on the continental slope and near the coastline. The mass of water on the shelf has some properties of a resonator. These properties can easily be illustrated by considering the multiple reflection of a long sinusoidal wave from the coast and the shelf under different conditions (resonance, anti-resonance, and intermediate situation). However, the non-sinusoidal wave form is more realistic. Consideration of the shelf effect influence upon the waves of different forms shows a variety of possible consequences. In particular, if a train of successively decreasing waves, of which the first is the highest, is approaching the coast, then the first crest recorded may not be the highest, but the second, or the third and so on. Thus, the mareogram will not correspond to the form of the initial wave train; the distortion will be definitely related to the shelf and continental slope features.

By means of the shallow-water theory of long waves (one-dimensional case, the coast being a total reflector), a number of numerical experiments have been made for some shelf types, schematic and real. The waves have been produced by an initial surface perturbation located at different places: in the deep region, on the continental slope, and on the shelf. The initial wave arising is of the solitary type and suffers a considerable transformation due to the shelf effects. After having admitted a positive solitary wave, the shelf radiates back a train of successively decreasing waves having both positive and negative fluctuations and with periods depending on the shelf width. Moreover, the steepness of the continental slope and the inclination of the shelf play a significant role because the continuous reflection, "spread" over the inclined part of the profile, may result in some relatively long periods in the reflected waves.

If the initial positive perturbation is located on the continental slope, then the initial wave is not of the solitary type, because from the start, there is a negative reflection of the wave directed to the deep region, and so the initial wave propagating to the coast consists of a crest and following trough. If the perturbation center is situated on the shelf, then, besides the parameters of perturbation and shelf, the distance of the perturbation center from the coast and the shelf edge is also of importance, influencing form and periods of the waves approaching the coast and radiating from the shelf to the open sea.

Some of the effects mentioned above might be detected in tsunami observations. For example, the periods of the Niigata tsunami, 1964, recorded on the coasts of Japan and U.S.S.R. may be explained by the dimensions and shape of the source and the characteristics of the Japanese shelf. It seems expedient to take shelf effects into account when classifying parts of the coast according to the danger of tsunamis.

INTRODUCTION

Many causes lead to the transformation of tsunami waves on the shelf, viz., reflection, shoaling effects, friction, processes of bore formation, breaking, and run-up. Here, some effects due to reflection of tsunami waves in the shelf region are considered. These effects are very simple by nature, but they are not always taken into account when analyzing real tsunamis.

As a rule, the most significant reflection occurs at two places: on the steep continental slope, and by the coast; the mass of water on the shelf thus acts as a resonator. To observe the characteristics of the resulting phenomena the penetration of a harmonic wave train into the schematized shelf (Fig. 1) is considered. If the reflection coefficients

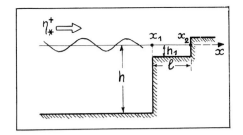

Fig. 1:
Simplified Geometry of Model for Continental Shelf and Coast.

at the shelf border (x_1) and by the coast (x_2) are r_1 and r_2, then the initial wave

$$\eta_*^+ = A \cos(\sigma t - kx)$$

after having reached x_1 generates a wave

$$\eta_*^- = r_1 A \cos(\sigma t + kx),$$

reflected from the shelf border and a wave

$$\eta_o^+ = (1 + r_1) A \cos(\sigma t - k_1 x),$$

enters the shelf. Here σ is the angular frequency, and k, k_1 are the wave numbers at the deep sea ($x < x_1$) and on the shelf ($x_1 < x < x_2$), correspondingly. The wave η_o^+, having passed the shelf, reflects at x_2, giving rise to the wave

$$\eta_o^- = r_2 (1+r_1) A \cos(\sigma t + k_1 x - 2\pi \frac{L}{\lambda_1}),$$

where $L = x_2 - x_1$ is the shelf width and $\lambda_1 = 2\pi/k_1$ is wave length on the shelf for given σ. The wave η_o^- suffers in turn the negative partial reflection at x_1, originating the new additional direct wave η_1^+ on the shelf. Reflections at x_1 and x_2 will continue, and after n pairs of reflections, there will exist an oscillation on the shelf that can be represented as a combination of two contrary progressive waves

$$\eta^+ = \sum_{j=0}^{n} \eta_j^+ = D \cos(\sigma t - k_1 x - \delta)$$

and

$$\eta^- = \sum_{j=0}^{n} \eta_j^- = r_2 D \cos(\sigma t + k_1 x - \delta).$$

The sum of these waves is a progressive-standing oscillation with amplitude F and phase f:

$$F = D\sqrt{1 + 2r_2 \cos^2(k_1 x) + r_2^2} \;;$$

$$f = \arctan \frac{(1 + r_2) \cos(k_1 x) \sin \delta + (1 - r_2) \sin k_1 x \cos \delta}{(1 + r_2) \cos(k_1 x) \cos \delta - (1 - r_2) \sin k_1 x \sin \delta} \;;$$

where

$$D = (1 + r_1) A \sqrt{\left[\sum_{j=0}^{n} q^j \cos(j\phi)\right]^2 + \left[\sum_{j=0}^{n} q^j \sin(j\phi)\right]^2} \;;$$

$$\delta = \arctan \frac{\sum_{j=0}^{n} q^j \sin(j\phi)}{\sum_{j=0}^{n} q^j \cos(j\phi)} \;;$$

$\phi = 4\pi L/\lambda_1$; and $q = (-r_1) r_2 = -r_1 r_2$.

The quantity q shows the decrease in amplitude of the new additional wave arising after each pair of reflections, and ϕ means the corresponding phase shift. The sign of q and the size of ϕ determine whether each additional wave will be in or out of phase with the preceding one, i.e., whether this addition will favor rise or fall of the oscillation amplitude within L. The coefficient r_2 is sometimes assumed to be equal to unity; however, in general, the process of interaction between wave and coastline is accompanied by some loss of energy, and so the reflection is not total, i.e., $r_2 < 1$.

The total reflection from the shelf region consists of reflected initial wave η^-_{*0} and "shelf radiation," arising from the nonperfect reflection of η^-_j-waves at x_1. The contribution of each such wave to the shelf radiation is

$$\eta^-_{*j} = r_2(1 - r_1^2)Aq^j \cos(\sigma t + kx - [j+1]\phi),$$

and the total shelf radiation has the form of a progressive wave

$$B \cos(\sigma t + kx - \beta),$$

where

$$B = A\, r_2(1 - r_1^2)\sqrt{\left[\sum_{j=0}^{n} q^j \cos(j+1)\phi\right]^2 + \left[\sum_{j=0}^{n} q^j \sin(j+1)\phi\right]^2},$$

and

$$\beta = \arctan \frac{\sum_{j=0}^{n} q^j \sin(j+1)\phi}{\sum_{j=0}^{n} q^j \cos(j+1)\phi}.$$

If $\phi = (2m+1)\pi$, where $m = 1,2,3,\ldots$, i.e., if $L = (2m+1)\lambda_1/4$, then the condition of resonance on the shelf takes place. In this case, each new addition $\eta^\pm_j$ is in phase with the preceding one, and the values D and B increase monotonically. On the shelf, there is an accumulation of energy, and so the total reflection from the shelf initially decreases. Under the resonant condition, the initial reflected wave η^-_*

is out of phase with the shelf radiation; this leads to a temporary reduction of the total reflection coefficient. Energy is accumulated on the shelf as radiation of the latter is intensifying; if r_2 is sufficiently large, the total reflection coefficient

$$R = r_1 - r_2 (1 - r_1^2) \sum_{j=0}^{n} |r_1 r_2|^j$$

changes sign and becomes negative (see Figure 2). Note that if $r_2 = r_1$, then the shelf is transformed from a reflector into an absorber, and the wave energy on the shelf is dissipated totally near the coastline after saturation has been achieved. The reflected wave is not sinusoidal but consists of pieces of sinusoids having different amplitudes and with phase changing by 180°. In the vicinity of the phase change, the component of oscillation is observed having double period equal to $T = 2\pi/\sigma$.

If the quantity $\phi = 2m\pi$, i.e., $L = m\lambda_1/2$, we have the condition of anti-resonance, when each new addition is out of phase with the preceding one. The change of amplitude on and in front of the shelf now has oscillatory character. After each new reflection of $n_{\overline{j}}$-wave at x_1, the shelf alternately emits or absorbs energy and, as a result, the quantity R may

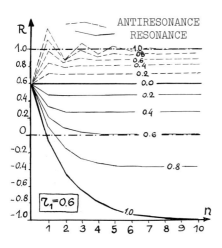

Fig. 2:
Value of Total Reflection Coefficient as a Function of the Number of Cycles, n, and the Reflection Coefficient at the Shoreline, r_2. The Reflection Coefficient at the Continental Shelf, r_1, equals 0.6.

exceed unity, i.e., in front of the shelf the reflected wave may at times have greater amplitude than the incident one.

When the shelf width L is neither equal nor multiple of $\lambda_1/2$ or $\lambda_1/4$, we have the situation intermediate between resonant and anti-resonant. The amplitude of oscillations on the shelf either increases or decreases at different stages depending on various combinations of q and ϕ.

If the initial wave train is not sinusoidal, which is more likely in nature, the repeating reflections at the coast and at the shelf border may lead to an extreme variety of results. In particular, the shelf effects may play an important role in determining which waves recorded at the coast will be the highest. This can easily be seen when considering the transformation of the exponentially modulated wave train under the resonant conditions. In this case, the form of the wave entering the shelf is

$$A (1 + r_1) \exp(-\frac{\gamma}{2\pi}[\sigma t - k_1 x]) \cdot \cos(\sigma t - k_1 x),$$

where γ is the decrement of the exponential curve enveloping the wave crests. Formation of oscillations on the shelf is determined, as previously, by reflections at the shelf border and at the coast. In the case of resonance the ratio of the height of the n-th crest approaching the coast to that of the first one is

$$q^{2n} e^{\gamma} \sum_{j=1}^{n} (q^{-2} e^{-\gamma})^j$$

In Fig. 3 the sequence of heights of waves approaching the coast one after another is shown for various values of q. One can see that though it is the first wave that has the highest crest in the initial train, near to the coast the maximum height may correspond to the second, third, or later waves. The number of the highest crest depends both on q and the form of the initial train and in this case, on the steepness of the enveloping curve (quantity γ).

In some cases the primary wave leaving the tsunami source seems to be similar to a solitary wave. Such a wave may be transformed by the shelf effects into a train of waves of which periods and lengths are determined by the shelf and continental

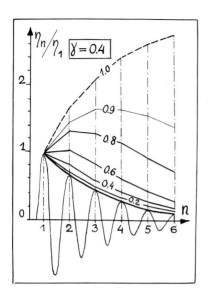

Fig. 3:
Normalized Wave Heights Approaching Coast as a Function of the Number of Cycles and the Response Parameter, q.

slope features. In this case the numerical experiments, based on numerical solution of hydrodynamical equations of the shallow water theory (Vol'tsinger and Pyaskovskiy, 1968) may serve as a convenient way of analysis. As an initial condition within the computation profile, the initial sea surface perturbation of a simple form is chosen (for instance, half a sinusoid). This perturbation generates two solitary waves travelling in opposite directions. In our calculations at the coast (at N in Fig. 4) the boundary condition has been set corresponding to the total reflection ($u = 0$, u being the total flow), whereas at the other end of the profile, at M, the boundary condition corresponds to absence of any reflection ($u = \eta\sqrt{gh}$); therefore, the wave approaching M had to go away from the profile without leaving any trace in it. The last boundary condition is physically proved if $h \gg \eta$. Note that in the case of partial reflection the corresponding boundary condition would be $u = [(1-r)/(1+r)] \cdot \eta\sqrt{gh}$, where r is the reflection coefficient. In Fig. 4 some results of calculation are shown. The initial perturbation has been positioned at various places relative to the shelf schematized by a section with constant depth and separated from the deep sea by either a vertical step or a rectilinear slope.

Case 1

The shelf width L is nearly resonant for the wave coming from the deep sea and, as a result, the process of radiating from the shelf is maximally stretched in time. The negative reflection at x_1 of the η_j^--wave generates the depressions; therefore, the shelf, admitting a solitary positive wave, emits a wave train with gradually decreasing double-humped cycles of contrary signs. If L is resonant, the radiating wavelength is 2d, where d means the length of the initial perturbation. The leading part of the train consists of two adjacent crests, of which the first is formed by the reflection of the initial wave from the shelf border, and thus here the feature of a double period exists. The sea surface variation by the coast presents a damped oscillation with the period $T = 2d/\sqrt{gh}$, with damping intensity depending on the value r_1, i.e. on the depth gradient on the shelf border.

Case 2

The shelf width L is less than resonant. Now the amplitude of oscillations by the coast decreases faster than in the previous case. The period of these oscillations and the length of the radiating wave decrease in the relationship $4L/\lambda_1$ as compared to the case of resonance. Amplitudes stay the same, so the radiating waves become steeper. Figure 5 shows that in each positive or negative half of a wave, the steepness of the forward slope increases considerably more than that of the back slope. The form of oscillations by the coast changes in the similar way -- the departure of sea surface from its nondisturbed position is more rapid than its returning back. In the leading part of the wave train travelling to the deep sea, there arises a long period constituent due to the confluence of two positive douple-humped cycles forming a continuous crest.

Case 3

The initial perturbation is located on the shelf and is nearly equal to L. The case would be "anti-resonant" if the perturbation were a progressive wave. However, the perturbation generates two opposite waves, which immediately begin to reflect from the coast and shelf border. Both the shelf radiation and the coastal oscillations include two principle components with periods T and 2T ($T = 2d/\sqrt{gh_1}$). The second period is due to the proximity of the tsunami origin to the

Fig. 4: The Normalized Wave Forms are Shown for the Deep-Water Position M and the Coastline Position N and the Variable

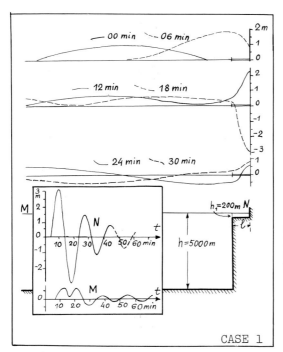

Case 1:
Deep-Water Position of Initial Wave: Shelf Width Nearly Resonant.

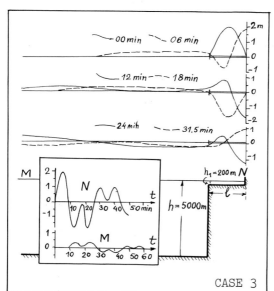

Case 3:
Shelf Position of Initial Wave: Shelf Width Nearly Resonant.

Shelf-Width Model in the Lower Right-Hand Side of Each Figure Due to a Half-Sinusoid Wave. The position of the initial wave is variable.

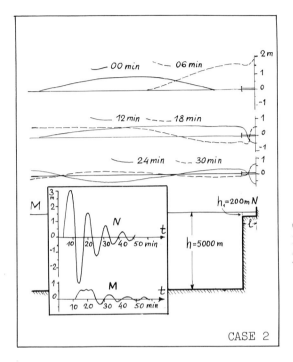

Case 2:
Deep-Water Position of Initial Wave: Shelf Width Less than Resonant.

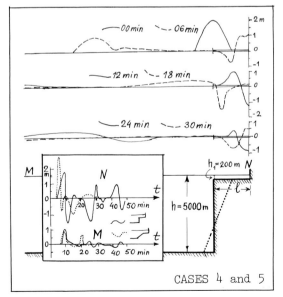

Cases 4 and 5:
Shelf-Border Position of Initial Wave: Shelf Width Nearly Resonant. Case 4 (---); Case 5 (·······).

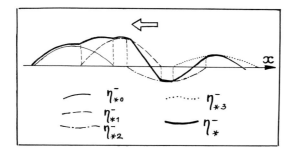

Fig. 5: Superposition of Multiple Reflections for a Step Model.

coast-reflector. The shelf oscillations are almost standing waves. The direct and return progressive waves, forming these oscillations, meet in the middle of the shelf now in-phase, now out-of-phase, and so the conditions here correspond alternatively to antinode and to node. The amplitude ratio of the opposite waves is also varying: if the waves are in phase, the amplitudes are equal to each other; otherwise the amplitude ratio is equal to r_1, the wave travelling to the coast being smaller.

Cases 4 and 5

If the initial perturbation is located over the shelf border (Case 4), the two oppositely directed waves have more complex forms. The energy contents in these waves are different (see Fig. 5) and, because of reflection at the shelf edge, the wave $\eta_{\overline{*}}^{+}$ has more energy than the wave η_o^{+}. Furthermore, the wave η_o^{+}, because of the negative reflection of the rear part of the primary wave travelling to the sea, is not solitary but presents a crest followed by a trough, the transition from crest to trough being rather rapid. The wave $\eta_{\overline{*}}^{-}$ consists of a main leading crest, followed by a positive tail. Thus, the depth distribution within the tsunami origin causes the additional periods in both the waves, this period being shorter in the η_o^{+}-wave and longer in the $\eta_{\overline{*}}^{-}$-wave than the initial fundamental period determined by the perturbation size. The coastal oscillations are characterized by "pulsations", in which the double-humped cycles of opposite signs have different durations. The shelf radiation consists of short phases having crest and trough of unequal length and separated by sections with relatively insignificant oscillations. In every other phase the crest and the trough exchange their seats and lengths.

The situation when the initial wave originates from the perturbation located over a slope (case 5) instead of a step approaches real conditions, but in principle it repeats the preceding case. The results of both cases are similar. The main difference is that the reflection effects are expressed not so sharply on the slope as on the step, because the reflection itself is "spread" over the slope. Sometimes such a continuous reflection on the slope or inclined shelf may result in relatively long periods in some portions of the back radiation, as occurs in the leading part of the wave going to the sea in cases 1 and 2. Some details of this phenomenon have been considered by means of a number of numerical experiments, which are not given here.

Some of the effects mentioned above should be detectable in real tsunami observation data. For example, the observations of the Niigata tsunami of 1964 show the presence of two principle periods: about 20 and 40 min. (Aida, et al., 1964). The tsunami origin is on the shelf close to the coast of Japan (Fig. 6) and corresponds roughly to case 3. The depth in this region is very variable but is approximately 200 to 300 meters (Soloviev and Militeev, 1968). The source dimensions in the direction normal to the coastline are about 30 kilometers. Taking this value as the half-length of the wave arising and using the formula $T = \lambda/\sqrt{gh}$, one can, putting h = 250 m., ob-

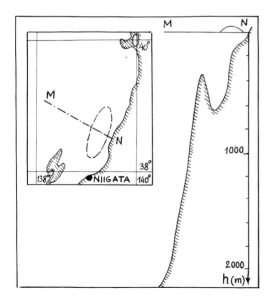

Fig. 6:
Tsunami Origin of the Niigata Tsunami in 1964. Map at Left: Cross-Section at Right.

tain for the principal initial oscillation T ≈ 20 minutes. The reflection from the coast and the shelf edge must contribute to generation of the second, double period. Therefore, the periods recorded during the Niigata tsunami of 1964 may be explained, first, by the parameters of the source itself and, second, by its disposition close to the coastline and the characteristics of the Japanese shelf.

From the practical point of view, it seems expedient to take the shelf effects into account when classifying parts of coasts by tsunami danger.

BIBLIOGRAPHY

Aida, I., Kajiura, K., Hatori, T., and Momoi, T. 1964. A tsunami accompany the Niigata earthquake of June 16, 1964. *Bull. Earthq. Res. Inst.* Tokyo Univ., Vol. 42, N. 4.

Soloviev, S. L. and Militeev, A. N. 1968. Proyavlenie Niigatskogo tsunami 1964 goda na Poberezh'e SSSR i Nekotorye Dannye ob Istochinke Voln. *Problema Tsunami*. Sbornik Statey. "Nauka."

Vol'tsinger, N. E. and Pyaskovskiy, R. V. 1968. Osnovnye Okeanologicheskie Zadachi Teorii Melkoy Vody. *Gidrometeoizdat*, Leningrad.

24. Experimental Investigations of Tsunami Waves

M. I. KRIVOSHEY
State Hydrological Institute
Leningrad, U.S.S.R.

ABSTRACT

Propagation of tsunami waves was studied on two spatial models and in a hydraulic flume. The first spatial model of a sector of the Pacific Ocean (horizontal scale 1:65,000; vertical scale 1:12,500) was used for the investigation of the influence of refraction and location of tsunami epicenter upon the angle of wave approach to the coast and the character of wave height change during its propagation from the epicenter zone.

The second or large scale model of one of the U.S.S.R. regions of the Pacific coast (horizontal scale 1:5,000; vertical scale 1:350) was used to investigate boundaries and depths of flooding of coastal territory for run-up of tsunami waves of various heights. By this model, data on variation of wave height, length, and period in shallow water (from 200 m. isobath up to water edge) were obtained.

In the hydraulic flume (100 m. x 1 m. x 1.20 m.) the influence of scale distortion on the results of modeling and the effect of wave length on coast flooding and the laws of wave transformation at various bottom slopes of the shallow zone were investigated. Considerable attention was paid to the analysis of formation and development of secondary waves ("ondulations") on a long wave. The method for computation of the transform of ondulation wave profile in a plane problem at different zones of shallow water (before the breaking,

after the breaking, and during run-up onto a dry coast) is
described. Computation results are compared with experimental
data.

INTRODUCTION

Experimental investigations of tsunami waves were conducted at the River Bed Laboratory of the State Hydrological Institute during 1964-66. The main purpose of the investigations was to determine the boundaries and the depth of flooding caused by tsunami in one of the U.S.S.R. regions of the Pacific coast. Additional experiments in a hydraulic flume have been undertaken to investigate the transformation of tsunami waves on a beach with variable slope and to elaborate methods for estimation of tsunami wave parameters.

EXPERIMENTS ON SPATIAL MODELS

The practical problem of estimation of the boundaries of the tsunami flooding zone was solved by means of successive modeling on two spatial models and in a hydraulic flume. The modeling was based on the laws of gravitational similarity deduced by the assumption of the equality of Froude's number in nature and on model.

Small-Scale Model

The investigations on the first (small-scale) model of a sector of the Pacific ocean were conducted to determine the influence of the location of the tsunami epicenter and of refraction on the angle of wave approach to the shore. For the small model the horizontal scale was 1:65,000, and the vertical scale was 1:12,500. Tsunami waves were generated by a pneumatic wave generator 0.45 x 0.50 m. placed at the bottom; the area of the wave generator approximately corresponded (to scale) to the area of the zone of the tsunami epicenter.

Vertical water level fluctuations were registered at 25 points of the model by electrical resistance sensors and recorded on an oscillograph. Waves of the period τ = 1.5 to 2.0 sec. (corresponding to τ = 15 to 20 min. of actual waves) were generated on the model. The analysis of the experiments on the small-scale model with different positions of the wave generator permitted compilation of outlines of successive positions of the crest of the first wave (Fig. 1) and of isolines of its relative height (Figure 2). The figures suggest the following conclusions:

1. Due to the decisive effect of refractions, the position of tsunami epicenter has no influence on the angle of wave approach to the coast line of the studied gulf.

2. When moving from the epicenter of tsunami towards the shore, the wave height at first diminishes due to energy dissipation and then increases due to the decrease of depth.

Large-Scale Model

The purpose of the investigations on the second (large-scale) model was to determine the boundaries and the depth of flooding of the shore for different wave heights and to analyze long wave transformation within the maritime part of the model.

For the model the horizontal scale was 1:5,000 and the vertical scale was 1:350. Modeling of the maritime part was made up to the depth of 200 meters. Long waves were created by means of water displacement and corresponded to the (actual) period of 10 min. on the shore of the bay. The distance between the wave generator and the coastal part made two wave lengths. Boundaries of flooding and the distribution of surface velocities were determined by filming at the rate of 16 sequences per minute. The boundary of flooding was determined by washing sawdust away from the studied land area. Surface currents were marked by floats. More than 300 experiments were run on the model.

As a result of the analysis of experimental data from the second model some regularities of the transformation of the wave parameters were deduced. For example, the relative height of the wave crest (i.e., the ratio of the height of the wave crest at the given point to the height of the wave crest near the wave generator) decreases from 1.0 to 0.6 during the propagation of the wave along the stretch with a constant depth of 200 meters. At varying depths (from 200 m. up to the water edge) the ratio increases up to 1.10. Relative wave period increases constantly from 1.0 up to 2.2, with an insignificant decrease only in the interior of the gulf. The ratio of the height of the first wave crest to the total wave height depends on the wave height; the mean value of this ratio equals approximately 0.5.

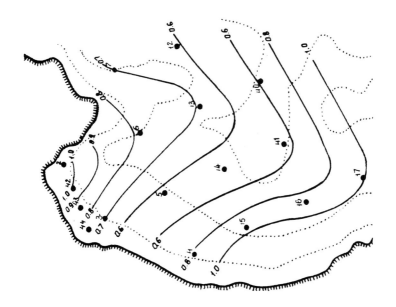

Fig. 2: Variations of Relative Heights of the First Wave Crest.

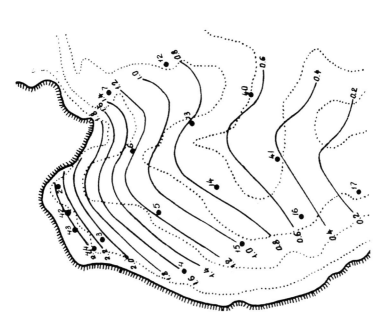

Fig. 1: Successive Positions of the Crest of the First Wave in 0.2 sec. The digits show numbers of levels gauges.

Experimental data from the large-scale model and field data on flooding boundaries were compared. The good agreement points to the reliability of experiments on a model with a distorted scale.

EXPERIMENTS IN A HYDRAULIC FLUME

Investigations in a hydraulic flume were aimed at:

1. Determination of the influence of scale distortion upon results of modeling with spatial models (Series I of experiments).

2. Investigations of a long wave transformation at a constant depth, on shallows in the zone of breakers and run-up (Series II-IV of experiments).

3. Development of a computing method for long wave transformation on shallows and in the zone of shore flooding.

The experiments were carried out in a 100 m. x 1.0 m. x 1.2 m. hydraulic flume.

The main results of Series I of the experiments, carried out in three profiles with equal horizontal and vertical scales (1:200), are as follows:

1. Modeling on spatial models with scale distortion has practically no effect on water level rise at the studied points.

2. Wave length is of considerable importance for flooding depth.

Series II and III of the experiments carried out on schematic profiles with changing bottom slope of the shallow zone indicated regularities of wave transformation in different zones of the flume. During wave propagation along the zone with constant depth, the height of the wave crest remains practically constant and the slope of the front part of the wave increases slightly. At profiles near the wave generator separation of the wave crest occurs; it bears the features of secondary oscillations. These oscillations were observed by Favre for the first time during experiments on unsteady flow and were called "ondulations". The formation of ondulations on a long wave is an interesting but little-studied phenomenon; the interaction of ondulations may influence considerably the depth and boundary of shore flooding; hence the phenomenon was studied in detail. The appearance of ondulations on a long wave is the first indication of wave breaking, i.e., the appearance of a point on the wave profile at which the tangent becomes vertical ($\frac{\partial \eta}{\partial x}$, where η is the water surface elevation above the undisturbed water level). Experimental data on the time of appearance of ondulations are compared with those computed by Vedernikov's formula. The formula is derived from Saint-Venant's equation for ideal and real fluids:

$$T = \pm \frac{2}{3} \sqrt{\frac{FB}{g}} \frac{W}{a_o} \bigg/ (1 - \frac{1}{3} \frac{F}{B} \frac{dB}{dH}) \tag{1}$$

where F = the cross section area of the stream;
W = the velocity of constant depth propagation;
B = the channel width;
g = the acceleration of gravity;
a_o = the rate of discharge variations at the initial profile.

The performed computation shows rather good agreement of the computation scheme suggested by Vedernikov with experimental data. This fact reveals the possibility of applying Saint-Venant's equations in cases when the curvature of free surface profile becomes considerable. The analysis of wave transformation along the constant depth zone indicated the existence of a reflected wave and allowed comparison of the propagation time of the beginning of the disturbance in still water calculated from experimental data with theoretical propagation time calculated by the Lagrange formula.

During wave propagation along shallow water, appreciable increases of height of the wave crest and of the steepness of the front part of the wave occurs. Experimental data on the increase of the height of the wave crest on shallows were compared with results of computation by Green's and Brovikov's formulas.

The progress of wave reflection along inclined bottom and in the zone of wave breaking was analyzed. The analysis was based on experimental data on water level fluctuations and on filming of stream velocity field at the beginning of the bottom rise. During the propagation of the wave along shallow zone with bottom slope i_o = -0.041 "concentrated" wave reflection occurs. It resembles the reflection of a wave from a vertical wall.

Results of computation of water level fluctuations on an electronic digital computer according to one of the experiments are given. The calculations were based on non-linear and linear equations of shallow water with the use of difference schemes of the first and second order of accuracy and with various friction coefficients. Computer results corresponded to experimental data (e.g., a "concentrated" reflection was obtained); this points to the reliability of the computation scheme used. The comparison of the experimental propagation time of the disturbance with the theoretical propagation time calculated by the relationship derived by the author from the Lagrange formula for shallow water with a given bottom slope showed an excess of experimental propagation time at the beginning of bottom rise and practically coincidence in the remaining part of the shallow zone.

Wave propagation after the wave breaking and run-up onto a dry coast is of quite a different character at direct and at inverse coast slope (Series II and Series III of experiments, respectively). The relationships for wave propagation velocity are:

(a) at wave propagation along a direct slope:

$$V = 1.8\sqrt{g\eta} \qquad (2)$$

where η is the wave elevation above the dry bottom; and

(b) at wave propagation along inverse slope:

$$V = 2.54\sqrt{g\eta} \qquad (3)$$

While considering Series IV of the experiments run at various initial water levels in the flume, attention was paid to the analysis of transformation of ondulations on a long wave. Besides, problems concerning relative height and the breaking of ondulations were considered in detail. In order to analyze the transformation of ondulations the following notions are introduced: ondulation period, crest height, hollow height, amplitude, slope, relative length, and relative height.

The analysis of four experiments run at various initial levels of undisturbed water surface in the flume and at equal boundary conditions (water level at the upper reach, the rate and duration of the opening of the shutter) suggests the following conclusions concerning the character of transformation of ondulations on a long wave:

1. When a wave propagates along a flume of constant depth, the crest height and the slope of ondulation remain practically unchanged and the hollow height decreases. The ondulation amplitude and period increase up to their maximum value at the beginning of bottom rise. Relative length of ondulation slightly increases. The decrease of hollow height (the deepening of hollows) and the resulting increase of amplitudes may be explained by gradual formation of ondulations. One can suppose that, due to hollow deepening, a long wave can break into a number of independent shorter waves. The increase of ondulation periods may be explained by wave flattening.

2. When a wave propagates along shallow water the steepness of ondulation, the crest height, the hollow height, and the relative length increase. Since the rate of increase of the height of crests and hollows is practically equal, the ondulation amplitude remains constant. Periods of ondulations retain their constant value equal to the period at the beginning of the bottom rise.

3. Filming of wave breaking shows that the moment of wave overturn is preceded by the appearance of a kind of jet on the crest, then the jet tears off, followed by a strong splash propagating up the slope as a roller (bore).

4. After breaking of the ondulation, sharp decrease of crest height and increase of hollow height take place -- resulting in the decrease of ondulation amplitude reaching its minimum (zero) value at the boundary of the flooding of the dry shore.

5. Intensive breaking is typical only for the first ondulation as breaking of subsequent ondulations is less intensive because it occurs at a higher level due to propagation of preceding ondulations and superposition of a reflected wave. Practically, only the first three to five ondulations break and every one of them breaks subsequently at deeper water.

6. In the zone of wave breaking, a reflected wave is formed. Concurrent with the above mentioned common characteristics of transformation of ondulation along the flume, there are some differences caused by various initial water levels in the flume.

a. With increase of initial water level in the flume the ondulation periods increase, the amplitudes decrease, and the intensity of their growth rises.

b. At higher initial water level in the flume, ondulations break at greater depth, so the ratio of the depth of water and the crest height at breaking profile increases from 0.86 (at h_o = 10 cm.) up to 1.33 (at h_o = 35 cm.) where h_o is the initial level of the undisturbed water level in the flume. The latter observation contradicts the assertion of some authors of the constant value of the ratio (h/η) break. For instance, Favre affirms that a long wave always breaks at the depth which equals 1.28 of the height of the breaking crest.

c. At the increase of the initial water level in the flume, and consequently at the extension of shallow zone, the character of wave reflection from an inclined bottom changes. When the shallow zone is not too large (h_o = 10 cm.), the reflection is of concentrated character with a clearly perceptible crest of a reflected wave at the profile of the breaking of the first ondulation. When the shallow zone is more extensive (h_o = 35 cm.), the reflection is dispersed along the entire inclined bottom, and the reflected wave crest is clearly seen at the flume profiles close to the filtering wall.

The problem of relative ondulation height at the profile of wave disturbance was studied. Some researchers, such as Favre, affirm that the first ondulation must be the highest. As a result of the analysis of these experiments performed, the governing factor of ondulation height at the disturbance profile has been found to be the initial wave slope which increases with a decrease in the duration of the shutter opening. The greater the wave slope at the initial profile, the greater the probability that the first ondulation will be the highest one. With a decrease of wave slope at the initial profile the second or the third ondulation will be the highest.

The detailed analysis of the breaking of ondulations suggests the following conclusions:

1. Ondulations in the zone of breaking may be regarded as independent waves; their breaking peculiarities depend on their slope in the deep zone.

2. With a decrease of the initial wave slope (in the beginning of the shallow zone) the breaking occurs in the

deeper zone. In this case, longer ondulations break before the shorter ones (Figure 3).

3. Earlier breaking of succeeding ondulations of shorter length than the preceding ones may be explained also by a counter current which accelerates the wave overturn.

4. In the zone of breaking, the ratio of ondulation length and depth of water is equal to 25 to 30 (Figure 4).

The analysis of transformation of ondulations on shallow water suggested the following working hypothesis: *Ondulations on shallow water may be regarded as independent long waves; their propagation is described by equations derived for long waves in shallow water (before the moment of breaking) and by equations derived for discontinuous waves (after breaking and propagation as a bore up to water edge).* The concept of ondulations as independent long waves is of both practical and theoretical interest; it is a new approach to the problem of ondulation profile computation. The transformation of ondulation profile was analyzed for the following reasons:

1. One may suppose that a series of successive tsunami waves running up on a shore are secondary oscillations on a long wave.

2. The process of shore flooding by tsunami waves is formed by a complex interaction of successive series of waves. Experiments prove that, after breaking, the ondulations propagate along a shore as superposing bores. For this reason the knowledge of methods for computation of ondulation transformations is of interest even if the hypothesis of ondulations as independent long tsunami waves is erroneous. In the case of run-up of longer waves, the character of their interaction will be comparable to superposition of shorter waves (ondulations). Computation of transformation of ondulation profile of a long wave under conditions of the one-dimensional problem was performed by different theoretical schemes for shallow zones.

COMPUTATION OF LONG WAVE TRANSFORMATION

Zone of Continuity of Motion

For computation in this zone non-linear equations of shallow water are used:

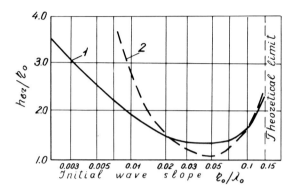

Fig. 3: Relationship of the Ratio Between the Depth of Undisturbed Water Surface at the Breaking Profile and the Wave Crest Height at Deep Water Beginning with the Initial Wave Slope at Deep Water: 1 - Curves by Sverdrup and Munk; 2 - The Author's Experimental Curve.

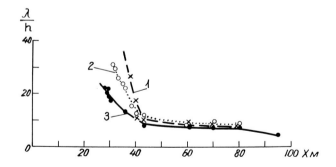

Fig. 4: Variation of the Ratio Between Ondulation Wave and the Depth of Undisturbed Water Surface Along the Flume (before the beach bottom): 1 - h_o = 10 cm.; 2 - h_o = 25 cm.; 3 - h_o = 35 cm.

$$\frac{\partial u}{\partial t} + u\frac{\partial u}{\partial x} = -g\frac{\partial \eta}{\partial x} - f\frac{u|u|}{h+\eta} + gi_o \qquad (4)$$

$$\frac{\partial [u(\eta+h)]}{\partial x} = -\frac{\partial \eta}{\partial x} \qquad (5)$$

where u = the current velocity,
η = the ordinate of free surface above the undisturbed water level,
h = the water depth,
i_o = the slope of the bottom, and
f = the friction coefficient.

It is possible to find integrals of these equations only in a limited number of cases; therefore, all the existing methods for integration of equation systems 4 and 5 are to some extent approximate. In this study the method of characteristics elaborated by academician S. A. Khristianovich during 1933-1937 was used. The method consists in the substitution of differential equations with total derivatives for equations 4 and 5 with partial derivatives.

Zone of Rupture of Continuity of Wave Motion
(Zone of Breaking) and the Propagation of the
Wave as a Bore up to the Water's Edge

The moment of rupture of continuity (breaking of wave) is characterized by the turning of one of the derivatives $\frac{\partial \eta}{\partial x}$, $\frac{\partial \eta}{\partial t}$, $\frac{\partial u}{\partial x}$, $\frac{\partial u}{\partial t}$ into infinity. In the wave plane, the characteristics intersect at that moment and discontinuous solutions appear.

In order to compute ondulation transformations at the moment of breaking and during the motion of ondulations as successive bores up to the water edge, the method for computation of discontinuous waves suggested by H. T. Meleschenko and S. M. Yakubov is used. The method is based on the equation of a discontinuous wave in a rectangular channel:

$$V = U_2 \pm \sqrt{\frac{gH_1(H_1+H_2)}{2H_2}} \qquad (6)$$

where V = the velocity of propagation of the front of the discontinuous wave,
U_2 = the velocity of the current ahead of the discontinuous wave front,
$H_1 = h_1 + \eta_1$ = the depth behind the front, and
$H_2 = h_2 + \eta_2$ = the depth before the discontinuous wave front.

The sign " + " corresponds to propagation of the discontinuous wave front with the current, the sign " - ", to propagation against the current.

Great attention is paid to consideration of conditions at the water's edge, where the velocity of wave front propagation becomes equal to the current velocity, the boundary condition for the computation of wave run-up onto a dry shore. Recommendations for evaluation of velocity of the current at the water's edge are given.

Zone of Wave Run-Up Onto a Dry Shore

After the run-up of the bore onto a dry shore, the water mass loses its wave properties and turns to a water jet. Due to the loss of wave properties, the notion of the velocity of wave front propagation loses meaning; the velocity becomes equal to the current velocity. The problem of computation of wave run-up on a shore is reduced to the determination of the boundaries, the depth of flooding, and water current velocity.

The method for computation of wave run-up on a shore is based on the integration of the equation of Newton's second law:

$$\frac{d^2x}{dt^2} = -gi_o \qquad (7)$$

This approach was first used by American scientists Ho and Meyer, who derived the equation of motion taking into account only inertia and gravity forces. Equations of motion taking into account friction forces have been derived:

$$\frac{d^2x}{dt^2} = -gi_o - \frac{(\frac{dx}{dt})^2}{C^2 \eta} \qquad (8)$$

In case of constant friction coefficient $K = \frac{1}{C^2\eta}$ (where C is Chézy's coefficient) within the design reach, the equations of current velocity $U = \frac{dx}{dt}$ and of overwash x become:

$$U = U_o - \sqrt{\frac{gi_o}{K}} \, tg \sqrt{gi_o K} \, t \qquad (9)$$

$$x = U_o t + \frac{1}{K} \ell n \cos\sqrt{gi_o K} \, t \qquad (10)$$

where U_o = the current velocity at water edge,
    t = the time.

In case of linear variation of friction coefficient K within the design reach, the motion equation is of the form:

$$U = U_o - \sqrt{(gi_o + \frac{U_1^2}{2C_1^2\eta_1})2C_2^2\eta_2} \, tg \sqrt{(gi_o + \frac{U_1^2}{2C_1^2\eta_1})\frac{1}{2C_2^2\eta_2}} \, t \qquad (11)$$

Two approximate relationships are suggested for determining flooding depth. The first one is derived from the equation of a discontinuous wave with the assumption that the current retains some wave properties:

$$\eta \approx \frac{U^2}{92.8} \qquad (12)$$

The second relationship is derived from the continuity equation when the values of η_1 and U_1 at the upper boundary of the reach are known:

$$\eta_2 = \frac{(\eta_1(U_1 - \frac{dj}{dt}))}{U_2 - \frac{dj}{dt}} \qquad (13)$$

where U_2 and η_2 are current velocity and flooding depth at the lower boundary of the design reach.

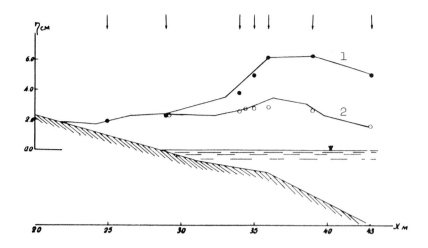

Fig. 5: Comparison of Experimental and Computed Values of Crest Heights (1) and Hollow Heights (2) of the First Ondulations at the Distance from the Bottom Rise Up to the Boundary of Coast Flooding: ● - Experimental Values of Crest Heights; o - Experimental Values of Hollow Heights. The arrows show water level gauges.

Computation shows that if the shore slope is small ($i_o \leq 0.001$), the neglect of friction results in an error in determining the inundation boundary which may cause an error of as much as 30% in the total area of the predicted flooded zone.

As calculations by the method of characteristics are rather labor-consuming, the computation of transformation was performed for only three ondulations.

Comparison of experimental and estimated parameters of waves showed their good agreement (Figure 5).

The methods used for calculation of transformation of a long wave with ondulations are quite suitable for practical purposes. The comparison of experimental and computation data has confirmed the working hypothesis of ondulations as secondary long waves on a tsunami wave.

25. A Laboratory Model of a Double-humped Wave Impingent on a Plane, Sloping Beach

JOHN A. WILLIAMS
University of Hawaii
Honolulu, Hawaii

JAN M. JORDAAN, JR.
105 Nassau Sunnyside
Pretoria, South Africa

ABSTRACT

The results of preliminary experiments on the generation of double-humped waves on a plane, sloping beach are presented. The experimental equipment and procedures are discussed. The generated wave form and the ensuing run-up are compared with the numerical results of J. P. Butler, which are based on the theoretical work of G. F. Carrier and H. P. Greenspan.

INTRODUCTION

Carrier and Greenspan (1958) used the inviscid nonlinear shallow-water equations to study theoretically the problem of waves on a sloping beach.

Butler (1967) applied the Carrier and Greenspan approach to waves having two crests, or so-called double-humped waves. These waves are represented analytically by the expression

$$\eta = 1024\varepsilon x^2 (a+8x)^2 \, e^{16px}, \qquad (1)$$

where the parameters a and p determine the relative height

and the spacing of the two crests, and where ε is a parameter which magnifies the vertical scale and is limited by the physical condition of wave breaking. The two variables η and x are dimensionless and represent the elevation of the free surface and the distance (negative) from the beach (x = 0), respectively. A characteristic length, l_o, which is determined by the parameters a and p and the location of the crests in the physical wave, is used to form the dimensionless variable x; this characteristic length multiplied by the beach slope, α, is used to form the dimensionless variable, η.

Using waves of this shape with their leading edges coincident with the beach and zero particle velocity at time zero, Butler calculated the elevation-time history at the beach for wave forms having several different values of the product ap with a = 0.5 and ε = 10^{-3}.

To validate the results of Butler, and hence the Carrier and Greenspan theory, an experimental study is necessary. Such experiments pose a unique problem -- namely, the generation of the correct double-humped wave form which reasonably satisfies the boundary and initial conditions of the theory. It is the purpose of this report to describe a technique for generating such wave forms and to present the results from an experiment which used this technique.

EXPERIMENTAL TECHNIQUE

The experiments were conducted in a lucite wave tank approximately 25 feet long by 1.0 feet wide with a maximum depth of 1.5 feet. The beach slope was 1 on 7.

The wave generator was constructed from strips of lucite 0.25 inch thick by 5.0 feet long, shaped to a profile given by equation 1 with a = 0.5 and ap = 1.3. Ten of these strips, spaced 0.948 inches apart, were connected by a series of six lucite crossbars placed across the plane edge of the strips. This assembly was then suspended from four points by cables leading up to a single cable which in turn passed around an axle placed across the channel. The generator was positioned such that its leading edge was coincident with the beach and the crests of the inverted wave form were about 0.10 in. above the free surface. A wave was generated by allowing the generator to drop vertically into the water until the trough of the inverted profile was on the same level with the undisturbed free surface, a distance of about 0.6 inches. The ratio of the change in water surface elevation to the depth of penetration of the lucite strips at the crests is of the order 0.1. A photo of the assembled wave generator is shown in Fig. 1 and the profile of the individual strips is shown in Figure 3.

Wave profiles were recorded by means of two-wire resistance probes in a bridge circuit with the output from the bridge passing through a rectifier and smoothing circuit and into a ten-channel Honeywell Visicorder. The sensitivity of the galvanometers was sufficient to preclude the use of amplifiers. Only seven galvanometers were operable during these preliminary runs and the seven two-wire probes were located at the following points: -0.75, -2.50, -4.25, -8.00, -11.75, -16.375, and -23.00 inches. These gauges and their identifying numbers are indicated along the abscissa of Figure 6.

Fig. 1: Double-humped Wave Generator.

Calibration of the gauge was carried out by raising and lowering the water surface in the tank and observing the corresponding change in the recorder trace. As the water depth at gauge No. 1 was only 0.107 inches deep, this gauge was placed in parallel with a second two-wire probe located in a separate container of water. This second resistance decreased the sensitivity of the gauge, which was no problem at a point so near the beach. At the same time, it produced a linear relation between water surface change and recorder trace deflection. Calibration constants in units of inches of water surface deflection per chart division (0.1 inch) for the several gauges are indicated in Figure 2.

The wave profile was determined by selecting time zero as that instant when gauge No. 3, located at the first crest,

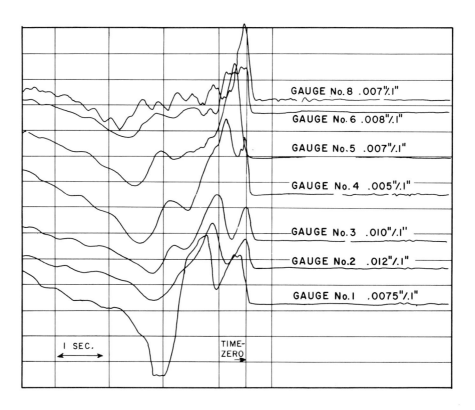

Fig. 2: Trace of Water Surface Time-Histories at the Several Gauges.

recorded a maximum as the wave generator penetrated the free surface. The corresponding water surface elevations at the remaining gauges were determined at this point in time and the wave profile plotted. Figure 2 shows the water surface-time history at the several gauge locations. The record from gauge No. 8 indicates that the generator dropped with a slight amount of rotation since the maximum deflection here occurs about 0.03 seconds later than it does at gauge No. 3.

In order to minimize the effect of surface tension at the beach, this region was brushed with a mixture of detergent and water prior to each run. This diminished the surface tension force by approximately 40 percent.

EXPERIMENTAL RESULTS

Figure 3 compares the initial wave form with the generator profile. The second crest is somewhat higher than it should be, probably because the generator deflected about 0.06 inches under its own weight and, consequently, more water was displaced than should have been. The generated profile conforms more closely to a wave with a value of ap = 1.2.

Figure 4 compares the generated wave form at time zero with equation 1 for a = 0.5 and ap = 1.2. The agreement with equation 1 is quite good except at the second crest, where the generated wave is about 15 percent too low. This implies that the generated wave has a value of ap slightly greater than 1.2. The parameter ε for the generated wave can be estimated by observing that at the first crest η/ε = 2.25 x 10^{-2} for ap = 1.2 and that $\eta^* = \eta \alpha l_o$ = 0.062 inches (from Figure 3). This gives ε = .12 for α = 1/7 and l_o = 14.0 feet.

Figure 5 compares the time history of the water surface elevation at gauge No. 1, where x = -.0046, (the circle symbols) with the predicted history at x = 0. Here the agreement is not as good. The run-up process appears to take place in a shorter time scale than that predicted. To determine the effect of the wave generator on the speed of propagation, a wave crest was formed at the seaward end of the wave generator, with the generator both in and out of the free surface. With the recorder speed and gauge spacing known, the average velocity between gauges of this crest was calculated and the results are presented in Figure 6. It is evident that the average velocity of the crest with the generator out of the free surface conforms very closely to the

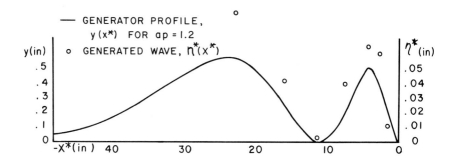

Fig. 3: Comparison of Wave Generator Profile with Water Surface Profile at Time Zero.

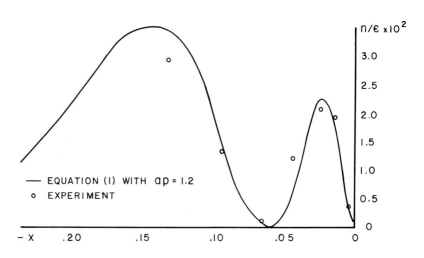

Fig. 4: Comparison of Water Surface Profile at Time Zero with Equation 1 for ap = 1.2.

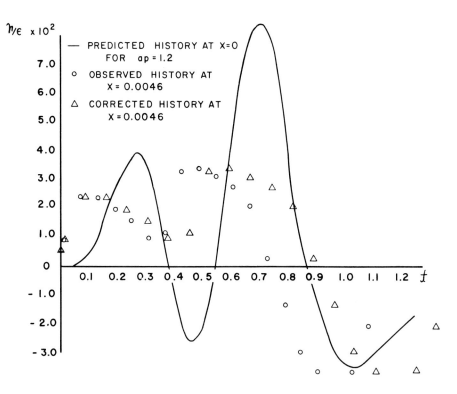

Fig. 5: Comparison of Observed Time-Histories at $x = -.0046$ with Predicted Time History at $x = 0$ for $ap = 1.2$.

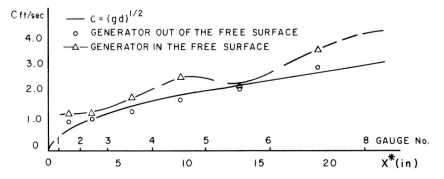

Fig. 6: Comparison of Wave Speeds with the Wave Generator In and Out of the Free Surface.

theoretical value of $(gd)^{1/2}$, while with the generator in the free surface the average velocity of the crest is as much as 20 percent greater than $(gd)^{1/2}$. This increase may be, in part, the result of a surface tension force along the intersection of the lucite strip and the water surface and, in part, the result of wave amplitude which increases or decreases as the ratio of water volume to submerged wave generator volume increases or decreases along the path of propagation.

From Fig. 6 the total travel time for a distrubance to reach gauge No. 1 from gauge No. 8 can be estimated from the relation

$$T = \int_{0.75"}^{23.0"} \frac{dx^*}{c} . \qquad (2)$$

This gives T = 0.82 seconds if the generator is in the free surface and T = 1.01 seconds if the generator is out of the free surface. A correction factor for the time scale can be estimated from these numbers, i.e., observed times should be decreased by 0.82/1.01 = 0.81. It is significant to note that the amount of water displaced by the generator weighs about 1.35 pounds while the total surface tension force on the generator for a static conditions is about 0.5 pounds.

This factor is applied to the time scale for the experimental wave and results in the second set of points (the triangle symbols) in Figure 5. The minimum and maximum points show better agreement with the theory. However, the first maximum still arrives about twice as fast as predicted and the first minimum and the second maximum arrive about 1.1 times faster than predicted. This may be, in part, the result of a failure to meet the initial condition of zero particle velocity. This condition is governed by the penetration time of the generator; it will not be met if the generator takes too long to make its maximum penetration after entering the water. The second minimum occurs about 1.15 times later than predicted. As the water line retreats beyond gauge No. 1, a film of water remains because of wetting of the beach. The thickness of this film decreases slightly during the time the beach line is seaward of gauge No. 1, thus causing the minimum point to occur somewhat later than it would otherwise.

Because the second crest is travelling faster than it would under gravity forces alone, it strikes the beach before the run-down of the first crest has had a chance to be completed. Thus, the first minimum point does not extend below the still water level. The two maximum points are about 40 to 50 percent lower than they are predicted to be. Again, this reduction is principally the result of surface tension. The advancing miniscus during run-up has a contact angle with the beach such that a component of surface tension force opposing the run-up is developed. Friction appears to be relatively unimportant here since the run-down of the second crest, when surface tension is unimportant, is of the predicted order of magnitude; the run-up, when surface tension is important, is about half its predicted value.

CONCLUSION

From the results presented above, the following conclusions have been drawn:

(1) A double-humped wave of the desired form can be generated using the wave generator described above. This generated wave satisfies the initial condition that its leading edge be coincident with the beach. The initial condition of zero particle velocity is approximated more closely as the penetration time for the generator becomes very small.

(2) The time-scale of the observed run-up agrees reasonably well with the predicted time-scale, except at the first maximum point, when the effect of the wave generator on wave speed is considered.

(3) The run-up is about 50 percent lower than predicted, primarily because of surface tension.

(4) An increase of three times in the length scale would produce waves for which the ratio of gravity to surface tension forces would be increased by nine times -- an increase of about one order of magnitude. This increase in the gravity forces should be sufficient to subordinate the surface tension to a second order effect. (Experiments at this increased scale have been completed by the Joint Tsunami Research Effort at the Hawaii Institute of Geophysics, and will be reported elsewhere.)

ACKNOWLEDGMENTS

We should like to express our thanks to Drs. William M. Adams and Gaylord Miller, the directors of the Joint Tsunami Research Effort at the Hawaii Institute of Geophysics, for their interest in and support of this research. We would also like to thank our student assistants Messrs Minoru Suzukawa and Ronald Siu for their help in constructing the experimental equipment and Eric Leu for his help in conducting the experiments.

BIBLIOGRAPHY

Butler, J. P. 1967. Double-humped waves on a sloping beach. Hawaii Institute of Geophysics *Report 67-16*.

Carrier, G. F. and Greenspan, H. P. 1958. Water waves of finite amplitude on a sloping beach. *Journal of Fluid Mechanics*. 4, 97.

26. Response of Narrow-mouthed Harbors to Tsunamis

G. F. CARRIER
Harvard University
Cambridge, Massachusetts

R. P. SHAW
University of Hawaii
Honolulu, Hawaii

ABSTRACT

The response of a narrow-mouthed harbor of general shape in a straight coast to a transient incident wave with the bulk of its energy spectrum at long periods (in the range of 3 - 60 minutes), e.g., a tsunami, is found as a Fourier synthesis of a corresponding time-harmonic solution which has been recently developed. The harbor basin is connected to the open sea by a narrow straight channel of arbitrary length (which may be zero) and a uniform depth is assumed for the basin, the channel, and the sea.

FORMULATION

The response of a narrow-mouthed harbor in a straight coastline to incident periodic plane waves has been found in approximate form by Carrier, Shaw, and Miyata, using small amplitude, linear, constant depth, shallow water wave theory. The present investigation is concerned with the analogous transient problem where an incident aperiodic wave, e.g., a tsunami, is the appropriate driving mechanism. The transient solution may be obtained by a Fourier synthesis of the time-harmonic problem. The analysis will consider, in particular, the effect of an entrance channel (see Figure 1).

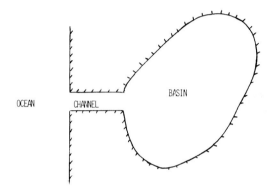

Fig. 1: General Geometry of Harbor Basin and Entrance Channel.

The normal modes $\phi_m(\bar{r})$ and natural frequency ω_m of the closed basin must be known to use this method. Then the steady state solution for the harbor response at a location $\bar{r}$ within the harbor to a unit amplitude wave of frequency ω is given as

$$\phi_{III}(\bar{r},\omega) = -4\pi k L H_1^{(1)}(kL)\ \sigma(\bar{r},k)/$$

$$\{[4kL\sigma(\bar{r}_o,k)H_o^{(1)}(kL)+\pi H_1^{(1)}(kL)]\sin(kW)$$

$$+[2H_o^{(1)}(kL)-2kL\pi\sigma(\bar{r}_o,k)H_1^{(1)}(kL)]\cos(kW)\} \quad (1)$$

where $k = \omega/c$ is the wave number, W is the entrance channel length, $c = \sqrt{gh}$ is the wave speed in the harbor (assuming that the constant depth h is shallow compared to the wave lengths involved in the problem), $2L$ is the entrance width, $\bar{r}_o$ is the midpoint of the entrance, $H_o^{(1)}$ and $H_1^{(1)}$ are Hankel functions of the first kind, and $\sigma(\bar{r},k)$ is a function containing information about the normal modes and natural frequencies of the closed harbor viz.,

$$\sigma(\bar{r},k) = \sum_{m=0}^{\infty} \frac{\phi_m(\bar{r})}{\int \phi_m^2} \frac{\phi_m(\bar{r}_o)}{dArea} \frac{1}{k^2 - k_m^2} \qquad (2)$$

($k_m = \omega_m/c$ is the corresponding natural wave number.)

For a monochromatic incident wave of amplitude $F(\omega)$, the response in the harbor is simply $F(\omega) \cdot \phi(\bar{r},\omega)$ and for a spectrum of incident waves, the response is

$$\psi_{III}(\bar{r},t) = \mathrm{Re} \left\{ \int_{-\infty}^{\infty} F(\omega) \phi_{III}(\bar{r},\omega) e^{-i\omega t} d\omega \right\} \qquad (3)$$

which can be considered an inverse Fourier transform. If ϕ_{III} were taken as unity (no shoreline or harbor), equation (3) would define the form of the incident transient wave at the origin, i.e., the location of the shoreline, as

$$f(t) = \eta_{inc}(0,t) = \mathrm{Re} \left\{ \int_{-\infty}^{\infty} F(\omega) e^{-i\omega t} d\omega \right\} \qquad (4)$$

implying that

$$F(\omega) = \frac{1}{2\pi} \int_{-\infty}^{\infty} f(t) e^{+i\omega t} dt \qquad (5)$$

The harbor response function ϕ_{III}, as obtained by Carrier, Shaw, and Miyata, is valid only for relatively long wavelengths compared to L; therefore, the range over which ϕ_{III} is expected to represent the harbor response accurately is limited to $|kL| \ll \pi$. Fortunately, most tsunamis, the driving forces of interest, contain the greatest portion of their energy at frequencies well within this range for most harbors, i.e., for most physically applicable values of L and h.

Since $F(\omega)$ will be real and symmetric for the input functions to be used, equation (3) may be expressed in terms of real values;

$$\psi_{III}(\bar{r},t) = \int_0^\infty 2F(\omega) \left\{ \text{Re}\,[\phi_{III}]\cos \omega t + \text{Im}[\phi_{III}] \cdot \sin \omega t \right\} d\omega \tag{6}$$

where $F(\omega)$ is found from the incident wave $\eta_{inc}(\bar{r},t)$ as

$$F(\omega) = \frac{1}{\pi} \int_0^\infty \eta_{inc}(0,t) \cos \omega t \, dt \tag{7}$$

INCIDENT WAVE FORMS

Consider first an incident wave form of a single crest with the bulk of its transform at long periods, e.g., at the origin (coastline)

$$\eta_{inc}(0,t) = A \,\text{sech}\,(bt) \tag{8}$$

From equation (7),

$$F(\omega) = \frac{A}{2b} \,\text{sech}\, \left(\frac{\pi\omega}{2b}\right) \tag{9}$$

and therefore from equation (8)

$$\psi_{III}(\bar{r},t) = \frac{A}{b} \int_0^\infty \text{sech}\,\left(\frac{\pi\omega}{2b}\right) \cdot \{\text{Re}[\phi_{III}]\cos \omega t + \text{Im}[\phi_{III}]\sin \omega t\} d\omega \tag{10}$$

At this point, it is convenient to nondimensionalize, using the length L and time L/c as bases. With $\kappa = kL$ and $\tau = ct/L$, equation (10) gives

$$\psi_{III}(\bar{r},\tau) = \frac{2\alpha A}{\pi} \int_0^\infty \text{sech}(\alpha\kappa)$$

$$\cdot \{\text{Re}[\phi_{III}]\cos \kappa\tau + \text{Im}[\phi_{III}]\sin \kappa\tau\}d\kappa \qquad (11)$$

as the response to

$$\eta_{inc}(0,\tau) = A \text{ sech}(\pi\tau/2\alpha) \qquad (12)$$

where $\alpha = \pi c/2bL$.

A representative time scale for this incident wave is $T = 3/b$ at which value the sech has been reduced to 10% of its maximum: this corresponds to $T_0 = 6\alpha/\pi$ in nondimensional units.

As a second incident waveform, consider an exponentially decaying sinusoidal wave

$$\eta_{inc}(0,t) = A e^{-q|t|} \cos(\omega_0 t), \qquad (13)$$

which gives

$$F(\omega) = \frac{A \cdot q}{2\pi} \left\{ \frac{1}{q^2+(\omega_0-\omega)^2} + \frac{1}{q^2+(\omega_0+\omega)^2} \right\} \qquad (14)$$

In nondimensional terms, using $\gamma = q \cdot L/c$ and $\kappa_0 = \omega_0 L/c$, this gives

$$\psi_{III}(\bar{r},\tau) = \frac{A \cdot \gamma}{\pi} \int_0^\infty \left\{ \frac{1}{\gamma^2+(\kappa_0-\kappa)^2} + \frac{1}{\gamma^2+(\kappa_0+\kappa)^2} \right\}$$

$$\cdot \{\text{Re}[\phi_{III}]\cos \kappa\tau + \text{Im}[\phi_{III}]\sin \kappa\tau\}d\kappa \qquad (15)$$

as the response to

$$\eta_{inc}(0,\tau) = A\ e^{-\gamma|\tau|}\cos(\kappa_0\tau). \tag{16}$$

The specific relationship $\gamma = \kappa_0/2\pi$ will be maintained in the numerical examples such that the incident wave form decays by a factor of e^{-1} in one period, although this restriction is not necessary. The time scale for this incident wave is $T = 2\pi/\omega_0$ or $\tau_0 = 2\pi/\kappa_0$. In both cases, A is taken as 1.0 in the calculations, i.e., all responses are for an incident wave having a peak amplitude of one.

BASIN GEOMETRY

The geometry chosen for the basin is rectangular with dimensions C by D and with an entrance channel of length W and half-width L located perpendicular to one of the sides of length C. The coordinates of the midpoint of the basin entrance are $(0,y_0)$ in a coordinate system located at a corner of the basin, as shown in Figure 2.

The normal modes are

$$\phi_{mn}(\bar{r}) = \cos\left(\frac{m\pi x}{D}\right)\cos\left(\frac{n\pi y}{C}\right) \tag{17}$$

and the natural wave numbers are

$$k_{mn} = \left[\pi^2\left\{\left(\frac{m}{D}\right)^2 + \left(\frac{n}{C}\right)^2\right\}\right]^{1/2} \tag{18}$$

with $\sigma(\bar{r},k)$ defined as a summation over all positive integer values of m and n. In practice, however, this summation must be terminated at a relatively small number of modes because the theory on which equation (1) was developed is valid only for small values of kL; i.e., the wave input must be such that the higher modes do not contribute appreciably to the harbor response.

Two sets of numerical values representing particular types of basins are chosen for use in the subsequent calculations. The first is a model representative of Barber's Point Harbor, a small barge harbor on Oahu, Hawaii. The model parameter values are C = 600 feet, D = 500 feet, h = 21 feet. The entrance half-width L is chosen to be 50 feet, although the actual harbor has a somewhat larger entrance and the entrance is located in a corner, i.e., y_o = L. The lowest natural period is 43 sec., corresponding to k_{o1} = 0.00524.

The second model is representative of a moderate-sized commercial harbor -- Duncan's Basin in Table Bay Harbor, Capetown, South Africa. The model dimensions are C = 6000 feet, D = 2200 feet, h = 43 feet, L = 200 feet, and Y_o = 1250 feet. The lowest natural period is 5.4 min., corresponding to k_{o1} = 0.000524.

These harbor models shall be referred to as B. P. and D. B., respectively, in the remainder of this discussion.

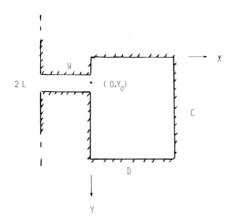

Fig. 2: Rectangular Harbor Geometry.

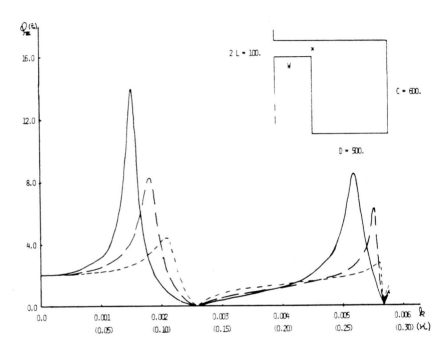

Fig. 3: Harmonic Amplification Factor for B. P. Model; C = 600 feet, D = 500 feet, L = 50 feet, Y_0 = 50 feet; W = 0 feet (----), W = 100 feet (— —), and W = 250 feet (———).

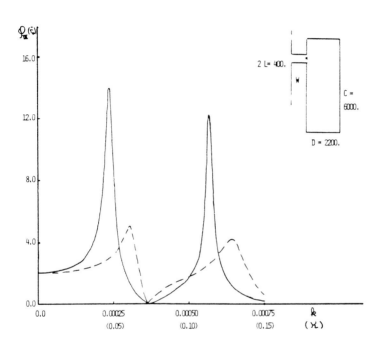

Fig. 4: Harmonic Amplification Factor for D. B. Model; C = 6000 feet, D = 2200 feet, L = 200 feet, Y_0 = 1250 feet; W = 0 feet (---) and W = 1000 feet (———).

NUMERICAL RESULTS

All calculations have been made for the field point $\bar{r}_o$. Although this is not necessarily the location of the maximum amplification, it should be typical of the harbor response. Figures 3 and 4 show the transfer function, $\phi_{III}(\bar{r}_o,k)$, for the two model basins chosen. Of particular interest is the reduction of radiation losses from the basin into the open ocean by the increase of channel length W, thereby increasing the amplification factor and reducing the resonance frequency.

The next group of figures show the calculated basin response to a single-crest wave with the form of equation (12) and having various values of α, the time scale of the incident wave, and W, the length of the entrance channel. For those examples with a non-zero length channel, there will be a delay of $\tau = W/L$ between the time at which the incident wave reaches the interface between the channel and the open sea ($\tau = 0$) and the time at which the incident wave would reach the field point, $\bar{r}_o$.

Apart from this delay, which is small for these examples, the response to a very long wave, $\alpha = 200$, is essentially a simple doubling, as shown in Fig. 5 for both basins and all values of W considered (up to 250 feet for the B. P. model and 1000 feet for the D. B. model, i.e., approximately, one-half of the width of the basin). This implies that the incident wave is so long that it essentially ignores the presence of the harbor. For the B. P. model, a value for α of 200 represents a time scale of 2070 sec. or 6.4 times the lowest natural period.

As α decreases, i.e., the time scale of the incident wave is shortened, the harbor is left with a damped oscillatory response after the incident wave has passed. The period of this oscillation is that associated with the maximum amplification factor shown in Figs. 3 and 4 and not that of the lowest natural period of the closed basin. The rate of decay of these oscillations lessens with increasing W, while their period increases with increasing W. Figures 6 and 7 show the response for B. P. with $\alpha = 20$, i.e., an incident time scale of 73 sec. which is 1.6 times the lowest closed basin period, and W = 0 and 250 feet, respectively. The periods of the response in these two cases are approximately 110 and 156 sec., corresponding to wave numbers of 0.00205 and 0.00150.

These response periods are obtained from a numerical

calculation with a step size of 5.0 units for τ, which may represent a substantial fraction of the period involved, and for a behavior which is primarily but not completely sinusoidal, i.e., other periods do play a role, and therefore the values calculated for the response periods may not be particularly accurate. Nevertheless, the change from one group of periods to another is quite noticeable.

The response at $\alpha = 20$ for D. B., shown in Figs. 8 and 9, indicates the influence of higher harmonics, but is dominated by fundamental wave numbers 0.00033 and 0.00024 for W = 0 and 1000 feet, respectively, while the corresponding values for the wave number of maximum harmonic response are 0.00031 and 0.00024. Intermediate values of α and W show similar results.

Finally, the maximum transient amplification factor is found to remain in the neighborhood of 2.0 for a range of α from 10 to 200 for both basins and various values of W, as shown in Figure 10.

The final group of figures give the transient response of the basins to the decaying sinusoidal input of equation (16). Figures 11 through 16 show the response of the B. P. basin to three types of waves: a relatively short-period input ($\kappa_0 = 0.25$), which is approximately the lowest closed-basin natural period; a longer-period input (0.10), which is close to the period of maximum harmonic response for W = 0; and a wave of period (0.07), which gives (approximately) the maximum transient amplification (for W = 250). Two channel lengths are used: W = 0 and 250 feet.

Again the effect of increasing the entrance channel length from 0 to 250 feet is seen to yield a decrease in the rate of decay of the long-time oscillatory harbor response. (Intermediate values of W give intermediate responses). As κ_0 decreases for W = 250, the maximum transient response may no longer occur at the time at which the incident wave is maximum but can occur at later peaks, indicating a build-up of amplitude over several cycles rather than an immediate response in one cycle. This does not appear to be directly the case for W = 0 although, for some values of κ_0, the magnitude of the highest peak is very close to that of its neighbors.

In all cases there is an oscillatory behavior in the harbor after the incident wave has died out. The dominant period of this response appears to be related to those periods at which the harmonic response was maximum but matches the incident period for very long period waves, as shown in Figure 17.

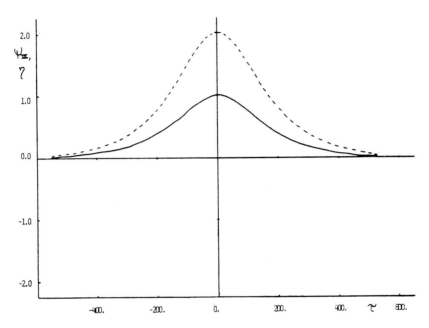

Fig. 5: Transient Response to Single-Crest Wave with $\alpha = 200$ and $W = 0$ for both B. P. and D. B. Models.

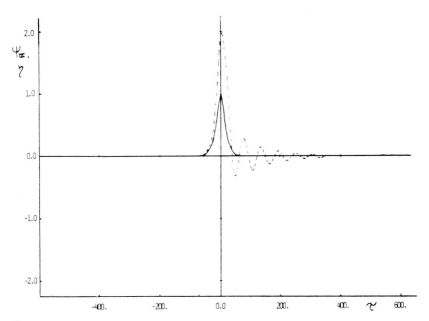

Fig. 6: Transient Response to Single-Crest Wave for B. P. with $\alpha = 20$ and $W = 0$.

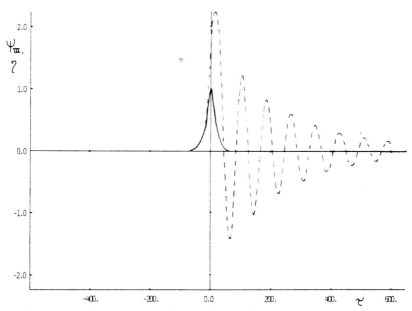

Fig. 7: Transient Response to Single-Crest Wave for B. P. with $\alpha = 20$ and $W = 250$.

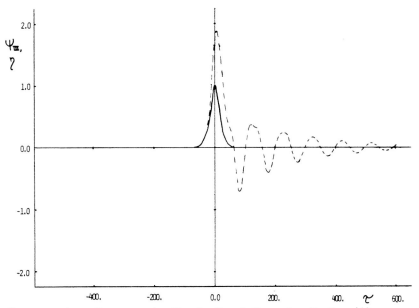

Fig. 8: Transient Response to Single-Crest Wave for D. B. with $\alpha = 20$ and $W = 0$.

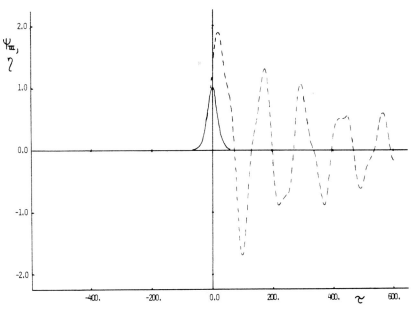

Fig. 9: Transient Response to Single-Crest Wave for D. B. with $\alpha = 20$ and $W = 1000$.

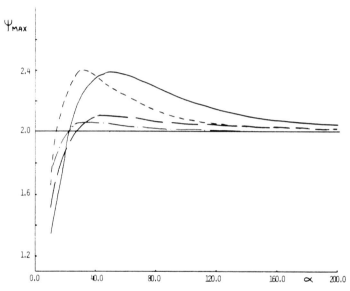

Fig. 10: Maximum Simplification Factor for Single-Crest Wave as a Function of α for B. P. with $W = 0$ (—·—), $W = 250$ (----), and D. B. with $W = 0$ (— —), $W = 1000$ (———).

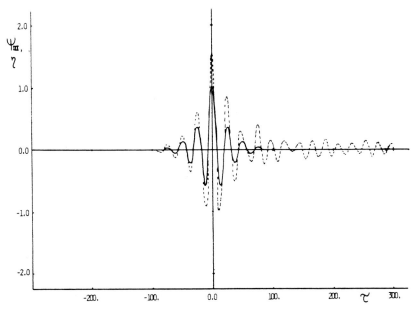

Fig. 11: Transient Response to Damped Sinusoidal Wave for B. P. with $\kappa_o = 0.25$ and $W = 0$.

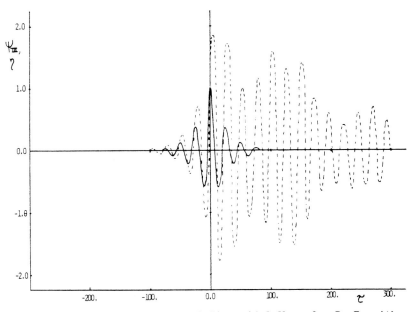

Fig. 12: Transient Response to Damped Sinusoidal Wave for B. P. with $\kappa_o = 0.25$ and $W = 250$.

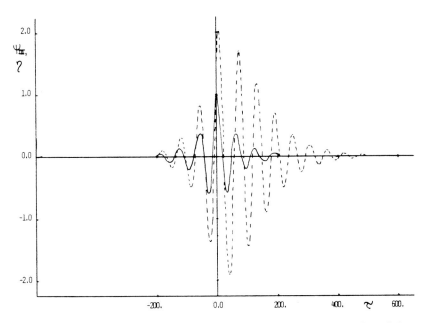

Fig. 13: Transient Response to Damped Sinusoidal Wave for B. P. with $\kappa_o = 0.10$ and $W = 0$.

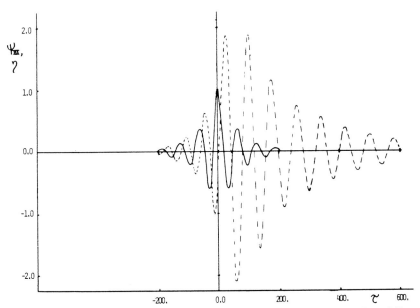

Fig. 14: Transient Response to Damped Sinusoidal Wave for B. P. with $\kappa_o = 0.10$ and $W = 250$.

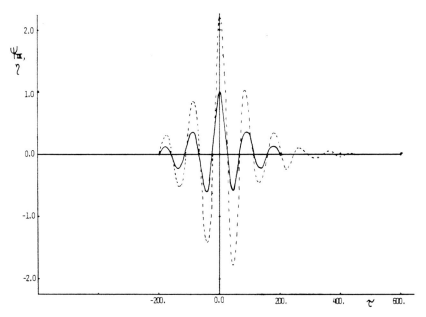

Fig. 15: Transient Response to Damped Sinusoidal Wave for B. P. with $\kappa_o = 0.07$ and $W = 0$.

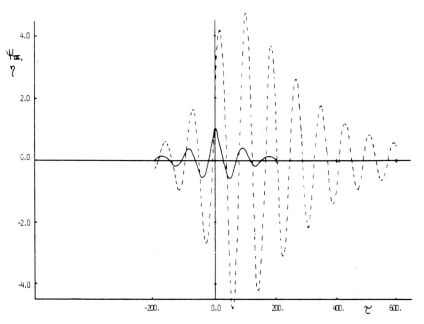

Fig. 16: Transient Response to Damped Sinusoidal Wave for B. P. with $\kappa_o = 0.07$ and $W = 250$.

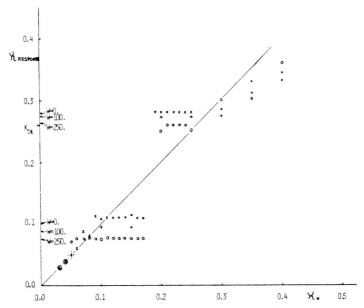

Fig. 17: Dominant Reduced Wave Number of Long-Time Response of B. P. with W = 0, (X) W = 100, (+) and W = 250, (O) to Damped Sinusoidal Waves.

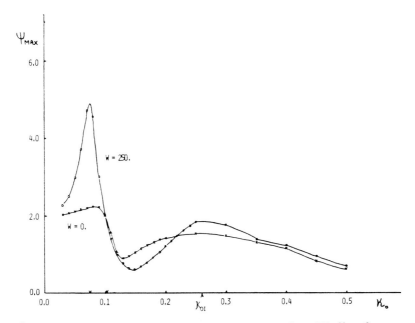

Fig. 18: Maximum Transient Amplification for B. P. with W = 0, (X) and 250, (O) for Damped Sinusoidal Waves.

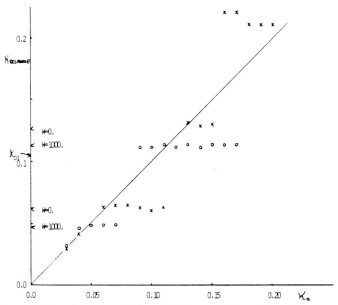

Fig. 19: Dominant Reduced Wave Number of Long-Time Response of D. B. with W = 0, (X) and W = 1000, (O) to Damped Sinusoidal Waves.

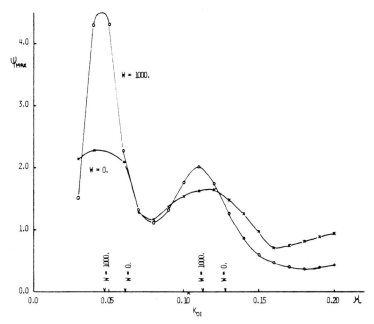

Fig. 20: Maximum Transient Amplification for D. B. with W = 0, (X) and W = 1000, (O) for Damped Sinusoidal Waves.

This behavior is further evidenced by the plot of the maximum transient amplification as a function of incident wave period shown in Figure 18. The local maxima occur very near to those values of κ_0 at which the harmonic amplification was maximum and the local minima occur near the zero's of the harmonic amplification response. The effects are more pronounced for the case of a non-zero length channel, e.g., W = 250, as would be expected from the decreased loss of energy due to radiation from the harbor entrance.

The results for the D. B. geometry are essentially the same as those for the B. P. geometry and are summarized in Fig. 19 showing the dominant period of the response as a function of κ_0, and Fig. 20, which shows the maximum transient amplification as a function of κ_0.

CONCLUSION

The transient response of harbors to long, aperiodic incident waves can be found based on a Fourier synthesis of solutions obtained for time-harmonic incident waves by a relatively simple, approximate procedure. For the two types of incident waves considered, a single crest and a decaying sinusoidal wave, the response of the harbor basins as calculated at one representative point was an initial amplification followed by a decaying oscillatory response which appears to consist of a dominant harmonic response with additional higher frequency terms.

Two particular observations can be made from these calculations. First, an increased entrance channel length acts to decrease the amount of energy lost per "cycle" and therefore to decrease the rate at which the harbor oscillations decay, to decrease the dominant frequency of this oscillatory response, and to magnify the basic harbor response, i.e., increase the maximum amplification. These conclusions would naturally be modified if frictional effects were included in the entrance channel.

Second, the long term response frequency of the harbor to the decaying sinusoidal incident wave appears to be "trapped" at those periods for which the harmonic response was maximum and to "jump" from one such group of periods to the next, corresponding to the next maximum, as the frequency of the decaying sinusoidal incident wave decreases.

Although the foregoing analysis is based on an approximate

theory, it seems advantageous to perform the relatively simple calculations to determine that range of incident frequencies at which the harbor response is predicted to be most severe and those frequencies at which the harbor is expected to oscillate after the incident wave has passed.

BIBLIOGRAPHY

Carrier, G. F., Shaw, R. P., and Miyata, M. The response of narrow-mouthed harbors to periodic incident waves. *Hawaii Institute of Geophysics Report* (in preparation).

27. An Inverse Tsunami Problem

R. O. REID and C. E. KNOWLES
Texas A & M University
College Station, Texas

ABSTRACT

The inverse problem investigated is that of attempting to estimate the deep-water input tsunami from recordings made at or near an island. This hinges on the ability to estimate the appropriate transfer function for the site in question for a given direction of input. A numerical integration of the wave equations is being investigated for the estimation of this transfer function. So far the method has been applied only to an island of simple geometry for which analytical solution of the scattered wave field is known. The test shows some promise for extension to islands of more realistic geometry.

INTRODUCTION

The problem investigated is that of estimating the deep-water signature of a tsunami based on an observed marigram in the immediate vicinity of an island. The basic assumption is made that the incident tsunami in deep water represents a plane wave but that its signature in time at a fixed point in deep water is unknown. This implies that the distance of the epicenter is large compared with the horizontal scale of the island at its base on the ocean floor. The observed tsunami signature near the island is, of course, distorted by scattering, refraction, diffraction, and possible resonance phenomenon created by trapping of wave energy by the island bathymetry. The present study is limited to the linear theory for long waves and accordingly its application requires that the observed water-level signatures be at locations where nonlinear effects and dispersion are minimal.

For a given direction of the input wave train in deep water and a given observation point P near the island, the solution of the problem as posed rests on the determination of the transfer function for the response at P. The transfer function is the ratio of the Fourier transform of the response to the Fourier transform of the input. In general it is a complex function of frequency. If it can be established from a known pair of input-output time sequences having a broad-band spectrum, then in principle one can estimate the deep-water input from other measured time sequences at the point P.

The transfer function at the selected point P of a given island and for given direction of the input plane waves can be estimated in at least three different ways:

(A) The transfer function can be estimated from field measurements in which both deep-water and island near-field data are obtained simultaneously for a given tsunami: this is the method referred to by Dr. Vitousek in another paper of this symposium.

(B) The transfer function can be estimated from laboratory model experiments with measurements analogous to those of (A), and for which the island bathymetry is duplicated in the model on an undistorted scale: this approach has been employed by Dr. Van Dorn in a previous unpublished study for Wake Island.

(C) The transfer function can be estimated from a numerical model in which the island bathymetry is represented on a discrete grid and the wave equations are solved numerically for given deep water input and the response at the point in question is obtained.

This paper is concerned with the latter approach. The study consists of several phases: the results of only the first few phases are reported on here. The other phases of the study are presently in progress. Specifically the different phases of the overall numerical study include:

(a) A test of the numerical model for islands of simple geometry for which analytical solutions are known and for which the input wave train is simple harmonic.

(b) A test of the numerical model for simple islands as in (a) but for an input sequence of long duration which possesses a broad band spectrum with respect to frequency.

(c) A test of the ability to reproduce the known input function from the output for the conditions of phase (a) and (b).

(d) Application of the numerical method to actual island bathymetry (such as that of Wake Island) to determine the transfer function for a given point P and given input wave direction.

(e) Application of the transfer function as evaluated in (d) to estimate the deep-water signature of a tsunami from actual recorded data at P for the island in question.

THEORY

Consider an island of given bathymetry within a circle of radius R_m. Outside of this circle it is assumed that the depth is constant (deep-water far field with respect to the island). Let X(t) represent the deep-water time sequence as might be measured at some point Q just beyond the circle of radius R_m for a tsunami arriving from a stipulated epicenter. The epicenter is considered far enough away from the island that the deep-water incident wave train can be considered as a plane wave of quasi-permanent form (at least in respect to its propagation in deep water past the outskirts of the island in question). Let Y(t) represent the response to the input wave at point P near the island. This is illustrated schematically in Figure 1.

Under the restriction of linear response at point P and the assumption of a plane progressive wave input at Q, the response Y can be expressed as an appropriate convolution of X or vice versa, i.e.,

$$Y(t) = \int_{-\infty}^{\infty} K(\lambda) \, X(t - \lambda) \, d\lambda \qquad (1)$$

or

$$X(t) = \int_{-\infty}^{\infty} G(\lambda) \, Y(t - \lambda) \, d\lambda \qquad (2)$$

where the integrations extend nominally over all λ. Actually the kernel function $K(\lambda)$ should vanish for $T < 0$, but this

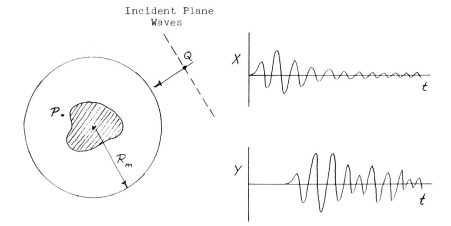

Fig. 1: Schematic Diagram of Plane Wave Incident on Cylindrical Island. Input is Trace x; Response is Trace y.

is not necessarily true of $G(\lambda)$. Both kernel functions depend upon the particular island bathymetry, the location of P, and the direction of the wave propagation, but not on the input function $X(t)$. Thus if $G(\lambda)$ can be estimated based on a known input and output information, then (2) could be used to estimate unknown input sequences X based on measurements at P for the same direction of wave propagation.

The Fourier transform of (1) yields

$$F_y(f) = R(f)\, F_x(f) \qquad (3)$$

where $F_x(f)$ and $F_y(f)$ are the Fourier transforms of $X(t)$ and $Y(t)$, respectively; and $R(f)$ is the Fourier transform of $K(t)$, or $R(f)^{-1}$ is the transform of $G(t)$. To be more specific:

$$F_x(f) = \int_{-\infty}^{\infty} X(t)\, e^{i2\pi ft}\, dt \qquad (4)$$

with similar relations for $F_y(f)$ and $R(f)$. In general, F_x, F_y, and R will be complex. If $X(t)$ and $Y(t)$ represent known input and output functions, then (3) can be employed to estimate the transfer function $R(f)$. Moreover $G(t)$ can be estimated as the inverse transform of $R(f)^{-1}$:

$$G(t) = \int_{-\infty}^{\infty} R(f)^{-1} e^{-i2\pi ft} df. \tag{5}$$

Clearly in application of these relations $G(t)$ will be only an estimate since the original sequences $X(t)$ and $Y(t)$ are generally discrete sequences at time step Δt and are of limited duration. The discrete time step imposes an upper limit in the frequency domain $(f_N = 1/2 t)$ and the finite duration of the record limits the resolution of the estimate of $R(f)$.

In the special case where the input $X(t)$ is simple harmonic, as can be realized in the numerical model, then the output will be simple harmonic - at least after the initial transient effects have subsided. In this case, the response R for the particular frequency of the input can be obtained without the necessity of Fourier analysis. However, this procedure does require many inputs at different frequencies in order to obtain adequate resolution of R versus f. Another possibility is to employ as input a stationary time sequence of long duration having a wide-band spectrum in terms of frequency range. In this case, the Fourier procedure can be used to estimate $R(f)$ for the range of frequencies involved in the input.

APPLICATION

So far application has been made only for simple island configurations for which the transfer function can be determined analytically. The purpose of these studies is to test the ability of the numerical solution of the wave equation to reproduce the analytical response for a broad range of frequencies. Also a numerical approximation of the convolution (2) is tested to see how well the known input record can be recovered from the output.

The results for the case of a simple circular cylinder are discussed here. The transfer function at two points on the cylinder for given direction of input is estimated by application of a numerical integration of the long wave equations with appropriate boundary conditions at the island and in the far field of the island. The numerical wave problem is discussed elsewhere by Vastano and Reid (1967). Moreover, the analytical solution for this case is contained in the above reference. The two points considered are the "wave side" and the "lee side" positions on a diameter parallel to the input wave propagation direction. For this case the response is influenced only by scattering at the cylinder and diffraction around the cylinder. There is no possibility of trapped waves, as is possible when bathymetry is included (Longuet-Higgins, 1967). As a consequence, the amplitude response $R(f)$ at any point on the cylinder is a monotonic function of f with no sharp resonant peaks.

The conditions used in the numerical solution of the wave equation are given in Table 1. The resolution in time for the input and output functions is 4 seconds. The input $X(t)$ was taken as monochromatic, and the wave program was run for the six different periods indicated in Table 2. The non-dimensional frequency T indicated in the table is employed in the analytical solution for the response (Vastano and Reid, 1967). From the output $Y(t)$ on the wave side ($0°$ azimuth) and lee side ($180°$ azimuth), the relative amplitude A and relative phase Φ can be evaluated by direct comparison with the input. Parameters A and Φ are the magnitude and phase of the complex response R; i.e.,

$$R = A\, e^{i\Phi} . \tag{6}$$

The phase is relative to the incident wave at the $90°$ azimuth position far from the cylinder. The results derived in this way are compared with the exact values derived from the analytical solution for both the wave side and lee side positions.

Table 1: Cylindrical Island Grid.

$D = 4$ km.	$r_o = 19$ km. (radius of cylinder)
$\Delta r = 1$ km.	$\Delta\theta = 11\ 1/4°$ (16 angular increments in $180°$)
$\Delta t = 4$ sec.	100 radial increments

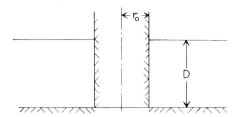

Table 2: Monochromatic Incident Wave -- Comparison of Numerical and Analytical Results

	Relative Amplitude						
	Period (T) (min.)	2	4	6	8	10	12
	Freq. (T) (non-dim.)	5	2.5	1.67	1.25	1.0	0.83
(0°)	Numerical	1.93	1.89	1.84	1.68	1.71	1.70
	Analytical	1.967	1.89	1.83	1.76	1.71	1.68
(180°)	Numerical	0.38	0.65	0.76	.85	.88	.92
	Analytical	0.483	0.67	0.78	.85	.89	.93

	Relative Phase Lag (degrees)						
	Period (T) (min.)	2	4	6	8	10	12
	Freq. (T) (non-dim.)	5	2.5	1.67	1.25	1.0	0.83
(0°)	Numerical	-298	-150	-108	-87	-72	-62
	Analytical	-291	-152	-107	-86	-69	-65
(180°)	Numerical	516	264	184	135	113	96
	Analytical	505	265	184	142	114	95

$$r_o = 19 \text{ km.}$$

$$c = \sqrt{gD} = 198 \text{ m./sec.}$$

$$T = \frac{2\pi f\, r_o}{c} = \frac{10}{T \text{ (min.)}}$$

Table 3: A Measure of Reproduction of Input from the Lee Side Output for the Cylindrical Island.

Period (T, mins.)	Standard Error of Estimate
2	0.313
4	0.126
6	0.124
8	0.129
10	0.154
12	0.145

The agreement for both A and Φ is seen to be reasonably good for every period on the wave side. A significant departure occurs in the amplitude on the lee side for the 2-minute period. Apparently the spatial resolution in the numerical model is not sufficient for proper rendition of diffraction into the shadow zone for periods of two minutes or smaller.

As a test of the ability to reproduce the input function from the output, the lee side response for the cylinder was employed to estimate $G(t)$ for discrete steps Δt based upon a numerical quadrature version of (5). In estimating $G(t)$ it was necessary to apply a window function to R^{-1} in the frequency domain such that R^{-1} is reduced to essentially zero at the maximum frequency f_N. The application of (2) with the estimates of the input function having standard error of estimate as indicated in Table 3. These are to be compared with the unit amplitude of the actual input functions.

Except for the lowest period input, the standard error of estimate of input based on the output is of the order of 13 per cent of the input amplitude for these experiments.

Part of the error of estimate is due to the window function that was applied to R^{-1} in estimating $G(t)$. This tends to give estimated values of $X(t)$ which are of smaller amplitude. A correction can be applied for this kind of error and would considerably improve the situation for the 2-minute period. Hopefully, improvement in the resolution of the derived transfer function R and its transform would further reduce the standard error of estimation for the input.

BIBLIOGRAPHY

Longuet-Higgins, M. S. 1967. On the trapping of wave energy round islands. *J. Fluid Mech. 29*, pp. 781-821.

Vastano, A. C. and Reid, R. O. 1967. Tsunami response for islands: verification of a numerical procedure. *Journal of Marine Science 25*, pp. 129-139.

28. Experimental Investigations of Wave Run-up under the Influence of Local Geometry

LI-SAN HWANG and ALBERT C. LIN
Tetra Tech, Inc.
Pasadena, California

ABSTRACT

This paper presents the results of three sets of experiments investigating the effect of local geometry on wave run-up. The first set was designed to investigate the effect of harbor oscillation--under periodic wave excitation--on run-up; the second investigated the effect of harbor oscillation--under nonperiodic wave-train excitation--on run-up; finally, the third set investigated run-up in a three-dimensional bay. A detailed discussion of the influence of wave period and geometric location on wave run-up is presented, and explanations are given for the scattering of results of field observation of tsunami run-up along a coast.

INTRODUCTION

Wave run-up on a straight beach has received considerable attention in the past years. Many theories have been developed (Miche, 1944; Keller and Keller, 1964 and 1965; Carrier, 1966; LeMéhauté and Hwang, 1967; and Hwang, et al., 1969) and verified experimentally to be reasonably satisfactory. However, field observations often give large deviations in run-up results. Theories developed for a straight shoreline with a uniformly sloping beach have not yet given satisfactory predictions.

Natural shorelines are seldom straight and uniform in slope; they usually contain small bays, curved beaches, canyons, and shelves. Such features cause wave reflection, diffraction, and refraction, resulting in large wave run-up at certain locations for some incident waves, or the reverse. In particular, when the period of the incoming waves matches the fundamental period or harmonics of the bay and refraction effects concentrate the waves at a certain location, very large run-up may be observed.

Figure 1 shows 254 run-up observations at the Hawaiian Islands from two Aleutian tsunamis (Van Dorn, 1965). The run-up heights vary by a factor of 10 from point to point along the coast in any particular region. Figure 2 shows run-up of the Sanriku tsunami (Horikawa, 1965). Such scattering may be partly due to local variation in bottom slope; however, a large portion may be due to variation of the local geometry other than beach slope.

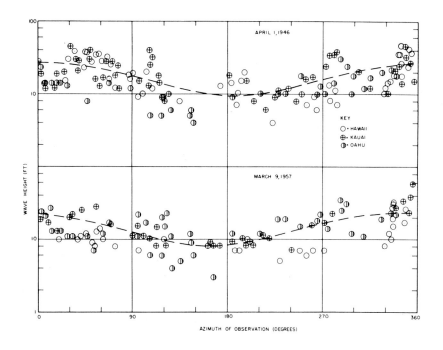

Fig. 1: Azimuthal Distribution of 254 Run-up Observations in the Hawaiian Islands for Two Large Aleutian Tsunamis. 0° azimuth corresponds to incident wave direction from the respective epicenters (from Van Dorn, 1965).

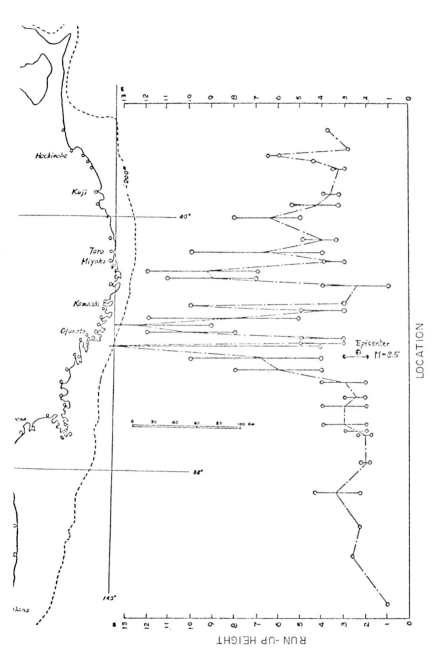

Fig. 2: Run-up Heights above Mean Water Level of Sanriku Tsunami, 1963 (K. Horikawa, 1965).

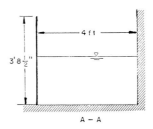

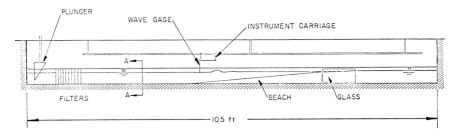

Fig. 3: Diagram of the Tetra-Tech Wave Tank.

To better understand the problem of tsunami run-up in a bay, this paper presents the results of experimental studies performed in the laboratory. The first set of experiments was designed to investigate the effect of rectangular-harbor oscillation (under periodic wave excitation on run-up. The second investigated the effect of rectangular-harbor oscillation (under nonperiodic wave-train excitation) on run-up, and the last set of experiments was performed to investigate run-up in a three-dimensional bay.

EXPERIMENTAL APPARATUS AND ARRANGEMENT

Wave Tank and Generator

The experiments were performed in the wave tank shown in Figure 3. The tank is 105 feet long, has a cross section approximately four-feet square, and includes a plate-glass window in the test section. The wave generator is of the versatile plunger type. A variable-speed electric motor operates through a clutch to drive the plunger, and the rotary motion is converted by an adjacent eccentric and an overhead linkage into a uniform, sinusoidal vertical displacement.

For generation of periodic waves, a plunger of right-triangular cross section slices into the water, producing periodic waves at its front inclined surface (hypotenuse) and creating negligible disturbance toward the rear since the rear surface is vertical. The wave period can be varied (by a motor speed control) through a range of approximately one to twelve seconds. The height can be varied from zero to several inches by adjustment of both the eccentric and the mean plunger immersion. Near the generator, a group of four-foot square, ten inch thick wave filters serves to absorb high-frequency components and noise from the waves. In addition, the filters function to attenuate reflections from the beach.

Nonperiodic wave trains are produced by a simple modification of the plunger and procedure. The triangular plunger is replaced by one of half-parabola cross section. Instead of periodic immersion and withdrawal, the plunger motion is controlled by the clutch between the motor and the eccentric to impact the water surface and suddenly stop. Alternative motions, such as initial immersion followed by sudden withdrawal or a cycle of first immersion and then withdrawal and so forth, may be used. These impulsive motions produce a variety of dispersive wave trains. The power transmitted to the waves, hence their maximum heights, can be varied by adjustment of both the speed of impact and the maximum depth of plunger immersion.

Wave Gauges and Recorder

The waves are measured and recorded by auxiliary equipment. An overhead carriage runs the length of the tank and is accurately positionable by means of a scale along its track and a brake. This carriage is used to support a point gauge mounted on a vernier, which permits resolution to one-tenth millimeter. In addition, the overhead carriage supports a resistance gauge which is also mounted on a vernier. This gauge consists of two stainless steel wires 0.01 inch in diameter, approximately 18 inches long, and spaced 3/16 inch apart. A high-frequency excitation voltage supplied by an external signal generator is impressed across these wires. Immersion in a conducting solution (the tank water) produces a voltage drop across the system that is proportional to the depth of immersion. This output is supplied to a two-channel Sanborn recorder, which continually monitors the surface elevation of the passing waves. In addition, an internal reference time marker permits accurate determination of the wave period.

Wave height calibration is achieved by immersing (and withdrawing) the gauge in still water by known increments with the vernier.

Experimental Arrangement

Of the three sets of experiments performed, the first two were performed in a rectangular model harbor; a detailed drawing is shown in Figure 4. The harbor size was adjusted for changing the characteristics of the harbor response. The walls on the two sides of the harbor were fixed and vertical. However, the wall at the back of the harbor was adjustable and was used to change the slope. At the entrance, an adjustable vertical wingwall was installed to permit change of the opening size.

The wave heights were measured with resistance-type wave gauges, as has been discussed in the previous subsection. The run-up on the back wall was also measured with resistance-type wave gauges, consisting of two parallel wires 1/32 inch in diameter and 1/2 inch apart, embedded in the wall.

The third set of experiments was performed in an idealized three-dimensional bay as shown in Figure 5. The half-bay was S-shaped with a sloping beach; the beach slope was 1/5, and everywhere perpendicular to the local shoreline. This arrangement, with a convex shoreline at the entrance and a concave region at the rear of the harbor, was adjacent to the tank wall and, therefore, represents, by symmetry, half of a bay with general features similar to many natural bays. With such an arrangement, refractive effects were introduced with convergence of orthogonals at the entrance and divergence at the rear; in this way, general three-dimensional effects could be observed.

EXPERIMENTAL STUDIES

Run-up from Periodic Wave Excitation

A large number of tests were run to measure the run-up of periodic waves along the beach installed at the back of the harbor; incoming wave period and height were varied. The experiment proceeded by setting the period and then running through a range of wave heights. For each wave period, several runs were made with both breaking and nonbreaking waves. In each run, the incident wave height was determined by taking the

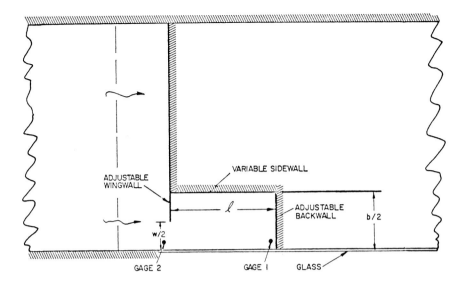

Fig. 4: Detailed Drawing of the Model Harbor.

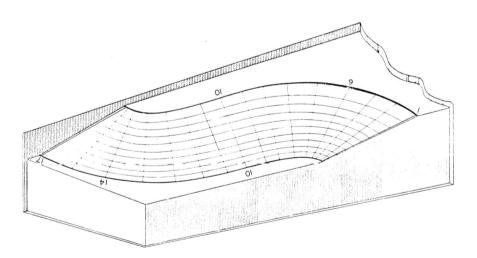

Fig. 5: Drawing of an Idealized Three-Dimensional Bay. This bay has a convex shoreline at the entrance and a concave region at the rear.

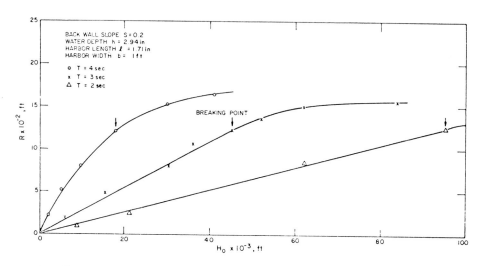

Fig. 6: Wave Run-up Inside a Harbor.

average of the heights at a node H_N and antinode H_A by

$$H_i = \frac{H_N + H_A}{2}$$

where H_N and H_A are taken at points located in the constant depth region of the tank to avoid any shoaling influence.

Figure 6 shows the results of three sets of experiments performed for different wave periods. The first group was performed with an incoming wave period of four seconds. The curve was obtained by changing the incoming wave height, as indicated in the figure. As the incoming wave height increases, the run-up increases until breaking occurs; the rate of increase of run-up is reduced considerably after breaking. The second group of experiments was performed with a period of three seconds. The slope of the curve in the nonbreaking region is less than in the first set; this is due to the fact that the harbor response factor for a four-second period is larger than for a period of three seconds. It is interesting to point out that the run-up of nonbreaking waves on a straight beach does not depend on the frequency of the incoming wave, while the run-up inside the harbor does depend on the period of the incoming wave. This fact has been pointed out by LeMéhauté, et al. (1968) in an analysis of run-up in a rectangular harbor.

For a small harbor, beaches may constitute a significant part of the harbor. These beaches will certainly modify the characteristics of harbor oscillation. Thus, the method proposed by LeMéhauté (1968) for calculating wave run-up inside the harbor -- multiplication of the response factor obtained for constant depth by the run-up evaluated as if the harbor were not present -- may be subject to considerable error. An accurate evaluation of the run-up has to consider the beach as a part of the harbor, solving the run-up and response of the harbor as a whole.

Run-up from Nonperiodic Wave-train Excitation

This group of experiments was also performed in the rectangular harbor. However, the waves were generated by an impulsive motion of the generator, forming a wave train. Table 1 lists the conditions under which the experiments were performed. Figures 7 and 8 show two samples of experimental results.

In each experiment, two measurements of wave height were performed: one for incoming waves and the other for the run-up inside the harbor. The incoming wave train was determined by taking half of the wave height measured at the harbor entrance when the harbor was closed. The reason for doing this can be clearly seen in Figure 7. In this figure, two wave records were taken at a model harbor entrance with conditions as shown in the upper left-hand corner of the figure. The first, with the entrance closed, is smooth and represents the incident system (one half of the measured height). The second, measured at the same point but with the entrance open, shows the strong interference arising between incident and reflected waves.

Figure 7 shows a distinct pickup of wave run-up at certain periods. This phenomenon is associated with harbor oscillation. Table 2 lists the comparison of the results of run-up of the maximum wave obtained from experiments and results obtained by use of Keller's formula (based on a straight shoreline):

$$\frac{R}{H} = \sqrt{\frac{\pi}{2\alpha}} \left\{ \tanh\frac{2\pi d}{L} \left(1 + \frac{4\pi d/L}{\sinh 4\pi d/L} \right) \right\}^{1/2}$$

where

    R = wave run-up   H = wave height at reference depth
    α = beach angle   d = reference water depth
                            L = wave length at reference depth

Table 1: Summary of Experimental Results on Harbor Oscillation Due to Dispersive Waves.

Run No.	Harbor Length ℓ in.	Harbor Width b in.	Harbor Opening Width d in.	Water Depth D ft.	Water Depth at Shelf h in.	Back Wall Slope S	Incoming Max.Wave Height at Station 62 H_m^*, cm	Max.Wave Height at Station 62 H_m^{**}, cm	Max. Wave Run-up R_m, cm	R_m/H_m
I - 1	7 1/8	6		2' 1"	3	0.4	1.40	1.20	5.60	4.00
I - 2	7 1/8	6		2' 1"	3	0.4	0.875	0.72	3.20	3.68
II - 1	6	6		2' 1"	3	Vertical	1.65	1.40	6.00	3.64
II - 2	6	6		2' 1"	3	Vertical	1.05	1.00	3.00	2.86
III - 1	12	24		2' 4 1/8"	6 1/8	Vertical	1.07	1.30	2.00	1.87
III - 2	12	24		2' 4 1/8"	6 1/8	Vertical	0.65	0.72	1.50	2.31
IV - 1	12	24	8	2' 4 1/8"	6 1/8	Vertical	1.05	1.00	1.28	1.24
IV - 2	12	24	8	2' 4 1/8"	6 1/8	Vertical	0.50	0.42	0.90	1.80
V - 1	18	24		2' 1 1/8"	3 1/8	Vertical	1.50	0.93	2.40	1.60
V - 2	18	24		2' 1 1/8"	3 1/8	Vertical	1.05	0.80	2.20	2.48
VI - 1	20 1/2	24		2' 15/16"	2 15/16	0.2	0.82	0.80	2.20	2.68
VI - 2	20 1/2	24		2' 15/16"	2 15/16	0.2	1.00	0.85	2.60	2.60
VII - 1	29 1/4	24	8	2'2 13/16"	4 13/16	0.2	1.10	0.80	2.36	2.14
VII - 2	29 1/4	24	8	2'2 13/16"	4 13/16	0.2	0.70	0.50	2.20	3.15
VIII - 1	20 3/8	24	8	2' 1"	3	0.2	1.10	0.72	1.38	1.25
VIII - 2	20 3/8	24	8	2' 1"	3	0.2	0.80	0.54	1.14	1.40

* Harbor Gate Closed ** Harbor Gate Open

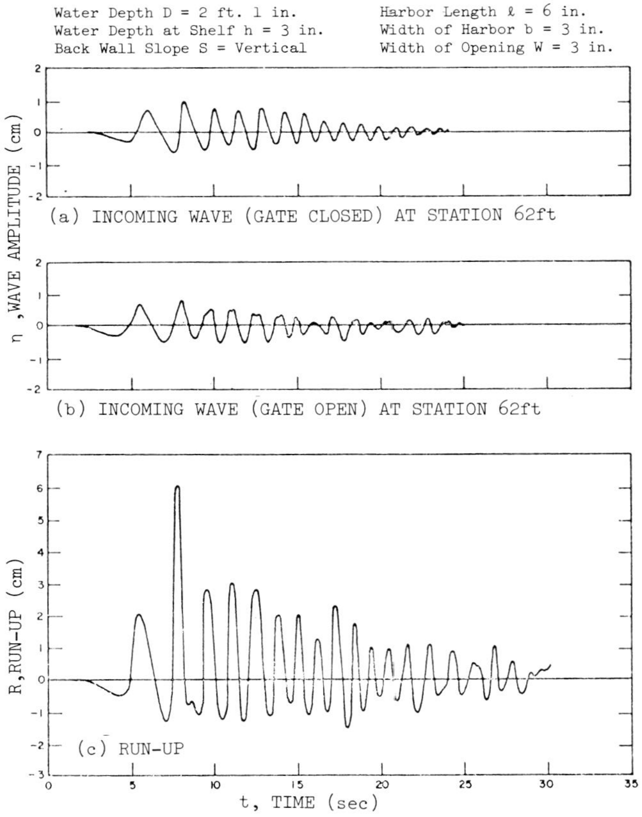

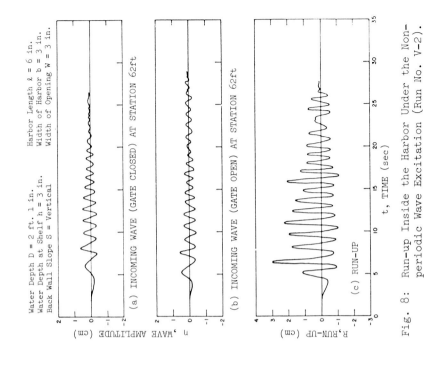

Fig. 7: Run-up Inside the Harbor Under the Non-periodic Wave Excitation (Run No. II-1).

Fig. 8: Run-up Inside the Harbor Under the Non-periodic Wave Excitation (Run No. V-2).

Table 2: A Comparison of Run-Up Results Obtained from Experiments and Keller's Formula.

Run No.	Period* (T sec.)	Max. Wave Height H_m cm	Max. Run-Up R_m cm	** Calculated $(R/H)_m$	Measured R_m/H_m
I-1	1.5	1.4	5.60	2.34	4.0
I-2	2.0	0.875	3.20	2.36	3.68
II-1	1.7	1.65	6.0	1.1	3.64
II-2	2.0	1.05	3.0	1.075	2.86
III-1	2.0	1.07	2.0	1.09	1.87
III-2	1.8	0.65	1.5	1.1	2.31
IV-1	3.3	1.05	1.28	1.01	1.24
IV-2	2.0	0.50	0.90	1.09	1.80
V-1	2.5	1.50	2.4	1.02	1.6
V-2	1.7	1.05	2.2	1.1	2.48
VI-1	1.7	0.82	2.2	3.1	2.68
VI-2	1.8	1.00	2.6	3.06	2.6
VII-1	2.5	1.0	2.36	2.9	2.14
VII-2	1.8	0.7	2.20	3.19	3.15
IIX-1	2.0	1.1	1.38	3.01	1.25
IIX-2	1.7	0.8	1.14	3.1	1.4

* Period used to calculate the wave length, L.
** Based on Keller's formula

Table 3: Experiments Performed in a Three-Dimensional Bay.

*Run No.	Period (T sec.)	Water Depth at shelf (d in.)	Incident Wave Height (H in.)	Water Depth in deep water (D in.)
R- I-1	2.0	2	9/16	2'
R- -2	4.1	2	1/4	2'
R- -3	6.2	2	1/4	2'
R- -4	8.0	2	1/4	2'
R- II-1	0.7	4	3/16	2'2"
R- -2	1.3	4	1/4	2'2"
R- -3	2.0	4	7/16	2'2"
R- -4	3.1	4	1/2	2'2"
R- -5	4.1	4	9/16	2'2"
R- -6	5.2	4	5/8	2'2"
R- -7	6.2	4	3/8	2'2"
R- -8	7.0	4	3/8	2'2"
R- -9	8.0	4	1/2	2'2"
R-III-1	2.0	6	3/8	2'4"
R- -2	4.1	6	5/8	2'4"
R- -3	6.2	6	11/16	2'4"
R- -4	8.0	6	13/16	2'4"

* All are Non-Breaking Waves

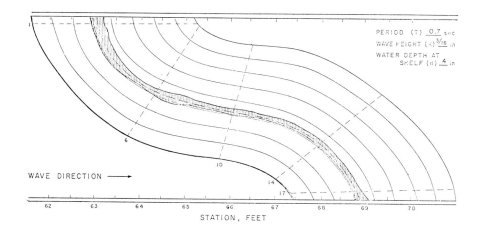

Fig. 9: Wave Run-up and Run-down on a Three-Dimensional Bay (Run No. II-1).

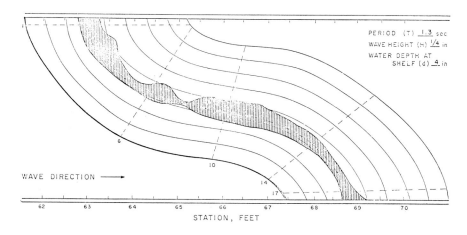

Fig. 10: Wave Run-up and Run-down on a Three-Dimensional Bay (Run No. II-2).

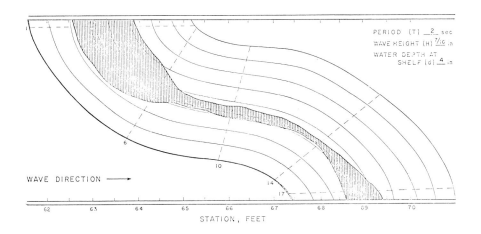

Fig. 11: Wave Run-up and Run-down on a Three-Dimensional Bay (Run No. II-3).

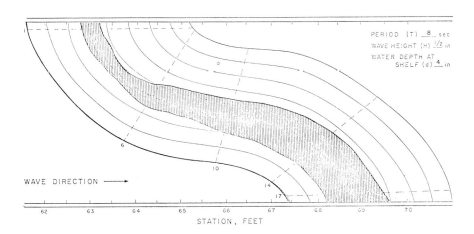

Fig. 12: Wave Run-up and Run-down on a Three-Dimensional Bay (Run No. II-9).

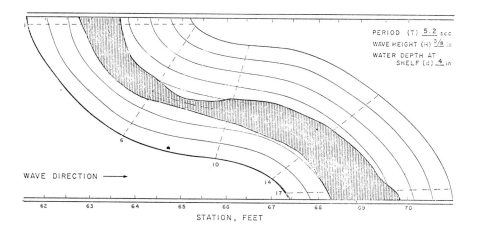

Fig. 13: Wave Run-up and Run-down on a Three-Dimensional Bay (Run No. II-6).

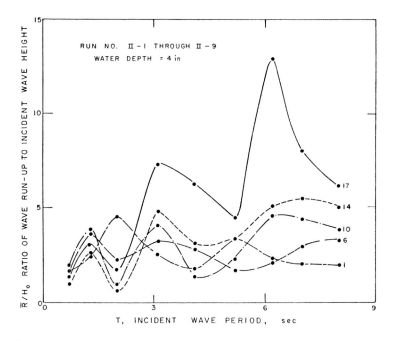

Fig. 14: The Effect of Wave Period on Wave Run-up on a Three-Dimensional Bay. The numbers on the right-hand side key the data to the locations shown in Figure 13.

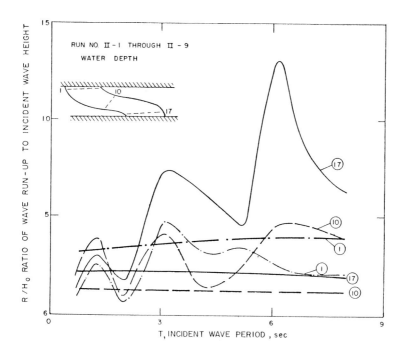

Fig. 15: A Comparison of Experimental and Calculated Results at Stations 1, 10, and 17. (* Experimental)

The wave period used to calculate the wave length L was taken from measurements. In most cases, the measured relative wave run-up results were considerably larger than those calculated (Table 2). It may be considered that the larger value of run-up is introduced by harbor oscillation.

Wave Run-up on an Idealized Three-Dimensional Bay

In this set of experiments, the incoming wave height was obtained as outlined previously. The run-up (and run-down) was measured by marking the maximum run-up (and run-down) on the beach at many locations to permit the drawing of as complete a run-up picture as possible. For example, in Fig. 9, the shaded area above the dotted line represents the run-up and below represents the run-down.

Three series of experiments were performed for three different water depths. The conditions for which the experiments were performed are listed in Table 3. The run-up and run-down results are given in Figures 9 to 13. From these figures, a large variation in run-up from one location to another can be seen; in some locations, there is hardly any run-up. Such variation of run-up results from wave refraction and bay oscillation.

From the figures, it can be concluded that the run-up appears to be larger inside the bay when the incident wave is long, and the run-up is larger at the entrance when the wave is short. In general, the run-up distribution tends to be more uniform for longer incident waves, while, for short waves, the run-up distribution tends to be very nonuniform.

The relative run-up, as a function of wave period, for the Run II series is plotted in Figure 14. The numbers on the right side of the figure indicate the location where the measurements were performed. The relative run-up tends to be much larger at longer periods. Figure 15 shows a comparison of experimental and calculated results from Miche's formula with correction for refraction:

$$\frac{R}{H_o} = K \sqrt{\frac{\pi}{2\alpha}}$$

where K is the refraction coefficient. These values are considerably smaller than those obtained from the measurements for long-period waves. Thus, it is clear that, when the shoreline is not straight, one has to take the bay oscillation into account in the calculation of wave run-up.

Figure 14 indicates that the relative run-up at different locations (with the same beach slope) may vary from a value of less than 0.7 to a value of 13, a factor of 20, while the variation of run-up at the same location for different periods may amount to a factor of 7. Such large variation is in agreement with field observations (Iida, 1961; Horikawa, 1965; and Van Dorn, 1965).

ACKNOWLEDGEMENT

This work was sponsored by the Atomic Energy Commission under contract AT(26-1)-289, M002.

BIBLIOGRAPHY

Carrier, G. F. 1966. Gravity waves on water of variable depth. *Journal of Fluid Mechanics*.

Horikawa, K. 1965. Distribution of tsunami run-up height along coast. Presented at *U.S.-Japan Cooperative Scientific Research Seminar on Tsunami Run-up*, Tokyo, Japan.

Hwang, Li-San, Fersht, S. and LeMéhauté, B. 1969. Transformation and run-up of tsunami-type wave trains on a sloping beach. Presented at *The 13th International Congress of the International Association of Hydraulic Research*, Tokyo, Japan.

Iida, K. and Yutka, Ohta 1961. On the height of the Chilean Tsunami on the Pacific coasts of central Japan and the effect of coasts on the tsunami; particularly on the comparison between tsunami and those that accompanied the Tonankai and Nankaido Earthquakes. *Report on the Chilean Tsunami, Field Investigation Committee for the Chilean Tsunami*.

Keller, J. B. and Keller, H. B. 1964. Water wave run-up on a beach. *Service Bureau Corporation Research Report, Vol. I*. Prepared for the Office of Naval Research under Contract Nonr 3828(00).

Keller, J. B. and Keller, H. B. 1965. Water wave run-up on a beach. *Service Bureau Corporation Research Report, Vol. II*. Prepared for the Office of Naval Research under Contract Nonr 3828(00).

LeMéhauté, B. and Hwang, Li-San 1967. Run-up of non-breaking waves. *Tetra Tech, Inc. Report TC-103*. Prepared for the Atomic Energy Commission under Contract AT(26-1)-289.

LeMéhauté, B., Koh, R., and Hwang, Li-San 1968. A synthesis on wave run-up. *Journal of the Waterways and Harbors Division, Proceedings of American Society of Civil Engineers*.

Miche, M. 1944. Mouvements ondulatoires de la mer en profondeur constante au décroissante. *Annals des ponts et chausées*, January to August.

Van Dorn, W. G. 1965. Tsunamis. *Advances in Hydroscience, Vol. III*. Academic Press, New York.

29. A Model Study of Wave Run-up at San Diego, California

R. W. WHALIN, D. R. BUCCI, and J. N. STRANGE
*U. S. Army Engineer Waterways Experiment Station
Corps of Engineers
Vicksburg, Mississippi*

ABSTRACT

Wave intrusion into San Diego, California (U.S.A.) harbor was investigated for prototype wave periods ranging from 40 to 186 sec. and deep-water wave heights varying from 12 to 45 ft. The wave heights and periods tested were chosen so as to be representative of intermediate period waves that might be associated with an impulsive source such as an offshore localized seismic disturbance, an explosion, a massive slide, or a meteoritic impact. Such waves are of considerable interest since they lie between wind-wave and tsunami-wave spectra.

The San Diego, California harbor model is distorted 5:1. It was thus necessary to perform two-dimensional flume run-up tests in order to extend Saville's run-up curves to waves of limit steepness and to aid in developing a methodology for adjusting wave periods in the model to obviate distortion effects.

The narrow, low-lying Silver Strand, the city of Coronado, California, and portions of North Island were completely inundated for most conditions tested. Wave heights in the restricted harbor entrance approached 26 ft., indicating the curtailment of ship traffic.

An extreme circulation pattern resulted from the large volume of water crossing the Silver Strand into the inner harbor. The return flow through the navigation channel measured as high as 10 fps. Oscillations, reflections, rip currents, and bores were observed in the smaller partially enclosed basins, bays, and creeks surrounding the harbor.

INTRODUCTION

Background

The destructive capabilities of long-period, low-amplitude, tsunami waves resulting from seismic disturbances have long been noted with apprehension. The setup, setdown and run-up of intermediate period water waves which could be produced by landslide, volcano eruptions, small earthquakes, explosions, or meteor impacts are of considerable interest since they fall between values associated with the wind-wave and the tsunami-wave spectra.

Objectives

The primary objective of this study was to assess the likelihood of intermediate-period wave intrusion into San Diego harbor. The point of wave-breaking is related to water depth, and the physical characteristics of waves and run-up are primarily a function of wave characteristics and beach slope. Therefore, a methodology was developed for conducting valid run-up tests in a distorted model.

Approach

The San Diego harbor model (Fig. 1) was utilized during early 1969 for limited wave intrusion tests. This harbor has significant commercial interests, is not protected by a broad continental shelf, and contains no wave reflector other than the narrow sand spit (Silver Strand). During previous tests, the model had been adjusted to reproduce prototype tidal fluctuations and appropriate inner-harbor circulations. The model is not altogether suitable for run-up measurements because of the low-lying Silver Strand; however, it is valuable for inferring inundation zones and reproducing the extreme circulation pattern resulting from the mass transport of water into the inner harbor and its return through the narrow neck at Point Loma.

The availability of the model and its capability for inferring long-period, wave intrusion damage at nominal cost justified the study. Also, the acquisition of run-up data in the wave flume extends the reliable prediction of wave run-up to limit steepness waves.

WAVE FLUME TESTS

Experimental Facilities

Two-dimensional run-up tests were conducted in a glass-walled wave flume about 97 ft. long, 1.5 ft. deep, and 1.0 ft. wide (Figure 2). The purpose of the initial tests was to extend Saville's run-up curves (Saville, 1956) for values of H'_o/T^2 greater than 0.4 for various beach slopes. A one-sided converging section was designed (Fig. 2) which narrowed the flume width from 1 ft. to 6 in., thereby increasing wave heights by a factor of $\sqrt{2}$.

Fig. 1: San Diego Bay, California Model -- Scale: 1:100 Vertical, 1:500 Horizontal. General View of Model with Existing Conditions. South San Diego Bay in Foreground, Point Loma in Background.

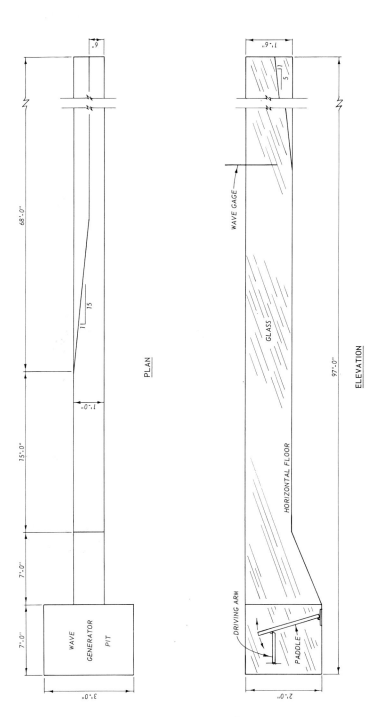

Fig. 2: Details of Wave Flume.

A wave gauge (printed-circuit type connected to a brush recorder) at the toe of the slope was used for wave-height measurements. Wave run-up was measured visually along a horizontal plane and converted to vertical run-up. The run-up was somewhat irregular and was measured after approximately four to six waves had arrived (thus a stable condition of setup resulted) and before reflections could return. Wave heights measured at the toe of the slope were converted to deep-water wave heights by the use of Wiegel's tables (Wiegel, 1954).

Similitude between the model and prototype was obtained by preserving geometric similarity and by adjusting hydraulic quantities in accordance with the Froudian relation, which assumes gravity as the dominant force.

Calibration

Calibration tests with the convergence section installed were conducted with water depths of 6 in. and 1.0 ft. for wave periods of 0.7 and 2.0 sec. (which encompassed the operational limits of the generator). These data simplified in a logarithmic plot (Fig. 3), making possible the setting of specific values for the testing program. All measured wave heights were corrected by linear shoaling theory to determine the deep-water wave height. This correction was small (about 2.4 percent) for 0.7 sec. waves in water 1.0 ft. deep.

Data

Tests were conducted on six different slopes. The slopes were 1:10, 1:20, and 1:30 and were divisible by a distortion factor of 5, i.e., 1:2, 1:4, and 1:6 because, in a distorted model, the bottom slopes are steepened. The data acquired are sufficient to extend Saville's curves of relative run-up as a function of deep-water wave steepness and beach slope. The composite run-up curves with the added H_o'/T^2 isolines of 0.5 and 0.6 are shown in Figure 4. The relative run-up for a limit-steepness wave (isoline for $H_o'/T^2 = 0.72$) is dashed to indicate its uncertainty.

METHODOLOGY FOR CONDUCTION RUN-UP TESTS IN A DISTORTED MODEL

Hypothesis

Since relative run-up data are now available for waves of steepnesses (H_o'/T^2) varying from 0.01 to 0.72 (limit steepness wave) a methodology can be proposed for conducting valid tests of wave run-up in distorted models. The following analysis assumes that the refraction coefficient is 1.0; that is, the orthogonals to the wave front are normal to the bottom contours. If run-up tests are to be conducted in a distorted model and if the wave height will be scaled in accordance with the vertical scale of the model, then a pertinent wave period must evolve which will produce the same relative run-up (R/H_o'). By determining the average beach slope and consulting Figure 4, the relative run-up can be obtained for any selected deep-water wave steepness.

EXAMPLE

Given: Average beach slope inside the surf zone, 1:20
Deep-water wave height, H_o', 10.0 ft.
Wave period (T) = 10.0 sec.; $H_o'/T^2 = 0.1$
Distortion factor 2:1

Find: Wave period in the distorted model (1 on 10 slope). From Fig. 4 the relative run-up on a 1:20 slope for a deep-water wave steepness of 0.1 H_o'/T^2 is $R/H_o' = 0.355$. Since the model is distorted 2:1, the wave steepness required to produce the identical run-up ($R/H_o' = 0.355$) on a 1:10 slope is $H_o'/T^2 = 0.43$ (Figure 4). Therefore, if the vertical scale of the model is 1:100, the model period should be

$$T = \left[(H_o')/(H_o'/T^2)_{\text{distorted}} \right]^{1/2} = 0.48 \text{ sec.}$$

Distortion Curve

In a similar manner, curves can be constructed to illustrate the proper deep-water wave steepness to use in the distorted model to produce the same run-up as would obtain in an undistorted model. Figures 5, 6, 7, and 8 depict the result of this analysis for distortion factors ranging from 2:1 to 5:1.

Applicability of Distortion Curves

Care must be exercised in conducting run-up tests in distorted models to assure that the results are reliable. The period which must be used in the distorted model is a strong function of the beach slope. The fact that waves have a limit steepness precludes the testing of run-up for a range of wave steepnesses depending on the distortion factor. The larger the distortion factor the more limited the range of wave steepnesses which can be tested.

Additional problems are encountered when the beach slope is steep (although, in reality, this is encountered only in rare instances). In applying this analysis a range may be reached where a rapid change in H_o'/T^2 produces a small change in R/H_o' and it would be difficult to construct curves similar to Figs. 5 through 8 accurately. This range is represented by the steeply descending portions of the curves. However, since R/H_o' changes slowly in this range, the run-up will probably be reliable even if some error is incurred in selecting the proper wave period.

Additional Problems Encountered
in Three-Dimensional Distorted Models

The primary problem encountered in performing run-up tests in three-dimensional hydraulic models is the need to compensate for wave refraction effects. As previously noted, if the wave orthogonals are normal to the bottom contours, the refraction coefficient will be 1.0 regardless of any differences in period between the distorted and the undistorted models. However, this will not be the case in most instances. The refraction coefficient should be plotted at some point near the shoreline for both the prototype wave being tested and the wave used in the distorted model to produce the correct run-up. If the difference in the two is uniform along the shoreline ($\pm$ some constant percentage), then it appears feasible to adjust the wave height accordingly in the distorted model to compensate for this difference. Without some experimentation and trial and error in both an undistorted and a distorted model, it is practically impossible to select the proper wave height because of the complex dependence of run-up on wave height and direction. In general, for a given wave period, beach slope, and angle of incidence, the run-up increases as the deep-water wave height increases, reaches a maximum and then may decrease or remain relatively constant. The method developed for performing run-up studies in distorted models is limited to those in which the refraction coefficient remains the same (within $\pm 10\%$) for both the prototype and the model waves being considered.

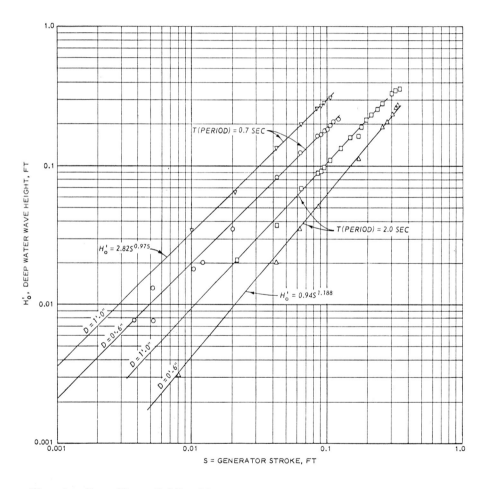

Fig. 3: Wave-Flume Calibration.

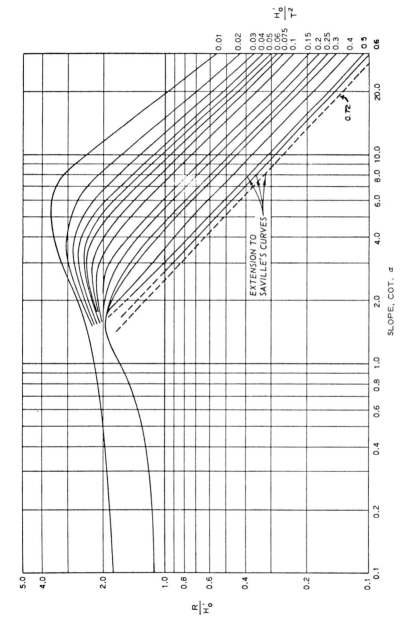

Fig. 4: Composite plot of Relative Run-up as a Function of Deep-water Wave Steepness and Beach Slope.

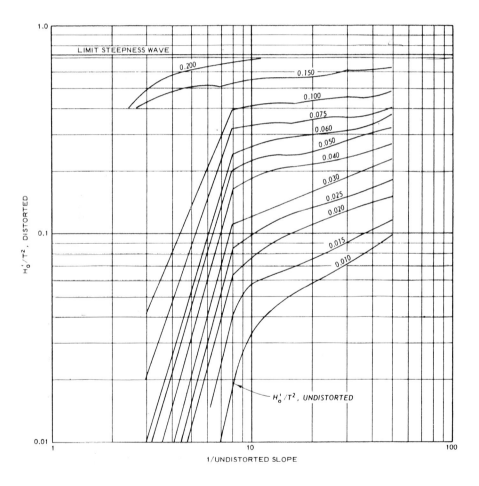

Fig. 5: Wave Steepness in Distorted Model as a Function of Undistorted Wave Steepness and Beach Slope: 2:1 Distortion.

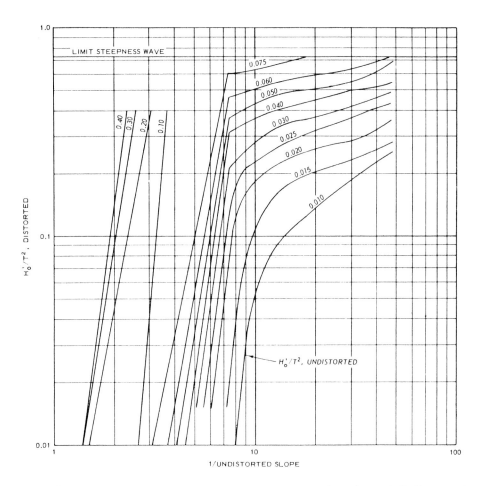

Fig. 6: Wave Steepness in Distorted Model as a Function of Undistorted Wave Steepness and Beach Slope: 3:1 Distortion.

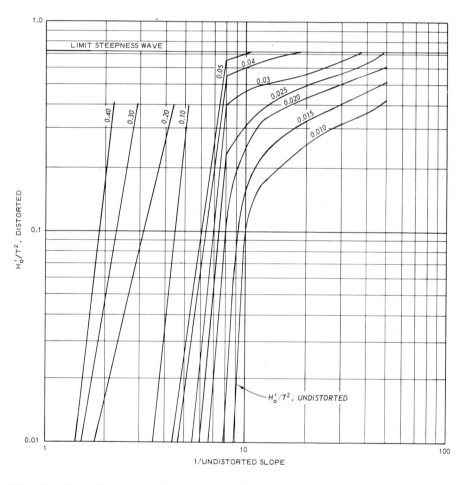

Fig. 7: Wave Steepness in Distorted Model as a Function of Undistorted Wave Steepness and Beach Slope: 4:1 Distortion.

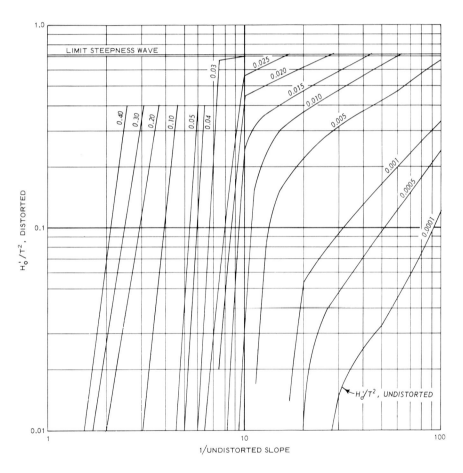

Fig. 8: Wave Steepness in Distorted Model as a Function of Undistorted Wave Steepness and Beach Slope: 5:1 Distortion.

EXPERIMENTAL PROCEDURES -- SAN DIEGO, CALIFORNIA, MODEL

Description of Model

The 5:1 distorted model (Fig. 1) was constructed to a horizontal scale of 1:500 and a vertical scale of 1:100 and reproduced the outer harbor to the 100-ft. depth contour, the Silver Strand, North Island, Point Loma, and a portion of the shore to either side of San Diego, California. Model topography was reproduced to the limits shown in Figure 9. Distorted models are designed so that the velocity scale is equal to the square root of the vertical scale, and the time scale is equal to the horizontal scale divided by the square root of the vertical scale. General scale relations for transferring model data to prototype equivalents are presented in Table 1.

Instrumentation

The wave generator (Fig. 9) was 40 ft. long and was positioned for wave intrusion from the S60°W. direction. A hydrographic map check of offshore bathymetry and islands indicated this to be an optimum placement for impulsively-generated wave intrusion tests. A sloping, 1-ft.-high float actuated by vertical push-pull pistons having a maximum stroke of 9.25 in. comprised the wave generating mechanism that produced periodic waves. Wave heights were measured by printed-circuit wave rods connected to a Brush recorder. The wave generator and the recorder were started simultaneously with the model pool still at +6.3 ft. mllw. Certain elements of the wave data were recorded photographically.

Intermediate period waves, by definition, fall between those of wind waves (<20 sec.) and those of tsunamis (>10 min.) and thus well below those of tidal cycles ($\approx$12 hr.). It was decided that the wave intrusion tests would encompass periods varying from 40 to 186 seconds. Wave rods were placed as shown in Figure 9. Wave rod No. 2 at the 75-ft.-depth contour (water depth 75 + 6.3 = 81.3 ft.) was utilized as a calibration standard.

Wave Intrusion Test Results

The model was not intended as an example to verify the run-up method proposed herein. Most of the waves generated in the study over-topped the low-lying Silver Strand (Fig. 9) and spilled into the inner harbor.

Wave gauges were installed at sixteen locations both inside and outside the harbor (Figure 9). Gauges 1, 2, and 3 provided data on offshore wave heights; gauges 4, 5, 6, 9, and 12 measured wave heights immediately offshore of the Silver Strand, City of Coronado, and North Island while wave rods 13 and 14 monitored wave heights in the critical narrow navigation entrance. Wave rods 7, 10, and 11 rendered wave heights in the inner harbor while wave rods 15 and 16 were used to study the bores that propagated up two small creeks which enter the inner harbor. Wave rods 6 and 8 were located to either side of Coronado, California.

Table 1: General Scale Relations for Transferring Model Data to Prototype Equivalents*

Dimensions	Scale Relations	
	General	Numeric
Length, horizontal	L_r	1:500
Length, vertical	Y_r	1:100
Time	$L_r(Y_r)^{-1/2}$	1:50
Velocity	$Y_r^{1/2}$	1:10
Weight	$(L_r)^2 Y_r$	25,000,000

*Notation used in this paper:

    d   Water depth, ft.
    H   Wave height, ft.
    H_o' Deep-water wave height, ft.
    L   Wave length, ft.
    L_o Deep-water wave length, ft.
  mllw. Mean lower low water, +6.3 ft.
    R   Vertical wave run-up, ft.
    S   Generator stroke, ft.
    T   Wave period, sec.
    ∅   Slope of beach

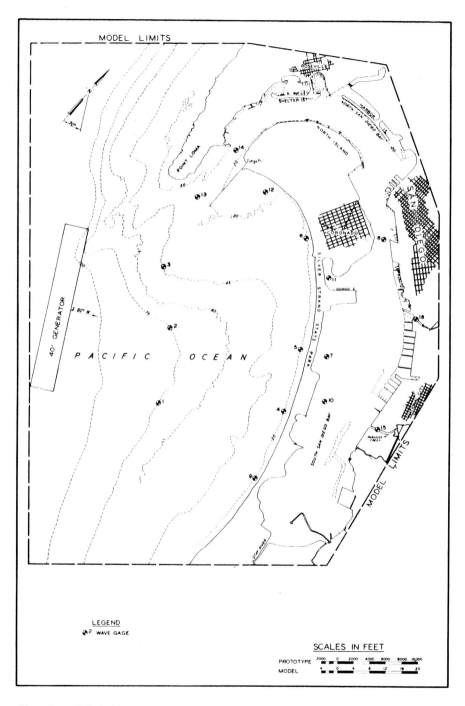

Fig. 9: Model Layout for San Diego Harbor, California.

Fig. 10: Wave Intrusion from S60°W into San Diego Bay (Wave Period 65 seconds; Wave Height at 81.3-ft. Depth, 56 feet; Still-Water Level +6.3 ft. mean low low water).

Fig. 11: Wave Intrusion from S60°W into San Diego Bay (Wave Period 85 seconds; Wave Height at 81.3-ft. Depth, 53 feet; Still-Water Level +6.3 ft. mean low low water).

Fig. 12: Wave Intrusion from S60°W into San Diego Bay (Wave Period 180 seconds; Wave Height at 81.3-ft. Depth, 33 feet; Still-Water Level +6.3 ft. mean low low water).

Fig. 13: Wave Intrusion from S60°W into Coronado (Wave Period 65 seconds; Wave Height at 81.3-ft. Depth, 56 feet; Still-Water Level +6.3 ft. mean low low water).

Fig. 14: Wave Intrusion from S60°W into Coronado (Wave Period 85 seconds; Wave Height at 81.3-ft. Depth, 53 feet; Still-Water Level +6.3 ft. mean low low water).

Fig. 15: Wave Intrusion from S60°W into Coronado (Wave Period 180 seconds; Wave Height at 81.3 ft. Depth, 33 feet; Still-Water Level +6.3 ft. mean low low water).

Fig. 16: Wave Intrusion from S60°W into the Entrance to San Diego Bay (Wave Period 65 seconds; Wave Height at 81.3-ft. Depth, 56 feet; Still-Water Level +6.3 ft. mean low low water).

Fig. 17: Wave Intrusion for S60°W into the Entrance to San Diego Bay (Wave Period 85 seconds; Wave Height at 81.3-ft. Depth, 53 feet; Still-Water Level +6.3 ft. mean low low water).

Fig. 18: Wave Intrusion from S60°W into the Entrance to San Diego Bay (Wave Period 180 seconds; Wave Height at 81.3-ft. Depth, 33 feet; Still-Water Level +6.3 ft. mean low low water).

Wave heights at the aforementioned sixteen sites for wave periods varying from 40 to 186 sec. are listed in Tables 2 and 3. The curves drawn in Fig. 8 were used to obtain the prototype value of H_o'/T^2 and consequently T from the model data. In the offshore area (wave rods 1, 2, and 3) wave heights (approaching 45 ft.) were far too severe to allow safe movement of even large craft. The waves in the navigation entrance (rods 15 and 16) indicate wave heights in excess of 26 and 13 feet, respectively; then it is unlikely that navigation in either direction would be possible without great hazard. Wave heights within the inner harbor (rods 7, 8, 10, and 11) approaching 10.0 ft. for the conditions tested can be expected. Wave heights in excess of 30 ft. were obtained immediately in front of the low-lying city of Coronado, California (wave rod 6). Docked craft in the inner harbor would be subjected to extreme buffeting. Generally, tranquil surface conditions resulted in the inner navigation channel (vicinity wave rod 8); however, the extreme circulation pattern developed (described later) should preclude any movement of small craft in this region. Large craft already under power, could possibly find a temporary haven to the lee of North Island and Coronado, California.

Table 2: Wave Heights off San Diego, California

Wave Generator Stroke in.	Model Period T, sec	Prototype Period T, sec Adjusted	Prototype H_o', ft	Wave Heights (H) in ft, Trough to Crest, at Indicated Wave Rod							
				75-ft Contour		45-ft Contour	20-ft Contour			Inside Harbor	
				Gage 1	Gage 2	Gage 3	Gage 4	Gage 5	Gage 6	Gage 7	Gage 8
3.00	0.53	39.7	19.7	20.6	22.4	19.0	14.9	20.5	14.5	0.9	0.9
	0.69	57.7	27.2	26.1	33.8	24.1	23.7	23.7	18.4	1.8	0.8
	0.79	76.8	21.7	20.3	20.1	21.6	21.8	19.0	18.5	1.8	0.5
	0.88	91.7	17.9	16.4	14.9	16.5	23.2	10.7	12.3	1.3	0.4
	1.00	117.0	17.7	16.1	17.3	16.4	20.0	13.3	14.2	2.0	0.9
	1.36	186.0	11.7	11.1	13.1	20.7	10.0	8.7	14.0	1.6	0.6
5.00	0.69	51.3	34.5	33.1	32.7	28.2	27.4	23.8	16.4	1.5	0.9
	0.79	66.7	30.3	28.3	34.4	33.4	22.1	22.2	20.5	3.2	1.0
	0.88	78.6	33.8	31.0	31.9	31.2	25.1	31.1	16.5	4.9	1.0
	1.00	97.5	25.5	23.3	27.7	29.1	26.8	25.4	17.0	4.0	1.0
	1.36	185.0	18.0	17.0	20.1	23.7	21.2	15.4	23.0	2.5	1.0
7.00	0.79	63.5	31.4	29.3	40.0	27.6	18.1	25.0	19.3	2.7	0.9
	0.88	71.8	37.0	34.0	49.0	33.6	19.8	26.6	18.4	3.2	1.0
	1.00	96.7	32.5	29.7	34.2	32.6	23.5	33.5	15.8	4.7	1.0
	1.36	180.0	22.9	21.7	24.3	22.6	24.2	24.6	25.6	6.0	1.3
9.25	0.88	69.2	44.1	40.5	51.2	29.3	24.3	23.6	25.3	3.8	1.6
	1.00	86.3	45.4	40.4	48.5	36.1	26.6	28.5	18.7	4.5	1.2
	1.36	165.0	32.8	31.1	31.0	34.9	30.3	29.3	22.8	6.8	1.6

Note: Data obtained with +6.3 ft mllw still-water level in model.

Table 3: Wave Heights off San Diego, California

Wave Heights (H) in ft, Trough to Crest, at Indicated Wave Rod

Wave Generator Stroke in.	Model Period T, sec	Prototype Period T, sec Adjusted	Prototype H_o', ft	20-ft Contour Gage 9	Inside Harbor Gage 10	Inside Harbor Gage 11	10-ft Contour Gage 12	Channel Entrance Gage 13	Channel Entrance Gage 14	In Creeks Gage 15	In Creeks Gage 16
3.00	0.53	39.7	19.7	19.8	0	0	8.8	9.6	6.6	0.5	0.5
	0.69	57.7	27.2	19.6	0.7	1.2	10.1	7.5	3.7	0.7	1.1
	0.79	76.8	21.7	8.5	0.8	1.6	5.1	7.4	2.9	0.6	1.2
	0.88	91.7	17.9	6.4	0.8	2.4	7.4	8.1	4.4	0.7	1.5
	1.00	117.0	17.7	6.3	1.5	2.9	4.7	7.2	4.8	0.8	1.6
	1.36	186.0	11.7	7.7	0.9	2.3	4.6	3.0	3.3	0.8	1.5
5.00	0.69	51.3	34.5	21.8	0	0.9	8.2	15.5	6.3	0.4	0.9
	0.79	66.7	30.3	19.5	1.2	2.0	9.5	13.5	11.7	0.7	1.7
	0.88	78.6	33.8	16.9	1.7	3.6	9.8	13.2	7.0	1.1	2.0
	1.00	97.5	25.5	10.3	1.9	6.3	9.6	11.8	6.6	1.4	3.5
	1.36	185.0	18.0	12.6	3.2	4.9	9.0	6.4	4.7	2.1	4.1
7.00	0.79	63.5	31.4	20.0	1.5	2.4	8.0	19.3	7.0	0.6	1.8
	0.88	71.8	37.0	23.4	1.6	3.7	13.1	23.0	8.0	1.0	2.8
	1.00	96.7	32.5	25.5	2.2	5.7	11.3	23.0	6.5	1.6	3.8
	1.36	180.0	22.9	16.4	4.3	9.0	10.1	11.6	4.8	2.9	6.0
9.25	0.88	69.2	44.1	25.2	2.2	6.0	14.3	26.3	13.3	0.8	2.7
	1.00	86.3	45.4	36.5	2.4	5.8	12.1	24.2	9.4	1.4	5.6
	1.36	165.0	32.8	21.3	4.4	9.4	12.7	15.6	6.9	1.9	6.2

Note: Data obtained with +6.3 ft mllw still-water level in model.

The relative severity of offshore wave heights for waves of medium and high steepness is shown in Figs. 10, 11, and 12. Note the reformed waves, reflections, diffraction, etc., occurring in this area. The setup which occurs from the mass transport of water in the surf zone increases water depths at the shore of the Silver Strand and results in a considerable amount of water entering the bay. Wave intrusion into Coronado, California, for the same wave periods and amplitudes as before (T = 65, 85, and 180 sec., H_o' = 56, 53, and 33 ft., respectively) is depicted in Figures 13, 14, and 15. The vulnerability of this city to inundation is very evident. Not much of North Island was subjected to wave run-up, generally to a maximum contour elevation of +15 ft. mllw (Figures 16, 17, and 18). Note the extreme waves in the entrance channel which have a maximum amplitude of about 26 feet.

After the propagation of 15 to 20 waves over the Silver Strand (Figs. 10, 11, and 12), a sufficient amount of water was deposited in the inner harbor to build up a hydraulic head. This buildup in head resulted in a flow through the navigation entrance. The velocity resulting was checked at gauge 8 (Fig. 9) in the narrow neck and the surface velocities approached 10.0 fps. Such an extreme velocity in an enclosed area results in eddies and vortices near the sides of the channel with consequent disturbance to both docked ships and small ships in transit.

The bore in the small creek (gauge 18, Fig. 9) was approximately 10 ft. high and propagated around a 45-degree bend in the creek with little attenuation. In fact, oscillations were observed in all the smaller partially enclosed basins, bays, and creeks surrounding the harbor.

CONCLUSIONS

Vulnerability of San Diego Harbor
to Wave Intrusion

Impulsively generated waves emanating from the deep ocean could result in waves, in the form of wave trains, intruding into harbor areas for some time. In the first several wave trains, it is conceivable that several hundred waves would be encompassed in each envelope. Therefore, a spectrum of waves with steepnesses approximating those resulting from model tests would be obtained. Since a finite number of waves of any particular steepness would occur, that portion of the wave train may be considered periodic. Thus, model data presented herein should be a valid quantitative indication of the severity of the problem.

A methodology has been proposed for conducting reliable wave run-up tests in a distorted model as long as the refraction coefficients for the prototype and model wave periods are negligibly different. The method utilizes the run-up curves developed by Saville (1956) and additional data obtained for large steepness waves in this study (H_o'/T^2 = 0.5, 0.6, and 0.72).

The wave periods utilized herein (general model) were determined from empirically constructed curves where the distorted deep-water wave steepness is given as a function of the prototype deep-water wave steepness and beach slope. The model wave height was determined by the vertical scale factor.

Extensive inundation of the Silver Strand, the city of Coronado, and part of North Island will result from wave intrusion similar to that reproduced herein. Offshore wave heights approaching 60 ft. in height can be anticipated in the surf zone. Wave heights of 26 ft. in the entrance to the harbor can be expected and wave heights within the harbor as high as 10 ft. can be expected. An extreme circulation pattern resulting from flow crossing the Silver Strand into the inner harbor builds up a head which causes a return flow of 10 fps through the navigation channel to the sea. This velocity is indicative of a velocity head (drawdown) approaching 1.6 feet.

It is unlikely that any craft could survive in the surf zone, and navigation during the wave intrusion would be impossible. A fairly tranquil zone exists in the navigation channel immediately behind North Island and Coronado in which even anchored small craft might ride out the wave environment.

Sufficient tests were not conducted to allow extrapolation of results. Therefore, the conclusions reached above are based on model results and observations.

BIBLIOGRAPHY

Saville, T., Jr. April 1956. Wave run-up on shore structures. *Journal of the Waterways and Harbors Division*, Vol. 82, No. WWZ, Paper 925, American Society of Civil Engineers, Ann Arbor, Michigan.

Wiegel, R. L. February 1954. Gravity waves, tables of functions. *Council on Wave Research, Engineering Foundation*. Berkeley, California.

30. The Numerical Simulation of Long Water Waves: Progress on Two Fronts

R. L. STREET and R. K. C. CHAN
Stanford University
Stanford, California

J. E. FROMM
IBM Research Division
San Jose, California

ABSTRACT

   Two numerical, finite-difference models have been developed and programmed for the study of long water waves, with particular application to aspects of tsunami propagation. The study has been confined to plane flows and the numerical results have been verified by comparison with experiments. Specific objectives of the work were the exact simulation of finite-amplitude waves near the shore and simulation, based on approximate equations, of long waves in moderately shallow water, e.g., on the continental slope and shelf. In the latter simulation, the effect of the bottom hydrography on waves is of prime interest. In the exact simulation, the results include the details of the wave motions such as breaking inception, pressure distributions, and water particle motions.

   The results presented illustrate the agreement between experiments and simulations, as well as potential application of the simulation methods.

INTRODUCTION

   The motion of large, long waves on and near the shore is a vital consideration in the planning, design, and protection of near-shore submerged structures, surf-zone facilities, waterfront buildings, and other equipment including harbors and vessels. In addition, the near-shore character of the large, long waves generated by storm or seismic disturbance is determined in large measure by the bottom hydrography of the continental slope and shelf near the concerned shore area. The importance of such considerations has been recently documented by Wilson and Tórum (1968), while the mathematical problems have been delineated by Mei and LeMéhauté (1966) and LeMéhauté, Koh, and Hwang (1968).

A review of the literature of coastal engineering (see, e.g., the Proceedings of the Tenth Conference on Coastal Engineering, 1966) indicates that the action of water waves on the continental slope and shelf and in the near-shore shoaling region or the surf zone is known only in terms of the results of approximate theories and of experiments and model tests on specific configurations (LeMéhauté, et al., 1968, and Van Dorn, 1966). In addition, the difficulties of making experimental measurements under transient conditions prevent more than a superficial knowledge of the total wave action. The consequences of the true or nonlinear properties of propagating and deforming water waves are neither fully understood nor accounted for in design practice. A study by Vanoni and Raichlen (1966) offers an excellent example.

Following preparation of a preliminary design for the nuclear plant island site, Vanoni and Raichlen ran a series of tests in the laboratory on a model of the proposed breakwater-berm cross-section for the island. This cross-section had been designed in accordance with the usual design wave, breaking, and run-up criteria. However, the tests showed that, as has been predicted by Camfield and Street (1968), the large, long design waves for this case grow much larger than expected before breaking and run up much farther. The Vanoni and Raichlen report indicated that the proposed design was inadequate. A major factor here was the general lack of knowledge and insight on nonlinear effects in the propagation of unsteady, finite-amplitude waves.

Similarly, Wilson and Tórum (1968) state, with regard to a city on the northern California coast, "We conclude then that Crescent City's susceptibility to large wave response from major tsunamis is, by its very name, related to its crescent-shaped coast and bowl-shaped continental shelf. Because of its dimensions, it will forever be a responsive echo-chamber for great tsunamis since their periods will be always capable of exciting full or partial resonance." While this is a tentative conclusion, to which not all subscribe, the fact remains that Crescent City is damaged by major tsunamis. It was apparent at the March 1969 Earthquake Engineering Conference at Berkeley, where tsunamis were discussed, that the nonlinear and bottom hydrography effects on tsunami propagation are not fully understood.

Of course, much evidence exists as to the importance of finite amplitude in the behavior of water waves. Dean (1965) gave an excellent theoretical method for dealing with steady-state nonlinear ocean waves. Similarly, Monkmeyer and Kutzback (1965) gave a higher-order theory for deep water waves. However, when one examines the nonlinear theory available for propagating and deforming waves, he finds that serious limitations, e.g., horizontal velocity uniform at any cross-section and pressure distribution hydrostatic, have been introduced in order to achieve a tractable mathematical model (see, e.g., Amein, 1964, Stoker, 1957; Peregrine, 1966 and 1967; and Carrier, 1966).

There is a need for comprehensive tools to provide both insight into and practical answers for the deformation of and forces caused by finite-amplitude water waves in the nearshore zone and for the character of waves as influenced or determined by continental slope and shelf hydrography.

SIMULATION AND EXPERIMENT

There are coherent, mathematically complete, and computationally feasible analytical methods currently available that can be modified and implemented as the needed tools in water-wave analysis. These methods can be developed to provide accurate, numerical simulation of a broad spectrum of significant water-wave problems. This development should provide not only the design answers sought, but also insight into a range of physical processes extant in water-wave mechanics. The use of these methods on simple problems to establish the credibility of the numerical schemes is reported here. These methods are based upon representation of the motion of incompressible fluids in terms of the Navier—Stokes or Euler equations in Eulerian coordinates. The flow boundary conditions are derived from physical requirements and the governing equations at the boundaries. The mathematical models thus obtained are then transformed to numerical, finite-difference models for the purposes of computation.

The numerical simulation can be likened to a physical experiment and is correctly called a series of numerical experiments. The essential difference between numerical and physical experiments is the character of the results obtained. In a physical experiment, the number and type of measurements that can be made are limited by instrumentation capabilities, interference by measurement devices in the flow, and the cost of sophisticated equipment. For example, it is difficult, if not impossible, to get a complete velocity map of the flow field as a function of time (cf., Miller and Zeigler, 1964, in which only the horizontal component was measured). On the other hand, from one of the simulation methods used here, one can obtain as either printed or graphical output the fluid velocity, pressure and acceleration fields, the free-surface configuration, and plots of the individual water-particle motions at every time step. Thus, an entirely different approach is necessary, in lieu of one point of achieving one type of data record; the numerical experiment should be designed so that a maximum significant use can be made of all the data obtainable. Simulation of generally useful configurations, as well as specific design configurations, is achieved. Dimensionless variables are used throughout the work.

The goal of this work is to produce numerical models (and their associated computer programs) that reproduce the essential natural conditions in given water-wave processes and, hence, can be used as engineering design tools. The work focuses on long, large water waves whose particular applications are to storm-generated waves and tsunamis. The specific objectives of this research were:

1. Exact simulation of finite-amplitude waves near the shore. This involved development and innovation with the Stanford-University-Modified Marker and Cell (SUMMAC) numerical finite-difference technique for analysis of two-dimensional, nonlinear wave motions.
2. Simulation, based on approximate equations, of long waves in moderately shallow water, e.g., the region of the continental slope and shelf. In this case, the equations proposed by Peregrine (1967) were used to derive a numerical model to study the two-dimensional motion of waves over complex bottom hydrography.

The following sections describe the concepts, computational features, present results, and proposed plans for the two numerical models and their related programs. The view was adopted that theoretical results were available but that implementation of these results into viable design and research tools for water waves was lacking. Therefore, two well-documented theoretical analyses were adopted that had never been optimized

for numerical application to design and research or adequately
verified against experiments. An essential consideration was
that each method must reproduce in its numerical results the
essential physical phenomena observed in relevant experiments.

AN EXACT SIMULATION -- SUMMAC

The objective of an exact simulation is to provide detailed information about wave processes near the shore and at
the ocean-structures interface. Nonlinear and finite-amplitude
effects must be included.

The Stanford-University-Modified Marker and Cell (SUMMAC)
method is based on a method presented by Welch, et al. (1966)
for computing time-dependent, viscous, incompressible fluid
flows with a free surface in several space dimensions. The
Navier-Stokes equations are used to compute the fluid motions.
At present the method is suitable for analyzing two-dimensional
flows.

Welch, et al. (1966) gave a detailed description of their
original method (MAC), flow charts for computations and several
examples of numerical solutions. One feature that distinguishes
the MAC method from others is that its finite-difference equations (derived from the Navier-Stokes and continuity equations)
rigorously conserve momentum. Also, the calculational procedure is arranged in such a way that it is self-correcting. The
errors made at any time-step tend to be corrected in the next
step. Furthermore, the use of hypothetical "marker" particles
to mark the free surface location makes it possible to carry
out the computations in a well-defined domain.

Basically, the method provides the solution to a set of
finite-difference equations that were derived from the partial
differential equations and boundary conditions of viscous fluid
flow. The essential physical principles used were the conservation of mass and the conservation of momentum (the full
Navier-Stokes equations). Only non-turbulent flows are considered. Pressure (p) and velocity (u,v) -- the important
dependent variables for water waves -- are used directly as
the primary field variables. The horizontal and vertical
components of velocity are defined at the center of the vertical and horizontal edges of each calculation cell, respectively, while the pressure, density, and viscous coefficient
are defined at the cell centers. The flow boundary conditions
are derived from the finite-difference momentum equations.
Usually, the calculation begins with the introduction of a dis-

turbance to a fluid system at rest or in uniform motion. The pressure and velocity are specified throughout the fluid field initially.

The operation of the method is quite simple. First, at some initial time the fluid region of a problem and the values of the pressure and velocities in the fluid are set. At the same time, massless "marker" particles are identified along any free surfaces of the problem. Second, with the pressure and velocities known at a time instant, the "Navier-Stokes" finite-difference equations can be used to find velocities at the next time step (the Navier-Stokes equations have a single time-derivative term that is approximated by a forward time difference). At the same time, the "markers," i.e., the free surface, are advanced one time step to their new positions in accordance with the kinematic relation between the surface geometry and fluid velocity. Third, by cross differentiation of the Navier-Stokes equations and use of the continuity equation, one can derive for the ratio ϕ of the fluid pressure to its density a finite-difference Poisson equation whose source term is a function of the velocity field at a particular time. Therefore, using the new velocities and fluid boundary description, one now finds the new pressure field at the advanced time step. This step-by-step process of finding new velocities, moving particles, and finding new pressures is repeated until the desired results are obtained.

The major effort in this first implementation of the method has been in establishing the desired program procedures and bookkeeping processes (number of cells occupied by fluid, location of markers, etc.) and modification of the method so that it is adequate for water-wave analyses. The SUMMAC program is implemented for use on IBM 360/67 or 360/91 digital computer systems. At the present time 60,000 words of core storage are required. Each time-step of a typical calculation requires about 0.1 minute on a 360/67.

For the process of implementing, testing, and verifying the program and method, a simple physical situation was selected for which good experimental data, knowledge of the physical processes, and easily observable physical phenomena existed. The propagation of solitary waves in a horizontal channel filled with fluid to unit depth (in terms of non-dimensional variables) and with vertical end walls was simulated. By solving the steady-state, propagation problem for solitary waves numerically (Chan, *et al*., 1969), a set of highly accurate initial conditions was obtained. The solitary wave propagation problem possessed several key features:

1. The theories for the wave motion against the wall were not in agreement with experiments (Street and Camfield, 1966);
2. The solitary wave should propagate stably (without change of form) in zones not near the channel walls; and
3. Perfect reflection from the wall should occur.

Several difficulties arose during the first attempts to calculate the motion of large-amplitude waves by the original MAC method. The well-shaped initial waves became very irregular after a few time-steps of computation, and the velocity fields did not vary as smoothly as they should. Accordingly, a significant modification of the MAC method was undertaken to create a numerical scheme suitable for water wave simulation. The resulting code has been called SUMMAC (Chan, et al., 1969). The modification and features introduced in SUMMAC include:

1. Correction of the free-surface boundary conditions and use of a special technique employing irregular (mesh) stars to represent the free-surface pressure condition correctly.
2. Use of successive over-relaxation to solve the pressure field equation.
3. Use of Gregory-Newton backward extrapolation to extend field variable values outside the fluid regions where required by the numerical scheme.
4. Higher-order interpolation formulae for marker velocity and undefined field variable calculation.

Results of calculation with SUMMAC are shown in Figures 1, 2, and 3. Figure 1 shows initial steps in the propagation of a truly finite-amplitude wave (height H_0 = 50 percent of the water depth d in a horizontal channel with vertical end walls). Runs in which the wave was permitted to reflect from the wall and return to its original position showed that perfect reflection occurred and the wave was stable, in height and form, when not near the end walls. In sum, the SUMMAC simulation met all three criteria stated above. This work established empirically that the SUMMAC method is computationally stable for long wave propagation at low viscosity, although analyses by Hirt (1968) suggest that some instabilities might be expected (up to 1000 time-steps were executed).

The successful application of the SUMMAC method to the preliminary example above of the run-up of a solitary wave on a vertical wall indicated the possibility of employing the same technique to attack a wide variety of water-wave problems.

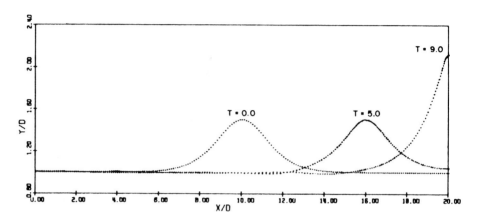

Fig. 1: Solitary Wave Deformation Against a Vertical Wall ($H_o/d = 0.5$).

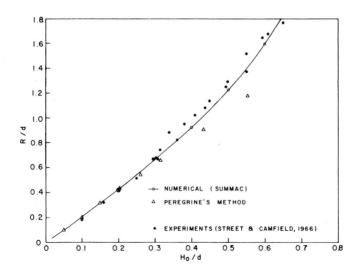

Fig. 2: Solitary Wave Run-up on a Vertical Wall (R/d).

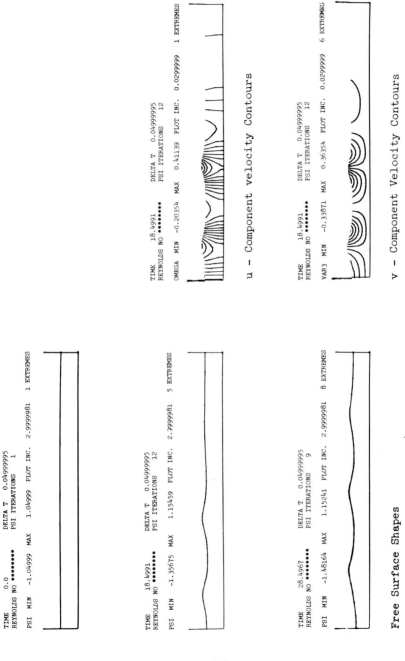

Fig. 3: Typical Graphic Display Results for Case with Sinusoidal Pressure Variation Applied in Zone near Left Wall.

Such extensions prove to be most valuable in problems where analytical methods are difficult, if not impossible. Completion of the implementation and verification phase of the work described above led to systematized data input and to graphical displays coupled to the output side of the program. Figure 3 shows a sample of the graphical output obtained from a movie made by photographing the face of the IBM B 2250 CRT device. Here, a wave train was generated by application of a transient pressure to a portion of the free surface. Again, a check shows that the waves of the train propagate at the correct speed and are approximately cnoidal when fully developed. The results pictured in Fig. 3 show the free surface at three times and one set of velocity contour plots. The plotting and graphic routines were developed by Schreiber (1968).

The water particle motions, the subsurface pressure distribution, the Eulerian velocity fields, the fluid mass transport, and the wave form deformation as functions of time, space, wave amplitude, bottom configuration, and the presence of a viscous stress term in the governing equations will be examined. The nonlinear nature of the motion will provide the most important and dramatic results in these studies and considerable insight into the above will be gained without even considering the effects of a viscous term (which can be set to zero in the present SUMMAC program). Also, the transfer relationship between wave surface shape and the subsurface pressure can be examined as affected by finite amplitude, bottom shape, and unsteady motion (cf., Ewing and Press, 1949, and Shooter and Ellis, 1967). It appears that no other method can give the wealth of basic and design information obtainable with the SUMMAC code.

AN APPROXIMATE SIMULATION -- APPSIM

The SUMMAC method described above provides an exact simulation of water-wave motion, yet the present speed and capacity of computers make it impractical to consider application of SUMMAC to three-dimensional problems. On the other hand, many (see, e.g., Carrier, 1966, Jordaan, 1965, and Wilson and Tórum, 1968) have pointed out that three-dimensional effects are an essential consideration in ocean and coastal design problems. Thus, while SUMMAC can provide insight to physical processes and useful design data in shallow water near many types of structures, some other technique is needed to treat the more general problem, e.g., long wave motions in moderately shallow water.

At the present time, model tests and some "recipes" that combine theory with empirical results (see LeMéhauté, et al.,

1968, and Van Dorn, 1966) provide some of the answers of the general nature needed to assess three-dimensional and unsteady effects. The essential point is that for design and planning purposes, numerical results must be produced. This requirement eliminates many theories. (One notable exception is the refraction theory based on geometric optics (Dobson, 1967); however, this theory is linear and does not include wave reflection or diffraction effects.)

In complex design problems where neither theory nor empirical curves can provide sufficient answers, model tests have been used extensively. However, large areas of ocean and shoreline where depths change by several orders of magnitude and the wave character is significantly modified cannot always be modeled. In some cases, quasi-full-scale model tests can be conducted. The Mono Lake tests are an example (National Marine Consultants, 1965). This type of test, like the model or laboratory test, is valuable in revealing basic physical processes or anomalies. However, not many runs can be made and the "model" is not easily modified to test various hypotheses or examine the consistency of results.

The continental slope and continental-shelf hydrography can have a significant effect on long, large waves such as tsunamis. Accordingly, three-dimensional numerical simulation is focused here on description of long-wave behavior in moderately shallow water (Mei and LeMéhauté, 1966) corresponding to conditions on the slope and shelf. The effects of bottom hydrography on the wave progress and character are of particular interest. The plan is: wave analyses based on a successive application of refraction theory in deeper waters where nonlinear and reflection effects are small, approximate simulation in moderately shallow waters, and SUMMAC simulation (perhaps implemented very locally in three dimensions) in the very shallow waters near shore.

The method used here, called the Approximate Simulation -- APPSIM, is based on an analysis by Peregrine (1967) and the perspective and justification outlined by Mei and LeMéhauté (1966). There are two key length ratios in shallow-water theory: the ratio ε of wave amplitude a to water depth d and the ratio σ of water depth d to wave length L. In their paper, Mei and LeMéhauté (1966) point out that nonlinear cnoidal and solitary wave theory, i.e., theory for nonlinear waves in moderately shallow water, is applicable when the Ursell parameter ε/σ^2 is of the order of unity. Peregrine's analysis is based on this assumed equality $\varepsilon = \sigma^2$. He derived equations of motion for long waves in water of varying depth. They were derived for three-dimensional motion but have only been used for two-dimensional examples.

Peregrine (1967) began with the Euler equations of motion and the continuity equation. Viscosity was neglected. His first step was to make a suitable choice of length scales for the space variables and to nondimensionalize the dependent variables so that an expansion in power series based on ε and σ (with $\varepsilon = \sigma^2$) could be carried out. Next, he showed that one can, as expected, solve successively for the values of the dependent variables at each order of approximation. However, Peregrine then made an important observation. Solutions at any order in this kind of process tend to be valid for only small times. In particular, the second-order terms have first-order effects over moderate times. For example, the first-order, nonlinear shallow-water wave equations (Airy equations) lead to the eventual breaking of any initial disturbance in the case of propagation over a horizontal bottom. Consider the propagation of solitary and cnoidal waves of <u>permanent</u> form over a horizontal bottom; clearly, the Airy equations are not a valid basis for the study of the unsteady propagation of finite-amplitude waves over moderate time.

Peregrine's unique and significant contribution was to utilize the equations and some solutions for the first two orders of expansion in his scheme to generate a set of new governing equations for the wave surface shape $\eta(x,y,t)$ and mean horizontal velocity $\vec{u}(x,y,t)$ where x and y are the plan form or horizontal plane coordinates. These new equations incorporated the following important features:

1. The effect of vertical accelerations and velocity at the second order, so that the non-hydrostatic pressure beneath steep waves is included.
2. The effect of the bottom hydrography, so that although only average horizontal velocity is computed the three-dimensional effects of the bottom can be incorporated.
3. The equations are valid for moderate time.

The new equations and their numerical results are then accurate to the second order in $\varepsilon = \sigma^2$.

If $H = H(x,y)$ describes the ocean bottom and dimensionless variables are used, Peregrine's equations take the form

$$\vec{u}_t + (\vec{u}\cdot\nabla)\vec{u} + \nabla\eta = \frac{1}{2} H \frac{\partial}{\partial t} \nabla [\nabla\cdot(H\vec{u})] - \frac{1}{6} H^2 \frac{\partial}{\partial t} \nabla(\nabla\cdot\vec{u}) \quad (1)$$

$$\eta_t + \nabla \cdot [(H + \eta)\vec{u}] = 0 \qquad (2)$$

Although these equations are nonlinear, a rather straightforward finite difference scheme allows detailed computations on wave propagation to be performed.

A program APPSIM has been implemented for two-dimensional studies on an IBM 360/67 digital computer system. The program requires about 33,000 words of storage (including 20,000 words for plot routines) and executes at a rate of about 200 timesteps per minute for the examples studied. As before, a simple physical situation is studied that was relevant to the overall goal. The propagation of solitary waves on a stepped slope which represents the configuration of the continental slope and shelf, i.e., long waves in moderately shallow water, was simulated. The key criteria to be satisfied were:

1. Solitary waves propagate stably on a horizontal bottom.
2. Solitary waves break down into undular bores when the waves propagate onto a stepped slope (Street, et al., 1968).
3. Wave heights must be in good quantitative agreement with available, relevant experimental data.

For two-dimensional (plane) flows, Equations 1 and 2 become

$$\frac{\partial u}{\partial t} + u\frac{\partial u}{\partial x} + \frac{\partial \eta}{\partial x} = \frac{1}{3} H^2 \frac{\partial^3 u}{\partial t \partial x^2} + H \frac{\partial H}{\partial x} \frac{\partial^2 u}{\partial t \partial x} + \frac{1}{2}\frac{\partial^2 H}{\partial x^2}\frac{\partial u}{\partial t} \qquad (3)$$

$$\frac{\partial \eta}{\partial t} + \frac{\partial}{\partial x}\left[(H + \eta)u\right] = 0 \qquad (4)$$

where $u = u(x,t)$ is the mean horizontal velocity, $H = H(x)$ describes the bottom configuration and $\eta = \eta(x,t)$ describes the free-surface shape. The finite-difference forms of Equations 3 and 4, with forward time differences and central space differences, lead naturally to a set of implicit equations for u and η because of mixed derivative terms in Equation 3. In particular, if an initial distribution of η and u is given or known at time t then:

1. A provisional new η at time t + Δt is computed from Equation 4.
2. Next, one finds a new u field at t + Δt by using the provisional η and Equation 3.
3. A final value of η is computed at t + Δt through the use of Equation 4 again and the results of step 2.

Because the scheme is implicit, no stability criteria is needed for calculations using the numerical forms of Equations 3 and 4. Equal time and space steps were used in the results presented here and about 400 node points were used in the horizontal direction in the finite difference equations. The implicit finite-difference scheme requires that a set of 400 simultaneous equations be solved for the horizontal velocity u at each node point at a given time. The coefficient matrix of these equations is tri-diagonal for the plane flow case and hence the equations are quickly and directly solved by Gaussian Elimination. If the matrix were not tri-diagonal, the solution processes would be slow and not practical for engineering design. The initial condition for solitary waves is obtained from the Boussinesq equation for η (Stoker, 1957) and Equation 4 which gives u for a wave propagation without change of form on a horizontal bottom. Linear wave theory (Stoker, 1957) was used for η in other cases with u given by Equation 4 as for solitary waves.

At the present the APPSIM program is set up to provide bottom hydrography as follows (cf., Figures 4 and 5):

1. First section: horizontal bottom from a left-hand vertical wall to beginning of variable zone. This flat section provides for simple boundary conditions at the wall -- u = 0 and ∂η/∂x = 0 -- and for easy prescription of initial waves. The bottom is held flat for at least one effective wave length.
2. Second section: variable bottom as desired. This section is for continental slope and simulation of the toe of the slope in work of Camfield and Street (1968). A cosine curve is shown in Figures 4-6.
3. Third section: bottom with constant slope to simulate continental shelf, for example.
4. Fourth section: horizontal bottom leading to right hand vertical wall, cf. (1) above.

As seen in Fig. 4, the APPSIM results do exhibit the wave breakdown phenomenon. A comparison with the experimental results of Street, et al. (1968) and Camfield and Street (1968) indicates that the APPSIM results are relatively good with

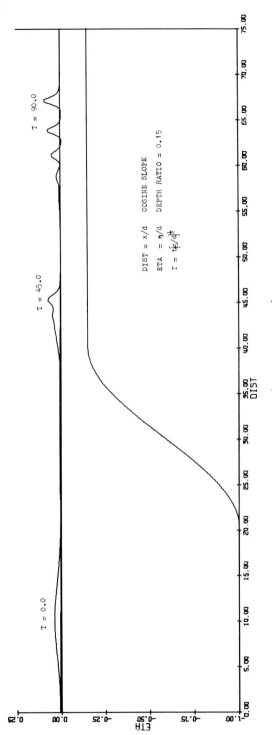

Fig. 4: Solitary Wave Shoaling on a Stepped Slope (Cosine Bottom).

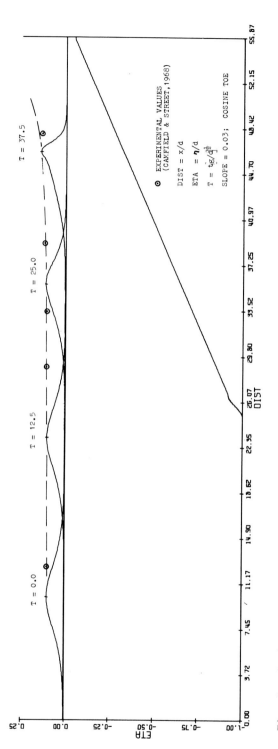

Fig. 5: A Simulation of Camfield and Street (1968) Experiment.

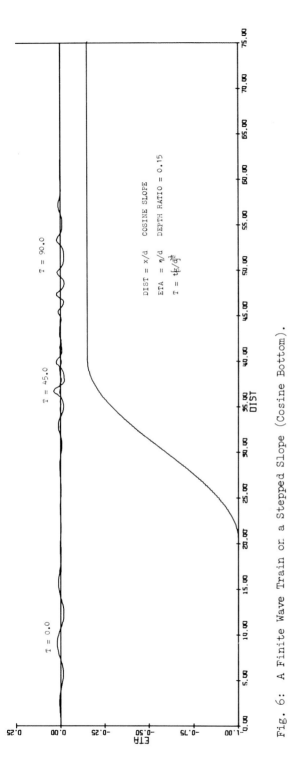

Fig. 6: A Finite Wave Train or a Stepped Slope (Cosine Bottom).

respect to modeling of physical phenomena. In APPSIM, solitary waves propagate stably on horizontal bottoms, as desired. Although an approximation such as used in APPSIM should not be expected to hold when ε (=H_o/d for solitary waves) is of the order of unity or in the presence of a vertical wall, Fig. 2 shows that APPSIM (Peregrine's method) agrees with SUMMAC and experiments for $\varepsilon \leq 0.4$. The other results agree quantitatively with experiments in this range, too.

Figures 5 and 6 illustrate the application of APPSIM for comparison with a specific set of experiments and for study of a finite wave train such as generated by a tsunami. Figure 6 has the same bottom configuration and maximum initial height as Figure 4.

The successful application of the APPSIM method to the preliminary examples for the plane motion of long waves in moderately shallow water indicated that the method can be applied to the three-dimensional motion of waves with consideration of the plan form variation of bottom hydrography. The result is a three-dimensional simulation (within the spirit of the original approximations). APPSIM is now being extended to handle general bottom hydrography and wave inputs.

The effects of specific idealized bottom configurations and wave inputs will be examined. For example, Dobson (1967) produced a wave refraction analysis program that is based on geometric optics and can be used for evaluation of the proposed work (or vice versa). The results cited by Dobson (1967) for circular islands are an interesting case, for instance. Also, Jones (1968) and Jordaan (1965) summarize extensive studies made in the large NCEL wave basin; these studies will be used to evaluate the present work.

After the three-dimensional simulation is implemented and verified against existing results, long-wave behavior and character over a number of specific, practical bottom hydrographies should be examined. In particular, idealized representations of well known coastlines near tsunami-affected harbors, e.g., Crescent City, California, and Hilo, Hawaii, should be studied. The final result of this work will be a completely documented computer program suitable for design and further research applications.

ACKNOWLEDGMENT

The preparation of this paper was supported in part by the Office of Naval Research (Contract NONR 225(71), NR062-320) and the National Science Foundation (Grant NSF GK-2506).

BIBLIOGRAPHY

Amein, M. 1964. Long waves on a sloping beach and wave forces on a pier deck. Contract Nby-32236, U. S. NCEL, Port Hueneme, Calif., DDC AD 451 244, Sept. (See also *J. Geophys. Res.*, Vol. 71, Pt. 2, Jan. 15, 1966).

Camfield, F. E. and Street, R. L. 1968. Shoaling of solitary waves on small slopes. *J. Waterways and Harbors Division, ASCE*, Vol. 95, No. WW1, Proc. Paper 6380, February, pp. 1-22.

Carrier, G. F. 1966. Gravity waves on water of variable depth. *Journal of Fluid Mechanics*, Vol. 24, Pt. 4, Apr., pp. 641-660.

Chan, R. K.-C., Street, R. L., and Strelkoff, T. 1969. Computer studies of finite-amplitude water waves. *Department of Civil Engineering Tech. Report No. 104*, Stanford Univ., Stanford, Calif., June.

Dean, R. G. 1965. Stream function representation of nonlinear ocean waves. *J. Geophys. Res.*, Vol. 70, No. 18, Sept. 15.

Dobson, R. S. 1967. Some applications of a digital computer to hydraulic engineering problems. *Department of Civil Engineering Tech. Report No. 80*, Stanford Univ., Stanford, Calif., June.

Ewing, M. and Press F. 1949. Notes on surface waves. *Annals*, New York Academy of Sciences, Vol. 59, pp. 453-462.

Hirt, C. W. 1968. Heuristic stability theory for finite difference equations. *J. Computational Physics*, 2, p. 339.

Jones, D. B. 1968. Wave-basin study of run-up on beaches from simulated underwater explosions near shore. *U. S. Naval*

Civil Engineering Lab. Tech. Report R604, Dec., Port Hueneme, Calif.

Jordaan, J. M. Jr. 1965. Feasibility of modeling run-up effects of dispersive water waves. *Technical Note N-691*, U. S. NCEL, Port Hueneme, California, May.

LeMéhauté, B. Koh, R. C. Y., and Hwang, L.-S. 1968. A synthesis on wave run-up. *J. Waterways and Harbors Div.*, Proc. ASCE, Proc. Paper 5807, Feb.

Mei, C. C. and LeMéhauté, B. 1966. Note on the equations of long waves over an uneven bottom. *J. Geophys. Res., Vol. 71*, No. 2, Jan. 15, pp. 393-401.

Miller, R. L. and Zeigler, J. M. 1964. The internal velocity field in breaking waves. *Tech. Report No. 1*, Fluid Dyn. and Sed. Trans. Lab., Dept. of Geophys. Sci., Univ. of Chicago, Chicago, Ill., Nov.

Monkmeyer, P. L. and Kutzback, J. E. 1965. A higher order theory for deep water waves. Chapter 13, *Coastal Engineering*, Santa Barbara Specialty Conference, Oct., ASCE, New York.

National Marine Consultants 1965. Mono Lake wave experiment: feasibility study. *Final Contract Report Nonr 21488(00)*, Anaheim, Calif., June.

Peregrine, D. H. 1966. Calculations of the development of an undular bore. *Journal of Fluid Mechanics, 25*, Pt. 2, pp. 321-330.

Peregrine, D. H. 1967. Long waves on a beach. *Journal of Fluid Mechanics, 27*, Pt. 4, pp. 815-827, Mar.

Proceedings of Tenth Conference on Coastal Engineering, Vols. I and II 1966. ASCE, New York.

Schreiber, D. E. 1968. A generalized equipotential plotting routine for a scalar function of two variables. *RJ499*, IBM Research, San Jose, Calif., May.

Shooter, J. A. and Ellis, G. E. 1967. Surface waves and dynamic bottom pressure at Buzzards Bay, Massachusetts. *Def. Res. Lab. Acoustical Report No. 292*, U. Texas (Austin), Austin, Texas, 18 Sept.

Stoker, J. J. 1957. *Water Waves*. Interscience Publications, Inc., New York

Street, R. L., Burges, S. J., and Whitford, P. W. 1968. The behavior of solitary waves on a stepped slope. *Dept. of Civil Engineering Tech. Report No. 93*, Stanford Univ., Stanford, Calif., Aug.

Street, R. L. and Camfield, F. E. 1966. Observations and experiments on solitary wave deformation. *Proc. Tenth Conf. on Coastal Engineering*. Tokyo, ASCE, Sept.

Van Dorn, W. G. 1966. Theoretical and experimental study of wave enhancement and run-up on uniformly sloping impermeable beaches, Univ. of Calif., Scripps Inst. of Oceanography., *Report SIO 66-11*. San Diego, Calif., May.

Vanoni, V. and Raichlen, F. 1966. Laboratory design studies of the effect of waves on a proposed island site for a combined nuclear power and desalting plant. *Calif. Inst. of Technology Report KH-R-14*. Wm. Keck Laboratory, Pasadena, Calif.

Welch, J. E., Harlow, F. H., Shannon, J. P., and Daly, B. J. 1966. The Mac Method, a computing technique for solving viscous, incompressible, transient fluid-flow problems involving free surfaces. *Report LA-3425*, Los Alamos Scientific Laboratory, Univ. of Calif., Los Alamos, New Mexico, Mar. (See also *Phys. Fluids, Vol. 8*, 2182 (1965) and *Vol. 9*, 842 (1965).

Wilson, B. W. and Tórum, A. 1968. The tsunami of the Alaskan earthquake, 1964; engineering evaluation. Coastal Engineering Research Center, U. S. Army, *Corps of Engineers, Tech. Memorandum No. 25*. Washington, D. C., May.

GENERAL REPORTS

31. Initiating an IBM System/360 Document Processing System for Tsunami Research Using an Existing KWIC Index Data Base

JAMES M. WALLING
University of Hawaii
Honolulu, Hawaii

DALE FREEMAN
IBM Corporation
Honolulu, Hawaii

WILLIAM M. ADAMS
Hawaii Institute of Geophysics
Honolulu, Hawaii
Contribution No. 302

ABSTRACT

In 1967 a KWIC (Keyword-in-Context) index of literature concerned with tsunamis was compiled (Adams, 1967). Abstracts have since been added to the bibliographic data which were punched into IBM cards to serve as input for the KWIC index. In 1969 this data base was selected as the first input for a System/360 Document Processing System. A PL/1 program converts the punched card data directly to tape in a format suitable for input to the Document Processing System.

Document Processing is a set of programs which convert machine-readable document data into searchable data sets, search these data sets, and produce various bibliographies and index listings. Information is stored in context and retrieval is by context-oriented search questions. Keywords and bibliographic data are extracted directly from incoming document records. Input records will include abstracts but may also include full text, if desired. Output will consist of document numbers, bibliographies and, optionally, a full-text printout.

The original documents, where available, are being microfilmed for storage in microfiche form, in a push-button, random access reader with five-second access time.

INTRODUCTION

For the scientists involved in tsunami research, and particularly for those concerned with tsunami warning services, ready access to the literature connected with specific events, geographic areas, or subjects is extremely important. To provide this quick access, the Tsunami Research Department of the Hawaii Institute of Geophysics is initiating an IBM System/360 Document Processing System. This system consists of a set of programs designed to convert machine-readable document data into searchable data sets, to search these data sets for Boolean combinations of keywords, and to manipulate elements of the data set to produce various forms of indexes. (IBM, 1967, 1969.)

In 1967, a KWIC (Keyword-in-Context) index of the literature concerned with tsunamis, from the earliest known references through 1966, was compiled (Adams, 1967). Abstracts have since been added to the bibliographic data for many items used for the KWIC index, and, in 1969, this data base was selected as the first input for Document Processing. This is a searchable, context-oriented information retrieval system which will be used in conjunction with a remote console and a Houston-Fearless CARD microfiche retrieval system to provide response to search questions and to display, within 5 seconds, any page of any document stored in the system.

The original bibliography issued with the KWIC index was arranged chronologically (Figure 1). Therefore, the present KWIC programs are designed to generate a new bibliography arranged by author and to include the abstracts which have been added (Figure 2). The 1967 index covered some 2,500 references, ranging from Herodotus' *History of the Peloponnesian Wars* through the events of 1966. The data exists in machine-readable form as EAM punched cards of four types: (1) author, (2) title, (3) source, and (4) abstract, in slightly modified KWIC format (IBM, 1964).

Document Processing is designed to convert machine-readable document data into searchable data sets, to search these data sets for keyword combinations, and to manipulate elements of the data set to produce various forms of indexes. Among the data sets produced are the following:

```
EXTRACT OF QUARTER·DECK  LOG BOOK OF FRIGATE DIANA.=                          55MS     0244
BLOCATION EARTHQUAKE IN  LOKRIS IN APRIL 1894.=+ THE LARGE DI                 95UA     0000
HE GREAT EARTHQUAKES OF  LOKRIS ON APRIL 20 AND 27, 1894.=    T               94VGE    0410
     THE EARTHQUAKES OF  LOKRIS WHICH OCCURRED THIS YEAR.=                    94VGF    0332
WING THE LAST TREMOR AT  LOKRIS.=      + CREVASS PRODUCED FOLLO               94CRAS   0380
     ON THE EARTHQUAKES IN  LOKRIS, GREECE IN APRIL 1894.=                    94CRAS   0112
        SPECTRAL ANALYSES OF  LONG-PERIOD OCEAN WAVES OBSERVED AT             62ERIB   0561
          AN INVESTIGATION OF  LONG-PERIOD OCEAN WAVES.=                      63IUGG   0026
                        A NEW  LONG-PERIOD WAVE RECORDER.=                    60JGR    1007
IGN AND CONSTRUCTION OF  LONG-PERIOD WAVE RECORDERS.=    DES                  62ERIB   0545
                              LONG-PERIOD WAVES IN JAPANESE PORTS.            57PIANC  0000
NS.=..                        LONG-PERIOD WAVES OR SURGES IN HARBO            53PASCE  0001
CONTINENTAL BORDERLAND+  LONG-PERIOD WAVES OVER CALIFORNIA'S                  62JMR    0031
CONTINENTAL BORDERLAND+  LONG-PERIOD WAVES OVER CALIFORNIA'S                  62JMR    0003
CONTINENTAL BORDERLAND+  LONG-PERIOD WAVES OVER CALIFORNIA'S                  62JMR    0119
RECONSTRUCTION ENDING.   LONG-TERM SCIENTIFIC PROGRAM IS                      64SCI    0038
A.= THEORY OF UNINODAL   LONGITUDINAL SEICHES IN LAKE YAMANAK                 32GM     0283
OSAKA BAY.=         ON THE  LONGITUDINAL SIMPLE NODE SEICHES OF               31KKI    0001
```

Fig. 1: KWIC Index (Reference Numbers from Cards).

```
05REMSM0294     ALFANI P G
                THE CALABRIAN EARTHQUAKE.=
                REV. FIS. MATH. SCI. NATUR.    1905            11  0294
07AHMM 0263     AMMON V
                ON THE EARTHQUAKE AND FLOOD ON THE 31 OF JANUARY, 1906 ON
                   THE SHORES OF COLOMBIA AND ECUADOR.=
                ANN. HYDROGR. MARIT. METEOROL. 1907 0035 0006    0263-0266
                   THE CONNECTION IS BROUGHT OUT BETWEEN EARTHQUAKES AND
                THE TSUNAMIS THAT FOLLOW AFTER THEM, THE FOCUS OF WHICH WAS
                IN THE SEA BETWEEN COLOMBIA AND ECUADOR  AND ALSO OF THE
2ORFT  0211     ABU'L FIDA
                ABU'L FIDA'S MUSLEM ANNALS II.=
                REISKE, THIELE, HAFNIAE          1320 0003         0211
                   THE SEISMIC SEA WAVES OF 1067 AND 957 WHICH CAUSED DAMAGE
                TO THE COASTS OF SYRIA, LEBANON, ISRAEL AND EGYPT ARE
                BRIEFLY DESCRIBED.  THIS WORK BY ABU L FIDA CONTAINS
                INFORMATION ON A LARGE NUMBER OF EARTHQUAKES IN THE MIDDLE
                EAST.
```

Fig. 2: Bibliography Arranged by Author (Reference Number Assigned by Computer).

1. Dictionary - contains one entry for each unique word in the data set. Each entry is accompanied by word frequency statistics, pointers to related information in other data sets, and an internally used identifier code.

2. Vocabulary - a series of document control number entries for each word in the dictionary. A set of entries identifies all documents containing a given word.

3. Master - a complete file of all document numbers, bibliographic data and text entries.

4. Synonym/Equivalent - two lists, compiled by the user, of substitution words to be utilized at search time at the user's option.

5. Text - contains the full text as supplied for each document.

In addition, the system produces a number of data sets for internal use.

Document Processing will store either abstracts, or keywords, or full text, depending on the availability of secondary storage and the requirements of the user. For the present application, abstracts were selected for storage. The full text, when available, will be stored in microfiche form in an automatic retrieval and display unit to be described later in this paper. Keywords are selected by the computer from each incoming document, not volunteered by the user or taken from a thesaurus. The location of each word with respect to its neighbors, to each sentence in its paragraph, and to each paragraph within the document, is stored. The system also lists the number of times the word appears in the entire file of documents and identifies each word with every document in which it has appeared. For each data base, a list of little-value words, also called "stop words", can be read into the system as "common words" to be automatically removed from incoming documents. In addition, the user has an editing option. When entering new documents, he can request a list of "non-found", or new, keywords found in the incoming document. He may choose to enter some, all, or none of these words into the system dictionary.

Bibliographic data is entered as fields of information, designated as "author", "title", "source", "year", etc., while text enters as unformatted data. Input Document Processing is accomplished in two phases. Phase I builds a data-base description and produces intermediate word data sets which are used in Phase II to update the searchable data sets. Phase I has the option of producing the list of new keywords for editing by the user. Phase II completes the updating of the permanent data sets and produces a listing of all words added to the dictionary and of all new words found but not added.

The Search function of Document Processing controls the interrogation of the data base and the resultant output, along with error indications. Search statements are made in near-English language and output may consist of document numbers, specified bibliographic data and/or full text as required. Keywords may be combined in such combinations as: (1) apples or pears; (2) apples and pears; or (3) apples and pears, but not peaches. Searches may be made for keywords in a certain order, or within the same sentence or paragraph, or within a stated number of words of each other. A typical request might be to list the text of all works published between certain dates containing the name "Hilo", and, also, the words "tsunami" or "earthquake". In this case, abstracts are entered as text and would be printed as output in such an example. If an abstract is not available, the title is repeated as text. When desired, keywords may be weighted in the Search statement.

The user has the option of expanding keywords at search time by the use of either, or both, of two substitution lists, rather arbitrarily called Synonyms and Equivalents, or by the use of truncated keywords. For example, suppose that equivalents and synonyms for the word "shore" were entered as follows:

Word	Equivalent	Synonym
shore	beach	bank
	seashore	coast
		seacoast

Search output could be as listed below, the last example demonstrating the use of a truncated keyword:

Search Statement	Output
SHORE	All documents containing the word "shore".
SHORE(E)	All documents containing the words "shore", "beach", or "seashore".
SHORE(S)	All documents containing the words "shore", "bank", "coast", or "seacoast".

SHORE(SE)   All documents containing any of the above words.
AGITAT($)   All documents containing any word beginning
               with the letters AGITAT, as agitate, agitates,
               agitater, agitated, agitation, etc.

The following examples demonstrate the use of the Boolean AND, OR and NOT statements:

Search Statement	Output
TSUNAMI, EARTHQUAKE	All documents containing the words "tsunami" or "earthquake".
TSUNAMI & EARTHQUAKE	All documents containing both the words "tsunami" and "earthquake".
TSUNAMI & EARTHQUAKE (NOT)	All documents containing the words "tsunami", but not containing the word "earthquake".

Other system functions provide for updating the Synonym/Equivalent word substitution lists and for file restructuring, or reorganizing the data base sets, in order to minimize storage requirements and access time. The Index List capability allows the user to obtain information about data base entries as a whole and provides the offline listing of such data as frequency of word appearance, documents associated with various dictionary words, and Synonym/Equivalent entries.

Document Processing requires that input documents be in machine-readable form and that the format conform to the following criteria:

1. Each document must consist of an ascending sequential document number, bibliographic data, and text or keywords, in that order.

2. The document number must be numeric.

3. The bibliographic data must be arranged in fields which are described in the data base description. Fields may be variable in length, but cannot exceed 255 characters. Maximum amount of bibliographic data is 1,636 characters.

4. A document may consist of one, or more, physical records, each of which must contain the document number.

5. The document number must be no larger than 2,147,483,647.

The existing KWIC input data is machine-readable, but does not fit the criteria listed above. A program is being written by Edith Fujikawa to convert the punched card data to tape and to accomplish the following:

1. Convert the alpha-numeric document number to numeric.

2. Move the document number to the beginning of the first record.

3. Compact the bibliographic data and move it into the fields selected in the data base description.

4. Eliminate unwanted data.

5. Repeat the document number at the beginning of the second, (text), record.

6. Enter the abstract as text. If there are no abstract cards, the title will be repeated as text.

In a few cases, where the title exceeds four cards, it will have to be shortened to fit within the maximum of 255 characters allowed for any field.

The original documents, when available, will be microfilmed and stored in microfiche form. The material will be filmed using a Recordak 35 mm microfilm camera with a 16 mm adapter. The processed film will be loaded into jackets and contact-printed to make the microfiche. Each fiche will have a capacity of approximately 75 pages.

The fiche will be stored in a Houston-Fearless CARD, (Compact Automatic Retrieval-Display), microfiche reader. This unit will store up to 750 microfiche and, according to the manufacturer's specifications, will provide access to any page of any fiche in less than 5 seconds. In our format, (similar to COSATI), this would allow storage of approximately 56,000 document pages in one unit.

The location key of the first page of each document may be included as a bibliographic field in the Document Processing input in order that it can be one output from any search question.

The use of Document Processing with automatic microfiche storage and retrieval provides a combination of a searchable information retrieval system with compact full-text storage, giving maximum flexibility of use with efficient utilization of computer storage.

Document Processing is designed for use with any data base that can be converted to machine-readable form. The use of near-English search questions, the lack of need for a thesaurus, and the ability of the user to use his own preferential search word relationships add to the value of Document Processing as a general purpose information system.

ACKNOWLEDGMENT

Financial support of this work has been provided by ESSA Contract No. E-22-131-68G.

BIBLIOGRAPHY

Adams, William Mansfield 1967. An index to tsunami literature to 1966. Prepared for the State of Hawaii, *Data Report No. 8. (HIG-67-21)*. Hawaii Institute of Geophysics, Honolulu, Hawaii.

International Business Machines Corporation 1969. *IBM System/360 Document Processing System: Application Description (H20-0315-0)*. 1st edition, White Plains, N. Y.

International Business Machines Corporation 1969. *IBM System/360 Document Processing System: Program Description and Operations Manual (H20-0477-1)*. 2nd edition, White Plains, N. Y.

International Business Machines Corporation 1964. *Keyword-in-Context (KWIC) Indexing Program for the IBM 1404 Data Processing System (1401-CR-02X)*. White Plains, N. Y.

32. Report of the International Union of Geodesy and Geophysics Tsunami Committee

B. D. ZETLER, Chairman

The Tsunami Committee met on 10 October 1969 following the Tsunami Symposium at Honolulu and prepared the following report.

The Tsunami Symposium was held at the East-West Center, University of Hawaii, Honolulu, Hawaii, 7-10 October 1969. The 59 registered attendants came from Australia, Canada, Federal Republic of Germany, Japan, New Zealand, the U.S.A., and the U.S.S.R. Seminars were held on (1) Seismic source and energy transfer, (2) Tsunami instrumentation, and (3) Tsunami propagation and run-up. A summary session was held on the last day with considerable emphasis on the inadequacies of the existing magnitude scale for tsunamis.

The following resolutions were adopted by the Tsunami Committee:

1. The proceedings of the symposium are to be published and sold at a price sufficient to cover the costs. The Committee respectfully requests that unused travel funds authorized for the symposium be made available by the International Union of Geodesy and Geophysics (IUGG) to Dr. William Mansfield Adams, Secretary of the Committee, for use in preparing the symposium proceedings.

2. The Committee proposes that an interdisciplinary symposium on tsunamis and earthquakes be arranged for the IUGG meeting planned for Moscow, U.S.S.R., during August 1971.

This symposium should have the following days, topics, and conveners:

DAY	TOPIC	CONVENER
First	Tsunami Committee Meeting	Zetler
Second	Redefining Tsunami Magnitude	Iida
Third	Prediction of Tsunami Inundation: Short Term	Miller
Fourth	Prediction of Tsunami Inundation: Long Term	Voit

The Committee respectfully requests that IUGG provide funds for necessary travel to this symposium.

3. The Committee endorses and encourages the use of offshore and deep-sea tsunami gauges (in particular, the re-establishment of the offshore gauge at Wake Island), intercalibration tests of these gauges, and studies involving the utilization of focal source mechanisms as significant steps toward more effective tsunami warnings.

4. The Committee expresses its deep appreciation to the East-West Center and the Hawaii Institute of Geophysics, University of Hawaii, for their efforts in arranging the symposium and for their hospitality, and to IUGG and the Office of Oceanography, the United Nations Educational, Scientific, and Cultural Organization (UNESCO), for the support of travel to the symposium.

5. The Committee wishes to clarify previous recommendations with reference to the compilation of tsunami records at the World Data Centers. Whenever a tsunami exceeds two (2) meters anywhere, records from representative stations should be submitted from all cooperating nations to the World Data Centers.

6. The Committee suggests that agencies distributing questionnaires relative to earthquake information include a request for information on sea state.

7. The Committee endorses the effort by the International Tsunami Information Center to compile a collection of photographic copies of marigrams showing tsunamis. It urges that this effort be continued and, if possible, accel-

erated. It defers a decision on publishing an atlas of these marigrams until the next meeting but all members are requested to investigate the possibility of publishing such an atlas, in whole or in part, by their respective countries.

8. The Committee commends Mr. G. Pararas-Carayannis, International Tsunami Information Center, for his fine effort in collecting the data for a catalog of Pacific tsunamis, as noted in his progress report of 10 October 1969. Inasmuch as his efforts and various preliminary and regional catalogs have largely exhausted the sources of data and information, it is therefore proposed that a final version of a Pacific catalog be published under the joint authorship of K. Iida, D. C. Cox, S. L. Soloviev, and G. Pararas-Carayannis. The Committee calls on all members to investigate in their own countries the possibility of obtaining some measure of support for the publication. The Committee requests any possible financial support for this project from IUGG. The total estimated cost is $12,000, of which $10,000 is for printing costs and computer time and $2,000 for travel.

The officers of the Committee were re-elected, to serve until the Moscow meeting. The Committee is presently constituted as follows:

Chairman:	B. D. Zetler, U.S.A.
Vice-Chairmen:	S. L. Soloviev, U.S.S.R.
	K. Iida, Japan
Secretary:	W. M. Adams, U.S.A.
	J. W. Brodie, New Zealand
	K. Kajiura, Japan
	C. Lomnitz, Mexico
	L. M. Murphy, U.S.A.
	G. L. Pickard, Canada
	R. O. Reid, U.S.A. (IAPSO representative)
	E. F. Savarensky, U.S.S.R.

Five additional scientists are being invited to serve on the Committee, viz.,

> R. D. Braddock, Australia
> E. Gajardo, Peru
> G. R. Miller, U.S.A.
> V. S. Moreira, Portugal
> S. S. Voit, U.S.S.R.

B. D. Zetler, Chairman, IUGG Committee on Tsunamis

33.   Ecumenical Tsunamigaku

DOAK COX
University of Hawaii
Honolulu, Hawaii

Tsunamigakusha and guests:

I feel that I must apologize to many of you who are old friends and colleagues, that until this Aloha dinner I have been able to show so little of the hospitality on which we Hawaiians used to pride ourselves. As some of you know, I have taken on new duties that permit me little time for work even on those few phases of tsunami study in which I may have some competence as well as a continuing interest. And during this week of your meetings I seem to have been more heavily loaded than usual with chores that could not be put off.

I suppose that I am expected this evening to aim toward the traditional goals of an after-dinner address in the course of a professional meeting: (a) that it should have some relation to the rest of the proceedings, but (b) that it should be not so weighty as to add significantly to the mental load from the regular sessions, and (c) that it should somehow at the same time be a little more broadly or philosophically oriented than the rest of the proceedings. My philosophical equilibrium is at the moment a little shaky. I have been thinking how apparent it may be to you that the only tsunami talk I felt I could hear this week was the one I am giving myself. There is, however, one further goal of an after-dinner speech toward which I can still aim; that is that it not be so lengthy as to postpone seriously our postprandial relaxation.

When Bill Adams first asked me to speak, and scheduled me for the first evening of the Symposium, he very kindly provided me with a title under which to speak: E Komo Mai! -- which means in Hawaiian essentially: Chotto, Ohairi kudasai! Hola, entrad! Pre vet stvah vaht, Fkhah-deet! Hello, come in! This no longer seems appropriate. You have now been in for several days -- not only in Honolulu but in the midst of the discussion of tsunamis, and I suppose if Bill were to choose a title for tonight it might goodbye, sayonara, adios, or do cvidaniya!

Before saying these things, however, I should like to discuss with you some aspects of tsunami research that have been of interest to me for a long time -- in particular, two groups of aspects that may seem rather distinct but that, I think, can be subsumed under the title, "Ecumenical Tsunamigaku."

Ecumenical, an English word of Greek etymology, means embracing the whole world and all its systems of thought. Tsunamigaku is Japanese, of course, for the science or study of tsunamis. The hybrid nature of the title is deliberate.

The first group of aspects is geographic. The distances you have travelled to attend this meeting are immediately one indication of the importance of geographical ecumenicity in the study of tsunamis. I have, myself attended tsunami meetings in Hilo, Crescent City, Fairbanks, San Francisco, Washington, Pasadena, Tokyo, and Helsinki, as well as Honolulu.

Since I enjoy travelling, I have greatly appreciated this physical-geographic aspect and I expect most of you do also. But there is another geographic aspect that I have appreciated even more. This is the politico-geographic aspect. It seems to me most unlikely that any endeavor other than tsunami research could have given me so great an opportunity to meet, to correspond with, and to greet again, so many people not only from such distant parts of the world, but from such diverse nations, among them some whose diplomatic relationships with my own nation have hardly been conducive to personal acquaintance.

All natural scientists share to some extent the advantages of world-wide and international acquaintance, because of the international commonality of interest in these sciences, but we tsunamigakusha have special advantages because, like those who study storms, insect pests, and epidemic diseases, we deal with disaster-bearing processes capable of international travel.

On this earth, where for decades we have been poised on

the brink of world-wide international chaos (or actually embroiled in it) as the result of national and ideological differences -- some no doubt real and significant, but many, also no doubt, imaginary or insignificant -- it is clear that we desperately need an enormous extension of opportunities for individual acquaintance across national and ideological boundaries, and especially for opportunities to live together and work together, even if only briefly, on an international basis. Yet think how few of the world's citizens have even a small fraction of the opportunities we have.

The operations of the IUGG Tsunami Committee have provided for often-repeated contacts among a few of us who now can genuinely consider ourselves old friends, and one of the deep satisfactions I have had in recent years has been the establishment of the International Tsunami Information Center in Honolulu, at which not only important scientific work of an international character can be done and now is being done, but international friendships can be developed as a most significant by-product. It seems to me absolutely crucial that this Center maintain and, if possible, even intensify its international character.

It is a most sobering thought that these extraordinarily fortunate opportunities we have stem from phenomena that brings such misfortune to others.

But the geographic aspects represent only one of the groups of essential ecumenical aspects of tsunamigaku. The other group can perhaps be described as disciplinary. It has, of course, been obvious for a very long time that the study of tsunamis involves at its very heart both seismology and oceanography. But consider the attendance of these meetings and the topics of the papers presented: tectonics, geology, instrumentation engineering, communication engineering, ocean and coastal engineering -- how many more aspects have been discussed? I suspect that the disciplinary ecumenicity of the field is one of the sources of interest, even enjoyment, for many of you as it has certainly been for me.

But the breadth of the interest of this meeting, appropriate though it may be for the International Union of Geodesy and Geophysics, is most certainly not enough for the full study of tsunamis as a source of human disaster, or of the capabilities of tsunami warning systems, because it does not include the study of the human part of the problems.

Obviously, the social sciences must be involved but further, and perhaps not quite so obviously, these must be part of the intimately concerted ecumenical effort. It isn't just the physical characteristics of tsunamis and the psychological and social characteristics of people that must be separately studied, but the characteristics of tsunamis and tsunami warnings in relation to the characteristics of the people affected, and vice versa. Similarly, the economics of warning systems and of all other sorts of tsunami control measures must be studied in intimate relation to the characteristics of tsunamis.

Further, as some of us have learned, the major source -- practically the only source -- of information bearing on one most important phase of the physical characteristics of tsunamis, their frequency of recurrence, is historical records. The necessary information can be abstracted from these records only by the use of techniques of this additional social science, history.

We have, I think, hardly begun on the extension of tsunamigaku ecumenicity to the social sciences.

As some of you know, some of my more recent tsunami research efforts have been related to the question of tsunami warnings in relation to human response. I am not well qualified to work on the response as such, but I have put some attention to those characteristics of the warning system that I was sure were of importance in controlling human response -- not the characteristics of the idealized system that we would all wish we had, but the characteristics of the system as it has actually functioned.

My study concentrated on certain phases of the warning system, those involving international, transoceanic, geophysical information collection and appraisal, and U.S.A. national dissemination of hazard information. It did not deal adequately with either local practices on U.S.A. coasts for hazard information dissemination to the intended public, or with the equivalent practices in other countries.

It is, of course, risky to generalize from conclusions reached or reinforced in my limited study to anything applicable to the broader system as a whole, which involves not only the international aspects, but the national aspects of all countries and the local aspects extending all the way to the people who are intended as the ultimate beneficiaries. However, if I may be permitted some opinions, based more on the rather sketchy studies to date of human behavior as such than on anything I have produced, they are that the success of a warning system cannot well be judged on the basis of single self-contained episodes, but that because of the effects of human individual and cultural memory (storage, if you like, or capacitance), there are cumulative effects which simply cannot be ignored. Further, such acceptance as the warning system has had on many coastlines in the Pacific has been based, in fact, on the expectations of capabilities not yet achieved though, hopefully, achievable.

To carry this a bit further, I conclude that, because of the differences not only in physical exposure but differences in cultural attitudes, the expectable success of a tsunami warning system must vary enormously from place to place, that successful operation of a tsunami warning system is simply not practicable on some coasts, and that if improvements do not keep pace with expectations, as the result of progress in both natural science and social science as applied to warning system operation, a system may lose its effectiveness.

I am afraid that a full explanation would involve a discussion far too lengthy and weighty for this occasion.

Let me wind up with just one more observation of a possible ecumenicity of approach to the two aspects of tsunamigaku ecumenicity that I have hitherto treated as fairly distinct.

It is, of course, taken for granted that the physical laws governing tsunami generation, propagation, and effects do not vary with politico- or ethno-geography. But what about the psychological and sociological laws? Are these really, as I have until now treated them, fundamentally distinctive from nation to nation? I think not, and furthermore I take it as a matter of faith that the ultimate goals of groups of men, no matter on which side of the ocean and under what set of national laws and cultural heritages they live, are the same. Opportunities for international and intercultural study of basic human goals are quite limited. It just might be that, because the study of tsunamis and warning systems on an international, intercultural, and world-wide basis is permitted and even encouraged in spite of the constraints introduced by our present disciplinary and natural divisions, such studies broadened to their humanistic aspects might provide significant insights as to these goals common to mankind.

With that let me say, not goodbye, sayonara, do cvidaniya, or adios, but, because of all of its additional connotations, ALOHA!

34. Names and Addresses of Authors and Conference Participants

(*Denotes author not in attendance.)

Ziyadin Abouziyarov
12 Pavlic Morozov D-3F6
Arctic-Antarctic & Marine Dept.
Hydrometeorological Service
Moscow, U.S.S.R.

Wm. Mansfield Adams
Tsunami Research Division
Hawaii Institute of Geophysics
2525 Correa Road
Honolulu, Hawaii 96822, U.S.A.

Richard P. Augulis
Scientific Services Division
ESSA/Weather Bureau, Western
  Region
Box 11188, Federal Bldg.
Salt Lake City, Utah 84111
U.S.A.

*L. M. Balakina
Institute of Physics
  of the Earth
Moscow, U.S.S.R.

*Lise Boilard
Dept. of Energy, Mines, &
  Resources
615 Booth Street
Ottawa 4, Ontario, Canada

R. D. Braddock
Dept. of Mathematics
University of Queensland
St. Lucia, Brisbane 4067
Queensland, Australia

J. W. Brodie
N. Z. Oceanographic Institute
P. O. Box 8009
Wellington, New Zealand

Robert R. Brownlee
Los Alamos Scientific Laboratory
P. O. Box 1663
Los Alamos, New Mexico 87544
U.S.A.

Don R. Bucci
Dept. of Army, Corps of
  Engineers
Waterways Experiment Station
Vicksburg, Mississippi 39180
U.S.A.

John N. Butchart
State Civil Defense
Building 24, Fort Ruger
Honolulu, Hawaii 96816, U.S.A.

Cohen B. Byrd
Weapon Safety Branch
Atomic Energy Commission
Albuquerque, New Mexico 87100
U.S.A.

*George Carrier
Harvard University
Pierce Hall
Cambridge, Massachusetts 02138
U.S.A.

*R. K. C. Chan
Stanford University
Stanford, California 94305
U.S.A.

*L. V. Cherkesov
Nautical Hydrophysical Institute
Lenina 28
Sevastopol, Crimea
U.S.S.R.

Niels Christensen
Dept. of Oceanography, HIG 321
University of Hawaii
2525 Correa Road
Honolulu, Hawaii 96822, U.S.A.

Doak C. Cox
Water Resources Research Center
University of Hawaii
2610 Pope Road
Honolulu, Hawaii 96822, U.S.A.

Peter DeMello
Honolulu Police Department
1455 S. Beretania Street
Honolulu, Hawaii 96822, U.S.A.

G. C. Dohler
Tides and Water Levels
Marine Sciences Branch
615 Booth Street
Ottawa, Ontario, Canada

*Robert A. Eppley
Coast & Geodetic Survey
6001 Executive Blvd.
Rockville, Maryland 20852
U.S.A.

Jean H. Filloux
Gulf General Atomic, Inc.
P. O. Box 688
San Diego, California 92112
U.S.A.

*Dale Freeman
IBM Corporation
1240 Ala Moana Blvd.
Honolulu, Hawaii 96814, U.S.A.

*J. E. Fromm
IBM Research Division
San Jose, California 95100
U.S.A.

A. S. Furumoto
Tsunami Research Division
Hawaii Institute of Geophysics
2525 Correa Road
Honolulu, Hawaii 96822, U.S.A.

Frans Gerritsen
Dept. of Ocean Engineering
University of Hawaii
Honolulu, Hawaii 96822, U.S.A.

Gordon Groves
Tsunami Research Division
Hawaii Institute of Geophysics
2525 Correa Road
Honolulu, Hawaii 96822, U.S.A.

James W. Hadley
K - Division
P. O. Box 808
Lawrence Radiation Laboratory
Livermore, California 94550
U.S.A.

Robert Harvey
Tsunami Research Division
Hawaii Institute of Geophysics
2525 Correa Road
Honolulu, Hawaii 96822, U.S.A.

Tokutaro Hatori
Earthquake Research Institute
University of Tokyo
Yayoi, 1-1, Bunkyo-Ku
Tokyo, Japan

Wilmot N. Hess
ESSA Research Laboratories
Boulder, Colorado 80302, U.S.A.

Li-San Hwang
Tetra Tech, Inc.
630 North Rosemead Blvd.
Pasadena, California 91107
U.S.A.

Kumizi Iida
Dept. of Earth Sciences
Faculty of Science
Nagoya University
Nagoya, Japan

C. W. Iseley
Engineering Dev. Laboratory, ESSA
6015 Executive Blvd., WCS
Rockville, Maryland 20852, U.S.A.

Rockne Johnson
Tsunami Research Division
Hawaii Institute of Geophysics
2525 Correa Road
Honolulu, Hawaii 96822, U.S.A.

*Jan M. Jordaan
105 Nassau Sunnyside
Pretoria, South Africa

Kinjiro Kajiura
Earthquake Research Institute
University of Tokyo
Bunkyo-ku, Tokyo, Japan

*C. E. Knowles
Dept. of Oceanography
Texas A & M University
College Station, Texas 77843
U.S.A.

*M. I. Krivoshey
State Hydrological Institute
2nd Line 23
Vassiliev Ostrov
Leningrad, U.S.S.R.

Jimmy C. Larsen
Tsunami Research Division
Hawaii Institute of Geophysics
2525 Correa Road
Honolulu, Hawaii 96822, U.S.A.

Theodore T. Lee
Look Laboratory of Oceanographic Engr.
University of Hawaii
811 Olomehani Street
Honolulu, Hawaii 96813, U.S.A.

*Albert C. Lin
Tetra Tech, Inc.
7730 Herschel Avenue, Suite M
La Jolla, California 92037
U.S.A.

Harold Loomis
Tsunami Research Division
Hawaii Institute of Geophysics
2525 Correa Road
Honolulu, Hawaii 96822, U.S.A.

Gaylord R. Miller
ESSA, Joint Tsunami Research
  Effort
Hawaii Institute of Geophysics
2525 Correa Road
Honolulu, Hawaii 96822, U.S.A.

Leonard M. Murphy
U.S. Coast and Geodetic Survey
Rockville, Maryland 20850
U.S.A.

T. S. Murty
Marine Science Branch
Dept. of Energy
615 Booth Street
Ottawa 4, Ontario, Canada

Alex I. Nakamura
ESSA, Joint Tsunami Research
  Effort
Hawaii Institute of Geophysics
2525 Correa Road
Honolulu, Hawaii 96822, U.S.A.

*A. V. Nekrasov
Leningrad Hydrometeorological
  Institute
Mallokhtinskiy 98
Leningrad K-196
U.S.S.R.

Gerd Niehaus
HAGENUK vorm. Neufeldt &
  Kuhnke GMBH
23 Kiel, Westring 431-451
West Germany

John T. O'Brien
Dept. of Ocean Engineering
University of Hawaii
Honolulu, Hawaii 96822, U.S.A.

Kenneth H. Olsen
Los Alamos Scientific
  Laboratory
P. O. Box 1663
Los Alamos, New Mexico 87544
U.S.A.

Robert Q. Palmer
Ocean Engineering Dept.
University of Hawaii
Honolulu, Hawaii 96822, U.S.A.

George Pararas-Carayannis
International Tsunami
 Information Center
P. O. Box 5214
Honolulu, Hawaii 96814, U.S.A.

G. L. Pickard
Institute of Oceanography
University of British Columbia
Vancouver 8, B. C., Canada

*G. S. Podyapolsky
Institute of Physics
 of the Earth
B. Gruzinskaya 10
Moscow D-242-U.S.S.R.

Fredric Raichlen
Keck Laboratory
California Institute of
 Technology
Pasadena, California 91109
U.S.A.

Robert O. Reid
Dept. of Oceanography
Texas A & M University
College Station, Texas 77843
U.S.A.

Robert E. Schank
State Civil Defense
Building 24, Fort Ruger
Honolulu, Hawaii 96816, U.S.A.

Margaret Elizabeth Schmitt-Habein
1700 Makiki Street
Honolulu, Hawaii 96822, U.S.A.

Richard P. Shaw
ESSA, Joint Tsunami Research
 Effort
Hawaii Institute of Geophysics
2525 Correa Road
Honolulu, Hawaii 96822, U.S.A.

*B. I. Sebekin
Institute of Oceanology
Sadovaya 1
Lublino
Moscow Zh-387
U.S.S.R.

Thomas J. Sokolowski
ESSA, Joint Tsunami Research
 Effort
Hawaii Institute of Geophysics
2525 Correa Road
Honolulu, Hawaii 96822, U.S.A.

S. L. Soloviev
Soviet Tsunami Commission
Soviet Geophysical Committee
Molodezhnaya 9
Moscow B-296, U.S.S.R.

Robert L. Street
Dept. of Civil Engineering
Stanford University
Stanford, California 94305
U.S.A.

W. Summerfield
Dept. of Earth and Planetary
 Sciences
The Johns Hopkins University
Baltimore, Maryland 21218
U.S.A.

Ziro Suzuki
Geophysical Institute
Tohoku University
Sendai, Japan

*Kazuhiko Terada
National Research Center for
 Disaster Prevention
6-1 Ginza-Higashi
Chuo-ku, Tokyo, Japan

Wm. G. Van Dorn
Scripps Institute of
 Oceanography
P. O. Box 109
La Jolla, California 92037
U.S.A.

Andrew C. Vastano
Dept. of Oceanography
Texas A & M University
College Station, Texas 77843
U.S.A.

Martin Vitousek
Tsunami Research Division
Hawaii Institute of Geophysics
2525 Correa Road
Honolulu, Hawaii 96822, U.S.A.

*S. S. Voit
Institute of Oceanology
Sedovaya 1
Lublino
Moscow Zh-387
U.S.S.R.

James W. Walker
Tsunami Research Division
Hawaii Institute of Geophysics
2525 Correa Road
Honolulu, Hawaii 96822, U.S.A.

*James M. Walling
Hamilton Library
University of Hawaii
Honolulu, Hawaii 96822, U.S.A.

Hideo Watanabe
Japan Meteorological Agency
Ote-machi, Chiyoda-ki
Tokyo, Japan

*Robert W. Whalin
Dept. of Army, Corps of
  Engineers
Waterways Experiment Station
Vicksburg, Mississippi 39180
U.S.A.

Syd O. Wigen
Canadian Hydrographic Service
512 Federal Building
Victoria, B. C., Canada

John A. Williams
Dept. of Civil Engineering
University of Hawaii
Honolulu, Hawaii 96822, U.S.A.

Basil W. Wilson
Consulting Oceanographic Engineer
529 South Winston Avenue
Pasadena, California 91107
U.S.A.

H. J. Wirz
Honolulu Observatory
Ewa Beach, Hawaii 96706, U.S.A.

T. Y. Wu
Thomas Building
California Institute of
  Technology
Pasadena, California 91109
U.S.A.

Robert E. Yoder
University of California
Lawrence Radiation Laboratory,
  L-517
Livermore, California 94550
U.S.A.

Lawrence T. Zerkel
American Red Cross
1270 Ala Moana Blvd.
Honolulu, Hawaii 96814, U.S.A.

Bernard D. Zetler
Atlantic Oceanographic
  Laboratories
901 South Miami Avenue
Miami, Florida 33130, U.S.A.

Index

Abouziyarov, Z. K., 271
Adak, 261, 268
Adams, W. M., 57, 63, 241, 265, 477, 478, 487
AEC, 425
Aftershock, 101, 105-108
Age hardening, 228
Aida, I., (University of Tokyo), 216, 349
Airy
  equations, 464
  integral, 38
Akita-oki, 101, 106
Alaska, 151, 158, 165, 286
  Command, 268
  Disaster Office, 268
  Gulf of, 35
Alberni Inlet, 165, 168
Alcock, E. D., 136
Alert Bay, 167
Aleutian
  Islands, 138, 157, 158
  Ridge, 254
  Trench, 34
Aloha Dinner, 489
Alsop, L. E., 130
Ambraseys, N. N., 152
Amchitka
  Island, 250
  station, 262
Amein, M., 455
Amphidromic oscillations, 169

Andrade, E. N., 228
Andrade's beta-creep law, 223
Andreanof Islands, 268
Antennas, dipole, 122
Anti-tsunami facilities, 87
Antofagasta observatory, 262
Aomori, 16, 85
Aperiodic incident waves, 397
Aperture, 257
Approximate simulation, 463
APPSIM, 463
Arc tectonics, 160
Arica, 294, 295
Arrival times, 40
Arthur, R. S., 286
Astoria, 294, 295
Atomic Energy Commission, 425
Attu station, 261
Australia, 296, 485
Avachinsk, Bay of, 276
Averyanova, V. N., 278
Ayukawa, 213
Azimuthal studies, T waves, 259

Background noise, 236
Balakina, L. M., 47
Banda Sea, 158
Barber, F. G., 186
Barber's Point Harbor, 383
Barkley Sound, 166, 167
Basins

closed, 378
  geometry of, 382
  rotation of, 307, 311, 320
  variable depth, 320
Batteries, in instrumentation, 246
Bay, idealized, three-dimensional, 412, 413
Bella Bella, 167
Ben-Menahem, A., 121, 127
Berkman, S. C., 292
Bimetal effect, 225
Block tectonics, 160
Bogert, B. P., 177
Boilard, L., 165
Bollinger, G. A., 9
Boolean algebra, 482
Bores, 362, 450
Born, M., 256
Boso, 8, 151
Boso-oki, 110
Bouguer's gravity anomaly, 109
Bourdon tube, 223
Braddock, R. D., 285, 286, 487
Breaking (waves), 356, 358, 362, 412
Brisbane, 295
Brodie, J. W., 283, 487
Brovikov's formula, 356
Brune, J. N., 121, 130
Bucci, D. R., 427
Buchbinden, G. G., 136
Buoy system instrumentation, 248
Butler, J. P., 367

Cable
  cutter, in instrumentation, 246, 251
  system instrumentation, 248
Calibration
  gauge, 370
  wave-flume, 434
California
  recurrence data, 158
  University of, 33
Call system, 210

Camfield, F. E., 454
Camp Cove, 295
Canada, 158, 165, 191, 485
Canton Island, 295
Capacitance-type meter, 219
Cape St. James, 167
Cape Shipunsky, 276, 280
Capetown, So. Africa, 383
Carrier, G. F., 367, 377, 455, 462
Catalogues
  tsunamigenic earthquakes, 4
  tsunamis, 149, 487
Central America, 158
Cepstrum analysis, 165
Chan, R. K. C., 453, 458
Channel equations, 166
Channels, rectangular, 362
Cherkesov, L. V., 319
Chézy's coefficient, 364
Chiba, 211
Chile, 85, 151, 158, 216
Closed basins, 378
Coastlines, straight, 377, 407
Coff's Harbour, 295
College station, 261
Collins, Curtis, 186
Colorado, University of, 268
Common words, 480
Computer programs
  adaptive pattern recognition, 136
  document processing, 478
Computers, digital, 357
Computing method development, 355
Conditional probabilities, 160
Continental shelf, effect of, 337, 341, 454
Convolution, 310
Coriolis force, 321
Correlogram, 256
Corvallis, Oregon, station, 138
Cox, Doak C., 59, 489
Creep, of Ni-Span-C, 228
Crescent City, 454, 470

Crest intervals, 40
Cross spectra, 177
Curvature, free surface profile, 356
Cylindrical island, 404

Dambara, T., 77
Dantzig, G. B., 288
Daughter stations, 210, 211
Dean, R. G., 455
Deep-sea tide gauges, 223
Deep-water signature, 399
Delta functions, 38
Density interface, 322
Depth of water, effective, 39
Dipole antennas, 122
Dip-slip
 earthquakes, 120
 faults, 48, 100
Discontinuous waves, 362
Discrete time step, 403
Dispersion equation, 23
Dispersive waves, 416
Displacement transducer, 41
Distorted models, 440
 wave steepness in, 436-439
Distortion curve, 432
Ditkin, V. A., 310
Diurnal constituents, 235
Dobson, R. S., 463
Document processing system, 477
Dohler, G. C., 186, 191
Doppler sounding technique, 119
Double-humped waves, 367
Duncan's Basin, 383
Duration of oscillations, 153
Dutch Harbor station, 261

Earthquake mechanisms, 47, 101-116, 119
Earthquake Research Institute, 69
Earthquakes
 geographic distribution, 100-108
 submarine, 58

Easter Island observatory, 262
Eaton, J. P., 259
Ebb tide, 345
Economics, warning systems, 492
Elastic
 liquid, 19
 solid half-space, 19
Electron density, 121
Ellis, G. E., 462
Energy flux, 88
Energy, Mines, and Resources, Dept. of, 165, 191
Eniwetok, 35
Enoshima Island, 91, 216
Ensenada, 294, 295
Entrance channel, 382
Environmental Science Services Administration, 484
 Coast and Geodetic Survey, 261
 Joint Tsunami Research Effort, 135, 239
Epicenters, geographic distribution, 4
Eppley, R. A., 261
Equations of motion, 20
Equivalent keywords, 480-481
Erimosaki, 87
Error, estimate of input, 406
Ewing, M., 462
Experimental investigations
 tsunami waves, 351-365
 wave run-up, 407

Fast Fourier Transform Technique, 174
Fault-scarps, 58
Favre, 355
Federal Aviation Administration, 268
Fedotov, S. A., 281
Ferchev, M. D., 281
Fermat's Principle, 286
FFTT, 174
Fiji
 Basin, 295

Islands, 157, 158
Filloux, J. H. 223
Finite amplitude, waves, 376
Finite-difference
  equations, 457
  models, 453
Flooding, shore, 360
Fluid, nonuniform, 320
Flume, hydraulic, 352, 355
Focal depth, 50, 112
Focal mechanism, 3, 62
Forecasting seismicity, 281
Forrester, J. W., 66
Fort Denison, 295
Fourier
  synthesis, 378
  transformation, 322
  Transform Technique, 174
Free-drop instrumentation, 242
Freeman, D., 477
Free periods, 169
Free surface, 322
Friction, 375
  coefficients, 362-364
  effect of, 172
  neglect of, 365
Fromm, J. E., 453
Fuess-type tide gauge, 213
Fukushima, 16, 77
Fukushimaken-oki, 107, 108
Fukuuchi, H., 183
Fulford Harbour, 167
Furumoto, A. S., 57, 64, 119, 121

Gajardo, E., 487
Galapagos Islands, 294
  station, 262
Gambier Island station, 262
Gauges
  deep-sea tide, 223, 486
  Fuess-type tide, 213
  GM, 275
  intercalibration tests, 486
  offshore, 486
  Scripps Institution, 235
Gaussian elimination, 466

Generation, tsunamis, 3, 19
Geographic distribution
  earthquakes and tsunami sources, 100
  epicenters, 4
Geostationary satellite, 269
Geostrophic relation, 169
Germany, Federal Republic of, 485
Gilmour, A. E., 286
Glass-sphere instrumentation, 245
Grade magnitudes
  classification, 100
  relationship between tsunami and earthquake, 112
Gradstein, I. S., 314
Gravity waves, 24
Green, C. S., 285
Greenberg, David, 186
Green's formula, 356
Greenspan, H. P., 367
Gregory-Newton extrapolation, 459
Greymouth, 295
Grid parameters, 289
Grigorash, Z. K., 280
Gulf General Atomic, Inc., 223

Haas, J. Eugene, 268
Hachinohe, 86
Hagenuk/Kiel, 192
Haleakala, Maui, station, 265
Hana, Maui, 126
Hankel function, 307, 379
Harbor response function, 379
Harbors, rectangular
  oscillations, 410
  run-up, 414
Hardy, W. A., 257
Harmonic
  amplification factor, 386-387
  input, 403
  wave trains, 338
Harvard University, 377
Hatori, T., 69
Hawaii

Institute of Geophysics, 57, 119, 132, 239, 253, 477
  recurrence data, 158
  seismic stations, 126
  University of, 132, 135, 239, 367, 377, 477, 489
  volcano observatory, 265
Healy, M. J. R., 177
Heat treatment, Bourdon tube, 228
Helicopter (installing instrumentation), 251
Hilaly, N., 297
Hilo harbor, 470
Hilo, Hawaii
  harbor, 470
  seismic station, 261
Hirasawa, T., 9
Hiratsuka, 219
Hirt, C. W., 459
Historical records, 492
Hiuganada, 69
Hobart, 295
Hodgson, J. H., 9, 47, 63
Hokkaido, 8, 16, 105, 157, 158
Honolulu seismic station, 261
Honshu, 157, 158
Honuapo station, 266
Horikawa, K., 408
Houston-Fearless CARD microfiche retrieval system, 478
Hwang, Li-San, 407, 454
Hydraulic
  head, harbor, 450
  flume, 355
  models, 3-dimensional, 433
Hydrodynamic models, 305
Hydrodynamics, tsunami waves, 319
Hydrometeorological Institute, 337
Hydrometeorological Service, 271
Hydrophone
  arrays, 256
  records, 258

Hyuganada, 109, 110, 111

Ibaraki, 16, 71
IBM Corporation, 453, 477
  Research Division, 453
  System/360 document processing, 478
Ichikawa, M., 9, 62, 116
Iida, H., 131
Iida, K., 59, 64, 120, 487
Ikonnikova, L. N., 278
Imabetsu, 86
Indonesia, 157
Input, estimate error, 406
Instrumentation
  deep-ocean, 239-252
  free-drop, 242
  in Japan, 207
  UBOPE (seismic), 275
  GM, 275
Intensity-frequency relations, 157
Intensity, tsunamis, 152
Intercalibration tests, 486
Interference, incident and reflected waves, 415
International Union of Geodesy and Geophysics, 491
Inundation, 486
Inverse tsunami problem, 399
Ionosphere, 119
Iriam, 158
Isaacs, J. D., 248
Isabelle, Maurice, 186
Isacks, B., 62, 116
Island-arc
  earthquakes, 100
  structures, 50
Island configurations, 403
Iterative procedure, 285
Ito, Y., 183
Iturup, 8, 16, 101, 105, 151
  foreshock 1968, 108
IUGG
  meeting, 1971, 485-486
  report, 1969, 485-487
Iwato Prefecture, 86

Japan
  earthquake mechanisms, 48
  Meteorological Agency, 85, 99, 100
  Sea, 158
  symposium participation, 485
Johnson Point, 167
Johnson, R. H., 253
Johnston Island station, 261
Jones, D. B., 470
Jordaan, J. M., Jr.,367, 462, 470
Junction Passage, 166

Kajiura, K., 35, 283, 487
Kalimantan, 158
Kamaishi, 86
Kamchatka, 48, 151, 157, 158
Kawasaki, 211
Ketoi Island, 281
Keyword, truncated, 481
Khristianovich, S. A., 362
Kinetic energy, 272
Kjawa, 158
Knowles, C. E., 399
Kodiak Island, 34
Koh, R. C. Y., 454
Koma Mountain, 213
Kona station, 126, 265
Krivoshey, M. I., 351
Kronoki Bay, 280
Kuril Islands, 48, 119, 157, 158
Kushiro, 16
Kutzback, J. E., 455
KWIC index data base, 477

Laboratory models, 367
Lagrange formula, 356
Lagrange-Eure formula, 280
Lamb, H., 306
Land damage, 101, 104
Landslides, submarine, 272
Langara Island, 193, 200
Laplace transformation, 322
Leading waves, 35
Least-squares analysis, 234

LéMehauté, B., 414, 454, 463
Lesser Sunda Island, 158
Lin, A. C., 407
Linear response, 401
Local geometry, influence of, 407
Lomnitz, C., 487
Long water waves, 453
Lord Howe Rise, 295
Loucks, R. E., 177
Luzon, 158
Lyttleton, 295

McKenney Island, 167
McMurdo Sound, 295
Macquarie Island, 295
Macroseismic
  effects, 152
  services, 160
Manzanillo, 262, 292, 293
Marigram, observed, 399
Marine observation tower, 219
Maritime Safety Agency, 216
Marker particles, 457
Marsden Point station, 262
Matua, 277
Maui, 126
Mauna Kea station, 265
Maximal
  distances of observation, 155, 156
  heights of inundation, 151
Mean heights of inundation, 151
Mei, C. C., 454, 463
Meleschenko, H. T., 362
Melline's integral, 314
Mera, 218
Merian periods, 172
Mexico, 157-158
Microfiche retrieval system, 478
Midway station, 261
Militeev, A. N., 349
Miller, G. R., 135, 239, 241, 475, 487
Miller, J. C. P., 40
Miller, R. L., 456

Miyagi
  Enoshima Tsunami Observatory, 70
  Prefecture, 86
Miyamura, S., 64
Miyazaki, 213
Model experiments
  Alaska tsunami generation, 33, 40
  nonstationary wave motions, 303
  three-dimensional hydraulic test, 433
  wave run-up, San Diego, 427
Mogi, K., 80
Monkmeyer, P. L., 455
Monochromatic
  incident waves, 379
  input conditions, 404
Mono Lake, 463
Monopolar uplift, 33
Montague Island, 38
Moreira, V. S., 487
Mortimer, D. H., 169
Moscow, IUGG meeting, 1971, 485-486
Mother station, 211
Mott, N. F., 228
Mud flows, 272
Multiturn helix, 224
Murphy, L. M., 1, 261, 487
Murty, T. S., 165

Naalehu station, 265
Nafe, J. E., 130
Nagasaki Marine Observatory, 209
Nagata (University of Tokyo), 216
Nagoya, 213
  University, 3
Nahmint River, 174
Nankaido, 111, 151
Narrow-mouthed harbors, 377
National Marine Consultants, 463

National Research Center for Disaster Prevention (Japan), 207, 219
National Science Foundation, 132, 471
Natural frequency, closed basin, 378
Nautical Hydrophysical Institute, 319
Naval Research, Office of, 44, 471
Navier-Stokes equation, 455
Nekrasov, A. V., 337
Nelson, 294, 295
Nemuro, Hokkaido, 131
New Britain, 157, 158
New Hebrides Islands, 138, 157, 158, 300
New Westminster, 167
New Zealand, 158, 262, 296, 477
  Plateau, 296
Niigata, 8, 101, 106, 108, 151
Niiyama-hama, 86
Ni-Span-C alloy, 223
Nodal-plane solution, 100
Nonbreaking waves, 412, 419
Nonlinear equations, 357
Nonperiodic wave-train excitation, 410
Nonstationary wave motions, 305
Nonuniform fluid, 320
Norfolk Island Ridge, 295
Normal-fault earthquakes, 15, 115
Normal modes, closed basin, 378
Norris, R. A., 254
North Arm, 167
North Island, 158, 440
North Solomon Island, 158
Northwestern Pacific, 47
Nuclear test, underground, 250
Nugent Sound, 167
Numerical
  method, application of, 401
  simulation, 453

Oahu, Hawaii, 383
Observed marigram, 399
Ocean-bottom seismometers, 269
Ocean Falls, 167
Oceanology, Institute of, 305
Ofunato Bay, 86
Ojika Peninsula, 213, 216
Okada, A., 77
Okinawa station, 262
Oliver, J., 116
Onagawa Bay, 91
Ondulations, 355
Optimal
  paths, local, 290
  problems, 285
Oscillations
  amphidromic, 169
  duration of, 153
  rectangular-harbor, 410
Oshika Peninsula, 87
Over-relaxation, 459
Overwash, 364

Pacific
  ridges, 157
  southwest, 294
  Tsunami Warning System, 64, 119, 261
    participants, 261
Pago Pago, 295
Pahoa station, 265
Palmyra station, 261
Pararas-Carayannis, G., 35, 487
Pattern recognition, 135
  adaptive, 136
Perception, tsunami hazards, 268
Peregrine, D. H., 455
Periodic
  run-up, 412
  wave excitation, 410
Peru, 158
Petropavlovsk, 276
Phase velocity, 127

Philippines, 157, 158
Physics of the Earth, Institute of, 19, 47
Pickard, G. L., 186, 189, 487
Pitt Lake, 167, 177
Plafker, G., 38
Plastic flow, 223
Platzman, G. W., 170
Plunger, 42
Pneumatic wave generator, 352
Podyapolsky, G. S., 19, 278
Point Atkinson, 167
Point Loma, 428
Port Alberni, 167, 177
Port and Harbour Bureau (Japan), 218
Port Moresby observatory, 262
Potential energy, 272
Press, F., 462
Pressure transducers, 41, 201
Prince Rupert, 167, 205
Probability, tsunami occurrence, 150
Proceedings, publication of, 485
Propagation, tsunami, 285
Prudnikov, A. P., 314
Public, education of, 268, 270
Pulsations, 348
Pulse-frequency system, 202, 204
Punta Arenas, 294, 295
Pyaskovskiy, R. V., 344

Queen Charlotte Island, 193
Queen Charlotte Sound, 193
Queensland University, 285

Radar, effect on instrumentation, 251
Radio beacon, 245
Radioscience Laboratory, 132
Raichlen, F., 454
Rao, D. B., 170
Rashuwa Island, 281
Rat Island, 143, 254
Rayleigh waves, 119
Real-time read-out, 242
Recall, acoustic, 246

Recordak microfilm camera, 483
Recorder, 411
Reflection
  coefficients, 338
  of waves, 356
Refraction
  diagrams, 88
  maps, 280
Reid, R. O., 399, 487
Relationship, tsunami generation and earthquake mechanism, 47
Remote stations, 210
Report, Tsunami Committee, 485
Residual noise spectrum, 237
Resistance
  probes, 369
  sensors, 352
Resonance
  characteristics, 166
  coastal, 343
Reverse-fault earthquakes, 15, 115
River Bed Laboratory, State Hydrological Institute, 352
Roberts, E. B., 66
Rotation, basin, 307, 311, 320
Run-up
  harbor, rectangular, 414
  model study, 427
  nonperiodic wave-train excitation, 415, 417
  periodic wave excitation, 412
  zone, 363
Rupture length, 58, 253
Ryukyu Islands, 158
Ryzhic, I. M., 314

Saint-Venant's equation, 356
Sakhalinsk Composite Research Institute, 275
Salina Cruz station, 262
Samoa-Tonga-Kermadec, 157-158
Samuseva, R. N., 280
San Diego, California, 224, 427
Sangihe Islands, 157, 158
Sanriku
  earthquake mechanism 107, 110
  inundation, 151, 409
  tsunamis, 16, 85, 216
Santa Cruz, 157-158
Satellite communications, 269
Savarensky, E. F., 278, 487
Saville, T., Jr., 429, 451
Scale distortion, 355
Scholte, J. G., 21
Schreiber, D. E., 462
Scripps Institution of Oceanography, 33
  tide gauge, 235
Sea-floor spreading, 62
Seaquakes, 271
Search statement, 481
Sebekin, B. I., 305
Seismicity, forecasting, 281
Seismic
  records, 135
  Sea Wave Warning System, 261
Seismometers
  ocean-bottom, 269
  UBOPE, 275
Seki, A., 82
Semidiurnal constituents, 235
Sendai Meteorological Observatory, 213
Shaw, R. P., 377
Shebalin, N. N., 153
Shelf radiation, 341
Shemya, 268
Shenderovich, I. M., 275
Shikoku, 77
Shimanokoshi, 71
Shiriyazaki, 87
Shooter, J. A., 462
Shore proximity, tsunami sources, 160
Shukeikin, V. V., 280
Signature, deep-water, 399
Silver Strand, 428

Simple harmonic input, tsunami problem, 403
Simple island configurations, 403
Simulation
  approximate, 463
  finite-amplitude waves, 453
Simushir Island, 281
Single-crest waves, 388-391
Single seismic records, 135
Sinusoidal wave, damped, 392-394
Sitka, 286, 294, 295
  station, 261
Skin friction coefficient, 170
Skott-Rassel formula, 280
Snodgrass, F. E., 246
Social sciences, 491
Sokolowski, T., 135
Solitary waves, 343
  propagation of, 458
Solomon Islands, 158, 302
Soloviev, S. L., 1, 50, 149, 156, 281, 349, 487
Somass River, 174
Sonogram, 122
Source
  mechanisms, 47, 49, 101-116, 119
  region, 135
South Japan, 157, 158
South Pole station, 138
South Solomon Island, 157, 158
Southwest Pacific, 294
Soviet Tsunami Commission, 149
Spaeth, M. G., 292
Spatial resolution, 406
Stanford University, 453
Stanford-University-Modified Marker and Cell, 456
State Hydrological Institute, 351
  River Bed Laboratory, 352
Statistical studies, 99

Stauder, W., 9
Stevens, Anne E., 63
Steveston, 167
Stoker, J. J., 455
Stoneley, R., 178
Straight coastline, 377, 407
Strait of Georgia, 177
Strange, J. N., 427
Street, R. L., 453, 454
Stressed plane, 115
Stretensky, L. N., 308
Strike-slip
  earthquakes, 120
  faults, 100
Submarine
  earthquakes, 58
  landslides, 20
  volcanoes, 272
Sulawesi, 158
Sumatera, 158
SUMMAC, 456
Superposition, multiple reflections, 348
Surface tension, 371
Surface waves
  analysis, 121
  attenuation, 237
Surf zone, 454
Suva observatory, 262
Suzuki, R. K., 121, 122
Suzuki, Z., 85
Sydney, 295
Sykes, L. R., 116
Synonym/Equivalent keywords, 480-481
Synonym keywords, 481
System effectiveness, 268

Table Bay Harbor, 383
Tacubaya observatory, 262
Taiwan, 157, 158
Takahashi, R., 154
Takahashi (University of Tokyo), 216
Talara, Peru, 262, 294, 295
Talaud Islands, 157, 158
Talcahuano, 294, 295
Tasman Sea, 296

Tasu, 167
Tateyama observatory, 218
Taylor, John, 186
Tectonics
 arc, 160
 block, 160
Tectonic structures, 50
Tele-announcer, 192
Telemetry, 191, 209
Telephone, 192
Telex, 192
Temperature measurement instrumentation, 248
Tendency change, 197
Terada, K., 207
Terdiurnal constituents, 235
Tetra Tech, Inc., 407
Texas A & M University, 399
Thermoelastic coefficient, 223, 225, 229
Thermoexpansion effects, 223, 229
Tide-gauge telemetry, 191
Tide gauges
 deep sea, 223
 Scripps Institute of Oceanography, 235
 Bourdon tube, 223
Timed-release instrumentation, 245
Tofino, 167, 193
Tohoku University, 85, 86
Tokachi-oki
 earthquake mechanism, 101, 105, 106, 108, 111
 earthquake, 1968, 85-97
 inundation, 151
 Rayleigh waves, 121
Toksoz, M. N., 127
Tokyo, University of, 69
Tonankai, 109, 110
Tonankaido, 151
Topography
 model, 440
 periods, 172
Tørum, A., 39, 454, 462
T-phase duration, 256
Transducers
 displacement, 41
 pressure, 41, 201
Transfer function, 396, 400
Transformation, long wave, 355
Transient
 amplification, 395, 396
 incident, 377
 response, 388-394
Transponder, acoustic, 246
Trapped waves, 404
Traveling storm, 183
Travel time
 between gauges, 374
 tsunamis, 286
Trevor Channel, 165, 168
Tri-diagonal coefficient matrix, 466
Troughs, oceanic, 50
Truk, 35
Tsunami Committee, 485
Tsunami
 control measures, 492
 detector, GM-30, 275
 energy, 154
 generation, 16
 hazards, 268
 inundation, 486
 magnitude, redefining, 486
 prediction, 486
 problem, inverse, 399
 propagation, 279
 recorders, 221, 275
 recurrence data, 159
 signature, 399
 sources
  area of, 101
  geographic distribution, 100, 102, 105-107, 113
  Japan, 99
  statistical studies, 99
 transformation, continental shelf, 337
 travel times, 280
 warning service, USSR, 271
 waves
  experimental investigations, 251

hydrodynamics, 319
zoning, 150, 281
Tsunamigaku, 489
Tsunamigenic earthquakes, geographic distribution, 99, 104
Tsunamis
 inverse problem, 399
 narrow-mouthed harbors, 377
Tuamotu Arch station, 262
Tucson, Arizona station, 138, 261
Tukey, J. W., 177
Tully, J. P., 174
T waves, 253

Uchucklesit Inlet, 168
Underground nuclear test, 250
Under-ship instrumentation, 246
Undulla Deep, 297
UNESCO, 486
United States
 Army Engineer Waterways Experiment Station, Corps of Engineers, 427
 Geological Survey, 265
 symposium participation, 485
Uppsala station, 156
Urakawa, Hokkaido, 131
Urup, 151
USSR
 IUGG meeting, 1971, 485-486
 symposium participation, 485
 Tsunami Warning Service, 271
Uthno-Sakhalinsk, 276
Utsu, T., 82, 136

Vancouver, 167
 Island, 193
Van-Der-Lind, 277
Van Dorn, W. G., 33, 39, 41, 178, 408, 454
Vanoni, V., 454

Variable depth
 in basins, 320
 with viscosity, 325
Vedernikov's formula, 356
Velocity
 map, 456
 wave, 127
Vera (typhoon), 213
Vertical step (ocean bottom), 324
Vibrotron system, 241
Victoria, 167, 193
Viscosity, influence on long waves, 325
Vitousek, M., 189, 239, 265
Voit, S. S., 305, 487
Volcanoes, submarine, 272
Vol'tsinger, N. E., 344

Wake Island, 35, 36, 40, 486
Walling, J. M., 477
Warning system
 effectiveness, 268
 USSR, 271
Watanabe, H., 64, 99
Water depth, 61
Wave
 gauges, 411, 431
 generators, 352, 368, 410
 interference, 415
 intrusion tests, 440
 motions, nonstationary, 305
 profiles, 369
 reflection, 356
 run-up
  experimental investigations, 407
  straight beach, 407
 steepness
  deep-water, 435
  in distorted model, 436-439
 tank, 410
Wave-flume calibration, 434
Waves
 aperiodic incident, 397
 breaking, 356, 358, 362
 breaking and nonbreaking, 412

damped sinusoidal, 392-394
  discontinuous, 362
  dispersive, 416
  double-humped, 367
  finite amplitude, 376
  gravity, 24
  nonbreaking, 419
  simulation of (finite-
    amplitude), 453
  single-crest, 388-389
  solitary, 343
  trapped, 404
Weaver, P. F., 121
Welch, J. E., 457
Wellington observatory, 262
Whalin, R. W., 427
White, W. R. H., 165
Wickens, A. J., 9, 47
Wiegel, R. L., 431
Wigen, S. O., 165, 186
Williams, J. A., 367
Wilson, B. W., 34, 35, 38, 454, 462
Wolf, E., 256
Words
  keywords, 480-481
  little-value, 480
  stop, 480
World Data Centers, 486

Yakubov, S. M., 362
Yaroshena, R. A., 280
Yokohama, 211
Yokosuka, 211
Yoshida, K., 178
Yuen, P. C., 121

Zeigler, J. M., 456
Zetler, B. D., 292, 484, 485

SIGNIFICANT DATES

1969
  August 11, 119
  October 2, 250

1968
  May 16, 121

1965
  February 4, 142, 254
  February 15, 143
  February 18, 143
  August 11, 144
  August 31, 144

1964
  January 12, 142
  March 28, 33, 294
  March 29-30, 143, 144, 165
  April 8, 142, 143, 144
  May 28, 144

1962
  May 7, 48, 49

1961
  February 12, 48, 49
  February 26, 48

1958
  November 6, 48

1952
  November 4, 46

1927
  March 7, 48